Lecture Notes in Computer Science 1106

Edited by G. Goos, J. Hartmanis and J. van Leeuwen

Advisory Board: W. Brauer

Springer
Berlin
Heidelberg
New York
Barcelona
Budapest
Hong Kong
London
Milan
Paris
Santa Clara
Singapore
Tokyo

Michael Jampel
Eugene Freuder   Michael Maher   (Eds.)

Over-Constrained
Systems

 Springer

Series Editors
Gerhard Goos, Karlsruhe University, Germany
Juris Hartmanis, Cornell University, NY, USA
Jan van Leeuwen, Utrecht University, The Netherlands

Volume Editors

Michael Jampel
City University, Department of Computer Science
Northampton Square, London EC1V 0HB, UK

Eugene Freuder
University of New Hampshire, Department of Computer Science
College Road, Durham, NH 03824, USA

Michael Maher
Griffith University, School of Computing and Information Technology
Nathan, Queensland 4111, Australia

Cataloging-in-Publication data applied for

Die Deutsche Bibliothek - CIP-Einheitsaufnahme

Over constrained systems / Michael Jampel ... (ed.). - Berlin ;
Heidelberg ; New York ; Barcelona ; Budapest ; Hong Kong ;
London ; Milan ; Paris ; Santa Clara ; Singapore ; Tokyo :
Springer, 1996
(Lecture notes in computer science ; 1106)
ISBN 3-540-61479-6
NE: Jampel, Michael [Hrsg.]; GT

CR Subject Classification (1991): D.1.m, D.3.2-3, D.1.3, I.2.8, I.2.3-4, D.1.6,
F.3.2, F.4.1

ISSN 0302-9743
ISBN 3-540-61479-6 Springer-Verlag Berlin Heidelberg New York

© Springer-Verlag Berlin Heidelberg 1996
Printed in Germany

Typesetting: Camera-ready by author
SPIN 10512423   06/3142 – 5 4 3 2 1 0   Printed on acid-free paper

Preface

Constraint-based systems are becoming more and more widespread because they allow many different problems to be expressed in a natural way and to be solved efficiently. However, most real-world systems contain inconsistencies and contradictory information. In this situation, many current techniques for constraints give no help to programmers and users. A number of research groups around the world are working to improve matters, using a variety of techniques at different levels of automation. In September 1995 a Workshop on Over-Constrained Systems was held as part of CP'95, the First International Conference on Principles and Practice of Constraint Programming, at Cassis near Marseilles. The aim of the workshop was to bring together researchers using all of these various techniques, and we especially encouraged cross-fertilisation between Constraint Logic Programming, Constraint Satisfaction, and other AI and logic-based techniques.

In addition to a selection of papers from the workshop, we have included some important background papers which have already been published elsewhere. The two theories represented by these papers, Hierarchical Constraint Logic Programming (HCLP) and Partial Constraint Satisfaction Problems (PCSP), are at the heart of most of the rest of the papers in this volume. We also commissioned a paper combining two existing general frameworks within which HCLP, PCSP, and other methods for relaxing constraints can be compared. We hope that the inclusion of these three papers, and also a bibliographic overview of the field, will make this book a very useful resource for all researchers and practitioners in the field of hierarchical, partial, and over-constrained systems.

Out of 24 papers submitted to the workshop, the programme committee selected 15 for full presentation and 6 as posters. Eleven of the 15 have been selected for inclusion in this volume. After the overview and three background papers, the eleven workshop papers are grouped into the following areas: CLP, CSP, and Alternative Paradigms.

Namur, May 1996

Michael Jampel
Workshop Chair

Programme Committee

Alan Borning
University of Washington, USA
borning@cs.washington.edu

Eugene Freuder (Co-Chair)
University of New Hampshire, USA
ecf@cs.unh.edu

Walter Hower
University College Cork, Ireland
walter@cs.ucc.ie

Michael Maher (Co-Chair)
Griffith University, Australia
M.Maher@cit.gu.edu.au

Thomas Schiex
INRA, France
tschiex@toulouse.inra.fr

Hélène Fargier
IRIT, France
fargier@irit.fr

Hans Werner Guesgen
University of Auckland, NZ
hans@cs.auckland.ac.nz

Michael Jampel
City University, UK
jampel@cs.city.ac.uk

Francesca Rossi
University of Pisa, Italy
rossi@di.unipi.it

Acknowledgements

The following two articles appear in the 'Background Papers' section of this book. They are reprinted with permission.

Constraint Hierarchies by Borning, Freeman-Benson, and Wilson originally appeared in the journal *Lisp and Symbolic Computation*, 5:223–270, 1992, and has since been reprinted elsewhere.

Partial Constraint Satisfaction by Freuder and Wallace originally appeared in *Artificial Intelligence*, 58:21–70, 1992, and has also been reprinted elsewhere.

Semiring-Based CSPs and Valued CSPs: Basic Properties and Comparison is printed here for the first time. However, it contains most of the following two papers from *IJCAI'95*:

S. Bistarelli, U. Montanari, and F. Rossi (1995). "Constraint Solving over Semirings." In Proceedings of the 14th International Joint Conference in Artificial Intelligence. Pages 624–630.

T. Schiex, H. Fargier, and G. Verfaillie (1995). "Valued Constraint Satisfaction Problems: Hard and Easy Problems" In Proceedings of the 14th International Joint Conference in Artificial Intelligence. Pages 631–637.

The above two papers are covered by the following copyright notice:

Contents

Alternative Paradigms

A Brief Overview of Over-Constrained Systems

Michael Jampel

Department of Computer Science, City University, Northampton Square, London
EC1V 0HB, U.K. Email `jampel@cs.city.ac.uk`.

Abstract. We survey previous work on resolving over-constrained systems, and previous theories and frameworks for expressing partial, preferential, and hierarchical information. We provide an extensive bibliography.

1 Introduction

In this paper we will attempt to provide a brief overview of the various different approaches to over-constrained systems (OCSs). Some work has been done on this problem directly, and a lot more has been done on what we call 'preference systems', i.e. frameworks and paradigms which allow the expression of partial, preferential, and hierarchical information. Using a preference system to model a problem may mean that over-constrained behaviour does not arise in the first place. Many of the papers in the rest of this book describe preference systems, as opposed to suggesting methods for relaxing OCSs once failure has occurred in a 'flat' (i.e. non-preference) theory.

We will not separate methods for resolving OCSs from general discussion of preference systems. Instead, we will survey the field according to classes of paradigm, such as Constraint Logic Programming, Constraint Satisfaction Problems, and so on.

An *over-constrained system* (OCS) is a set of constraints with no solution, caused by some of the constraints contradicting others. For example, $X < Y$, $Y < Z$, $Z < X$ is an OCS.

There are a number of different approaches to weakening OCSs in order to obtain a solution. One possibility is to change the *structure* of the problem, i.e. the relationship between the constraints. For example, we may decide that $X < Y$ and $Y < Z$ are more important than the constraint $Z < X$, which can therefore be discarded if necessary. This is the philosophy behind, among others, HCLP (Hierarchical Constraint Logic Programming, see below, Sect.2).

Another possibility is to change the *meaning* of the constraints themselves. This relaxation is achieved by adding extra values to the domain of the constraints, all of which are considered equally as candidates for relaxation. In the above example, if a pair (x, z) such that $z > x$ was added to the domain of the third constraint $Z < X$, its meaning would change and it would become consistent with the first two constraints. Optimisation functions can select which relaxations are preferred, perhaps minimising the number of additional variable pairs, or else trying to force the model towards some structure which allows specialised algorithms to be used. This general approach is embodied by PCSP (Partial Constraint Satisfaction Problems, see below, Sect.3).

Various frameworks exist which generalise the HCLP and PCSP paradigms, and which often also include other models, such as fuzzy constraints, as special cases. These are discussed in Sect.4.

There are a number of other paradigms, some of which use ideas originally developed in other parts of AI (Sect.5). They can be divided into those which prevent over-constrained behaviour from arising in the first place, and others which specifically assume that contradictions *will* occur, and which present mechanisms for resolving the resulting OCS. There is also the related problem of *diagnosing* the cause of the OCS; two methods are intelligent backtracking and Chinneck's work on infeasible systems in linear programming, which are discussed in Sect.6.

The bibliography for this paper also includes many references which are not specifically mentioned in the text.

2 Constraint Logic Programming

The work of Borning, Wilson, et al. on Hierarchical Constraint Logic Programming (HCLP) provided a new approach to including preference information in CLP languages; early references include [16, 19, 237], motivated by work on ThingLab [14]. This book includes in the 'Background Papers' section a reprint of [18], which has also been reprinted elsewhere. Algorithmic aspects have been discussed by Borning's students Freeman-Benson [85] and Sannella [192, 193]. More recent HCLP references include Wilson's PhD thesis [236] and an article in the Journal of Logic Programming [238].

HCLP is an extension of CLP where each constraint is given a strength or level of importance modelled by an integer, and 'hard' or *required* constraints are placed at level 0. 'Soft' or optional constraints are placed at levels 1,2,..., where the constraints at level 1, often called *strong*, are considered more important than those at level 2 (*medium*), and so on. Given such a constraint hierarchy, different comparators can be used to compare solutions of the required constraints, to decide which is the 'best'. A feature of the HCLP approach is that a constraint at level i can override arbitrarily many constraints at weaker levels (i.e. at levels $j > i$). Clearly it is also possible to define comparators that do not have this property, as a modification of the HCLP approach.

There is similar work by Satoh and Aiba [198, 199, 200], done in the context of the ICOT Fifth Generation Computer project. However, their approach involves the use of second-order logic, which is undecidable. Admittedly, first-order logic is only semi-decidable, but in the context of constraints that difference appears to be very important. Whatever the reason, Satoh and Aiba's work has not had as much exposure as that of HCLP, especially outside the field of CLP[1]. This might be considered to be a pity, especially as their ideas have been implemented

---

[1] HCLP is treated as the representative of CLP approaches to preference systems in many cross-domain works, including those discussed in Sect.4 below.

in the language CHAL, which fits into the framework of the various other implementations from ICOT. Satoh and Aiba's work [200], already mentioned, also considers the best solutions to constraint hierarchies in terms of circumscription (see below for more on circumscription and non-monotonic logic), as discussed in Jaffar and Maher's survey of CLP [129].

Almost all subsequent work in CLP has been affected by HCLP, usually using its terminology. Often, therefore, general practical issues have been considered in terms of implementing HCLP. Menezes, Barahona and Codognet have defined an incremental compiler for HCLP [163] — incrementality is a key issue for all CLP implementations. Menezes and Barahona have also considered defeasible constraint solving in the context of HCLP, as printed later in this book [162]. De Backer and Beringer have developed an implementation of HCLP with efficiency in mind [53]. There is also work by Wolf [239] which is presented later in this book: he treats the order of constraints in each level of the hierarchy as significant, again for efficiency reasons. Tsutsumi [219] is also concerned with efficiency.

Fages, Fowler and Sola have a different approach: instead of using strength labels they use optimisation functions which are local to each predicate invocation. They show how various HCLP-style problems can be modelled in their formalism [72]. Jampel et al. have attempted various semantic changes, especially concerning compositionality [132, 134, 135, 136], including a paper later in this book [131]. They have also defined transformations between HCLP and PCSP (Partial Constraint Satisfaction Problems) [137] — see below for more on PCSP.

A variety of algorithms have been designed for satisfying constraint hierarchies. One common class is *local propagation algorithms*, in which each constraint has a collection of 'methods' for initially satisfying the constraint, or resatisfying it after the system has been perturbed (e.g., in a graphics environment, by dragging one part of an object). For example, the $a + b = c$ constraint would have three methods to compute values for a, b, and c respectively to satisfy the constraint: $a := c - b$, $b := c - a$, and $c := a + b$. This basic idea can be extended to solve constraint hierarchies, and to produce incremental algorithms. Two examples by Sannella et al. are DeltaBlue [195] and SkyBlue [193]. Hosobe et al. [117], Bouzoubaa, Neveu and Hasle [23], and Borning and Freeman-Benson [17] extend these algorithms to accommodate constraint cycles (simultaneous equations).

Maher and Stuckey [154] combine several approaches for addressing OCSs arising from over-specified queries of CLP programs. They propose the use of objective functions and/or hierarchical constraints to find the 'best' solution to a query. In addition, they propose an almost spreadsheet-like interface which allows the user to make incremental modifications to a query, for example, to delete or weaken constraints in it.

There is quite a lot of work on Fuzzy Logic and CSPs, as discussed in the next section, but less on Fuzzy CLP. However, one exception is Matyska [160].

Optimisation in a CLP context is discussed by Van Hentenryck in his classic book [220]. The semantics of CLP extended with optimisation predicates is considered by Fages [74] and by Marriott and Stuckey [157]. However, optimisation and objective functions are more often associated with PCSP, one of a number of CSP approaches which we consider in the next section.

3 Constraint Satisfaction Problems

Early work on relaxation of CSPs includes the 1976 paper by Rosenfeld et al. [187] which applied fuzzy CSPs to vision, and also Shapiro and Haralick's 1981 paper on maximal satisfiability [213]. However, it is probably true to say that in the field of CSP, the central role of HCLP in CLP approaches is taken by Partial Constraint Satisfaction Problems (PCSP), albeit to a slightly lesser extent. The original PCSP article is by Freuder [89], and the longer journal article, reprinted in this book, is by Freuder and Wallace [91]. Hertzberg et al. address similar issues in [116].

We now provide a very brief characterisation of PCSP: often, a problem is modelled via a CSP, but then no solution is found for it because it is over-constrained. In most cases, a reasonable approach consists in relaxing some of the constraints, thus considering as solutions instantiations of the variables which satisfy only some of the constraints (the non-relaxed ones). Several different optimisation criteria (called 'metrics' in PCSP) may be chosen in order to decide which is the smallest set of constraints to be relaxed in order to find a solution.

Wallace and Freuder have other papers in the area of PCSP later in this book [229, 233], as well as publications elsewhere [228, 230, 231], mainly dealing with one particular PCSP metric called Maximal Satisfiability, or MaxCSP. MaxCSP selects solutions which maximise the total number of constraints satisfied, without trying to distinguish between ones which are more or less important.

Other authors have used PCSP as a basis for their work including Meyer auf'm Hofe and Tschaitschian [164], who also use hierarchical constraints in their scheduling example. Beck's masters thesis about schedule optimisation uses PCSP as an instance of a more general schema [4], and Hower has also worked in this area, especially with direct reference to inconsistent systems [118, 119, 120, 121].

Various other formalisms have been proposed, including Fargier and Lang's Probabilistic CSP [76] and Schiex's Possibilistic CSP [203]. Dendris et al. [61, 62] have developed Mohr and Masini's [167] ideas on Partial Arc Consistency, which is not directly related to PCSP; [62] is printed later in this book. As mentioned above, Rosenfeld et al. [187] did early work in the area of Fuzzy CSP, as have Fargier et al. [77], Ruttkay [188], Bowen et al. [24], and Dubois et al. [65, 66]. Godo and Vila have used fuzzy temporal constraints as a basis for possibilistic reasoning [100]. Martin-Clouaire considers fuzzy linear constraints in the context of CSPs [158], and has also worked on soft (optional) constraints in CSPs [159].

Work on Dynamics CSPs (DCSPs) was not originally motivated by OCSs, but many of the issues are similar. In fact, there are two different uses of the term 'Dynamic CSP' in the literature [204]. The first was proposed by Dechter and Dechter [58] and defines a DCSP as a sequence of CSPs, each instance differing from the previous one by the addition or deletion of one constraint. Many other articles have since been published on this topic, including work by Berlandier and Neveu [6], Bessière [7, 8], Luo et al. [149], Prosser et al. [182], and Schiex and Verfaillie [206, 207, 208, 227]. Van Hentenryck and Le Provost have considered similar issues in the context of CLP [221, 223].

The second theory called DCSP was introduced by Mittal and Falkenhainer [166]. It assumes that all the constraints are present at the start of computation, but that some of them may be 'switched off' depending on the values of certain variables. This is related to the idea of 'hangouts', as worked on by Neveu [175]. Hangouts are wildcard values added to the domains of variables to ensure that a constraint can be satisfied, but this is equivalent to using the value of a variable to switch the constraint off entirely.

This type of DCSP can easily be modelled in as a PCSP in the finite domain case [204]: add a hangout to the domain of each potentially inactive variable, augment all the constraint relations accordingly, to ensure that this new value is supported by all values of all incident constraints, and finally, forbid the hangout value with a soft (optional) unary constraint on each affected variable. DCSP 'activity constraints' can be modelled by binary constraints which forbid the hangout value in order to activate the variable, or which forbid all other values in order to make it inactive. Then define a metric which is appropriate for the problem. For example MaxCSP would minimise the number of hangouts actually used in a solution, i.e. it would maximise the number of active variables used.

Gelle and Smith have applied the second type of DCSP to bridge design, as printed later in this book [96], and Gherida [97] has explored the links between DCSP and PCSP. Jégou's work [139] allows the solution of PCSPs and also both sorts of DCSP; it is based on Binary Decision Diagrams.

4 General Frameworks

Recently two separate frameworks have been proposed which have HCLP, PCSP, and fuzzy, probabilistic, and possibilistic CSPs, as instances. The SCSP theory of Bistarelli et al. [11] is based on semirings, whereas VCSP, the theory developed by Schiex et al. [205], is based on ordered monoids.

Instead of assigning levels of importance to constraints as in HCLP, SCSP assigns a value to each *tuple* of each constraint (i.e. to each collection of values in the domain of the constraint). Because of this, the way constraints are combined becomes more complex (that is, it is not just the join of the relations, as in classical CSPs). In fact, SCSP requires the creation of a new method for combining shorter tuples and their corresponding values into a longer one, and then giving the correct value to the resulting tuple. This is modelled by defining the values assigned to the tuples to be taken from a set which is the domain of a semiring, where the two operations of the semiring ($\times$ and $+$) define how to combine constraints and how to compare the values assigned to different tuples, respectively. Classical CSPs, as well as fuzzy and partial CSPs, can all be cast in this framework.

In VCSP each constraint is assigned a value, which is taken from a monoid. The monoid operator $*$ defines how constraints can be combined. Again, classical, fuzzy, and partial CSPs are all instances of this framework, and a detailed discussion of algorithmic complexity is also provided.

These two theories have been combined and compared in a single paper by Bistarelli et al. [10], printed in the 'Background Papers' section of this book. It shows how it is possible to pass from a particular class of SCSPs to VCSPs and vice versa, without losing any information.

One of the most important contributions of both SCSP and VCSP is to define conditions for the applicability of arc-consistency and k-consistency algorithms. Specifically, the idempotence of the SCSP combining operator $\times$ is a sufficient condition for k-consistency enforcing to work. The work on VCSP suggests the converse is also true: the instances of this framework which have idempotent combining operators (specifically fuzzy and possibilistic CSPs) are the only ones for which efficient arc-consistency algorithms are known. Furthermore, the VCSP papers show that strict monotonicity, a property incompatible with idempotent combining operators, is sufficient to make arc-consistency NP-complete.

Jampel et al. have developed a similar general framework [132, 136], but with an emphasis on software engineering issues rather than the mathematical and algorithmic characterisations of Bistarelli et al.

5   Other Areas of AI

There are a number of paradigms for dealing with incomplete or imprecise information outside the domain of constraints but still in the general area of AI, which ensure that over-constrained behaviour will never arise. Other theories specifically assume that contradictions will occur, and develop mechanisms for resolving the resultant OCS.

One of the best-known examples of the first approach is Zadeh's Fuzzy Logic [242], which has already been mentioned in the context of CSP, above. Similar themes are addressed by Bundy's Incidence Calculus [30]. Fuzzy Logic applications relevant to constraints can be found in Wolfgang Slany's bibliography on fuzzy scheduling [214].

Consider the following statement: "All birds fly." This is true in England but not in the Arctic or Antarctic. This general rule, in conjunction with the fact "Penguins are birds but cannot fly," can be considered an over-constrained system. Among theories which assume that contradictions such as this will occur and try and resolve them is Reiter's Default Logic [185]. This allows specific information to override rules which are nonetheless generally true in the absence of exceptions, i.e. which are true only 'by default'. In the case of the example, default logic would say "All birds can fly, by default" which means "in the absence of a specific indication to the contrary." When penguins are introduced, one of the facts that must be mentioned is a specific contra-indication of this type. Default Logic is used by Sattar et al. later in this book [202].

A collection of papers on non-monotonic logic in general has been edited by Ginsberg [98], and Etherington's journal paper [70] provides a formalisation. There is related work by McCarthy on circumscription [161]. Brewka provides an extended logical framework for default reasoning [26] and has also worked with others on constraint relaxation as a form of non-monotonic reasoning [27], as mentioned above. Other work includes Descotte and Latombe's paper on compromising between antagonist constraints in planning [64].

Ryan's work on belief revision tries to extract as much as possible out of older information, while resolving contradictions with more recently learnt pieces of knowledge [191]. He also considers preferences and priorities in default logic [189, 190]. Ryan's work on belief revision would try to keep the universal statement 'as true as possible' in the presence of the fact about penguins, leading to a similar result to the models of the default logic formulation. We could also model this in, say, HCLP, by having general rules at a weaker level of the hierarchy than specific facts. However, unlike HCLP and PCSP, in belief revision the 'preference' is not based on the desires of the user, but on the changing state of the world.

This book has a section on 'Alternative Paradigms' in which OCSs are solved using various methods from other parts of Artificial Intelligence (AI). As already mentioned, Sattar et al. used default logic [202]. Jussien and Boizumault use ATMSs (Assumption-Based Truth Maintenance Systems) [140], Boltzmann machines are used by Weißschnur et al. [235], and Bouquet and Jégou use Ordered Binary Decision Diagrams [22]. Elsewhere, Brewka et al. [27] investigate the relationship between constraint relaxation and non-monotonic logic (see above for more on non-monotonic logic itself). Bakker et al. [2] apply techniques from model-based diagnosis to an over-constrained scheduling application. The difference between these papers and the others mentioned earlier in this section is that the ones mentioned here all *explicitly* consider the problem of OCSs modelled by constraints. As we have seen the other papers mentioned in this section consider similar problems, as well as the general issue of preference, but not specifically with reference to constraints.

6 Diagnosis of OCSs

In CLP, an OCS arises as a failure of the current execution path. It is the role of 'intelligent backtracking' to avoid other execution paths that will fail for the same reason. Thus diagnosing the reason why the constraint system is over-constrained is central to this technique. Consider the following query in a CLP language with left-to-right execution of goals:

```
?- p(X), q(Y), r(X).
```

If all the different rules for r(X) fail, backtracking occurs. This will cause the choice of a different rule for q(Y). But it is clear that q(Y) cannot be the cause of the failure of r(X), as it concerns a completely different variable. Intelligent backtracking would go straight to the choice of rule for p(X), which is in fact the cause of the problem. We can say that r(X) is 'independent' of q(Y). In terms of an OCS, this shows that any inconsistency arises either due to $r(X)$ or $q(Y)$, but not because of their interaction. This also means that $r(X)$ and $q(Y)$ can be executed in parallel, or in arbitrary order. See the work of Chen, Lassez and Port [33], De Backer and Beringer [54], and the PhD thesis of Kotzamanidis [144]. For more on independence in the context of CLP parallelism see García de la Banda et al. [3].

In linear programming, Chinneck has done some work on what he terms IIS's i.e. infeasible systems of linear equations and inequalities [34, 38, 39]. He relaxes each inequality i by adding a distinct new variable ϵ_i to it, and replaces the original optimisation function with one which minimises the sum of the ϵ's. Any non-zero ϵ in the answer indicates one member of the minimal infeasible subset. If that ϵ is then removed and the system solved again, another member of the infeasible subset is identified. When all the inequalities have been enforced in this manner, the next attempt fails. (In fact his method is slightly more complicated, in order to cover cases when there is more than one independently inconsistent subset, and in order to include equalities.) This work is interesting, and the ϵ's have some similarities with the notion of hangouts (see above), however it is computationally expensive — if the system as a whole contains n constraints and one of the minimal infeasible subsets contains s of them, it is necessary to solve $s + 1$ different problems each of size n. Solving one of these problems is hard enough, and yet the worst case is when $n = s$ and so one must solve $n + 1$ problems of size n. Chinneck addresses this issue in [35], and also discusses implementations [36, 37, 40].

Other work on infeasible linear programs has been published by Greenberg [103, 104, 105, 106, 107]. Although it is not clear to us how it might be used by the constraint programming community, it also appears to be relevant to the problem of diagnosis.

7 Conclusion

We have provided a brief survey of previous work on resolving over-constrained systems, and previous theories and frameworks for expressing partial, preferential, and hierarchical information. We hope that we have shown where the papers in the rest of this book fit into this work, and provided useful starting points for anyone wishing to enter this interesting and highly relevant field.

Note that the following bibliography includes many papers which have not been mentioned in the above text.

Acknowledgements

Thanks especially to Alan Borning, Michael Maher, Francesca Rossi, Thomas Schiex, Rick Wallace, and also the other members of the workshop Program Committee, for many helpful suggestions and comments.

References

1. E. Aarts and J. Korst. *Simulated Annealing and Boltzmann Machines.* John Wiley & Sons, Cichester, England, 1989.
2. R. R. Bakker, F. Dikker, F. Templeman, and P. M. Wognum. Diagnosing and Solving Over-Determined Constraint Satisfaction Problems. In *IJCAI'93: Proceedings International Joint Conference on Artificial Intelligence*, 1993.
3. M. García de la Banda, M. Hermenegildo, and K. Marriott. Independence in Constraint Logic Programs. In *ILPS'93: Proceedings 3rd International Logic Programming Symposium*, pages 130–146, Vancouver, 1993.
4. J. C. Beck. A Schema for Constraint Relaxation with Instantiations for Partial Constraint Satisfaction and Schedule Optimization. Master's thesis, University of Toronto, Canada, 1994.
5. F. Benhamou, D. Mc Allester, and P. Van Hentenryck. CLP(Intervals) revisited. In *ILPS'94: Proceedings 4rd International Logic Programming Symposium*, 1994.
6. Pierre Berlandier and Bertrand Neveu. Arc-Consistency for Dynamic Constraint Problems: A RMS-Free Approach. In Thomas Schiex and Christian Bessière, editors, *Proceedings ECAI'94 Workshop on Constraint Satisfaction Issues raised by Practical Applications*, Amsterdam, August 1994.
7. Christian Bessière. Arc Consistency in Dynamic Constraint Satisfaction Problems. In *Proceedings AAAI'91*, 1991.
8. Christian Bessière. Arc Consistency for Non-Binary Dynamic CSPs. In *ECAI'92*, 1992.
9. Stefano Bistarelli. Programmazione con vincoli pesati e ottimizzazione (in Italian). Dipartimento di Informatica, Università di Pisa, Italy, 1994.
10. Stefano Bistarelli, Hélène Fargier, Ugo Montanari, Francesca Rossi, Thomas Schiex, and Gerard Verfaillie. Semiring-based CSPs and Valued CSPs: Basic Properties and Comparison. In [133].
11. Stefano Bistarelli, Ugo Montanari, and Francesca Rossi. Constraint Solving over Semirings. In Chris Mellish, editor, *IJCAI'95: Proceedings International Joint Conference on Artificial Intelligence*, Montreal, Morgan Kaufman, August 1995.
12. Karl-Friedrich Böhringer. Using Constraints to Achieve Stability in Automatic Graph Layout Algorithms. In *CHI'90 Conference Proceedings*, pages 43–52, Seattle, Washington, April 1990. ACM SIGCHI.
13. Patrice Boizumault, Christelle Guéret, and Narendra Jussien. Efficient Labeling and Constraint Relaxation for Solving Time Tabling Problems. In Pierre Lim and Jean Jourdan, editors, *Proceedings of the 1994 ILPS Post-conference Workshop on Constraint Languages/Systems and Their Use in Problem Modeling : Volume 1 (Applications and Modelling)*, Technical Report ECRC-94-38, ECRC, Munich, Germany, November 1994.
14. Alan Borning. The Programming Language Aspects of ThingLab, A Constraint-Oriented Simulation Laboratory. *ACM Transactions on Programming Languages and Systems*, 3(4):353–387, October 1981.
15. Alan Borning and Robert Duisberg. Constraint-based Tools for Building User Interfaces. *ACM Transactions on Graphics*, 5(4):345–374, October 1986.
16. Alan Borning, Robert Duisberg, Bjorn Freeman-Benson, Axel Kramer, and Michael Woolf. Constraint Hierarchies. In *OOPSLA'87: Proceedings of the 1987 ACM Conference on Object-Oriented Programming Systems, Languages, and Applications*, pages 48–60. ACM, October 1987.

17. Alan Borning and Bjorn Freeman-Benson. The OTI Constraint Solver: A Constraint Library for Constructing Interactive Graphical User Interfaces. In Ugo Montanari and Francesca Rossi, editors, *CP'95: Proceedings of the 1st International Conference on Principles and Practice of Constraint Programming 1995*, LNCS 976. Springer, 1995.

18. Alan Borning, Bjorn Freeman-Benson, and Molly Wilson. Constraint Hierarchies. *Lisp and Symbolic Computation*, 5:223–270, 1992. (Reprinted in B. Mayoh, E. Tyugu, and J.Penjaam, editors, *Constraint Programming: Proceedings 1993 NATO ASI Parnu, Estonia*, NATO Advanced Science Institute Series, pages 80–122. Springer, 1994. Also reprinted in [133].)

19. Alan Borning, Michael Maher, Amy Martindale, and Molly Wilson. Constraint Hierarchies and Logic Programming. In Giorgio Levi and Maurizio Martelli, editors, *ICLP'89: Proceedings 6th International Conference on Logic Programming*, pages 149–164, Lisbon, Portugal, June 1989. MIT Press.

20. Annalisa Bossi, Maurizio Gabbrielli, Giorgio Levi, and Maria Chiara Meo. An Or-Compositional Semantics for Logic Programs. In J.-M. Jacquet, editor, *Constructing Logic Programs*, chapter 10, pages 215–240. Wiley, 1993.

21. Fabrice Bouquet and Philippe Jégou. Knowledge Base Revision Using Dynamic Weighted OBDDs. Technical report, Université de Provence, 1995.

22. Fabrice Bouquet and Philippe Jégou. Solving Over-Constrained CSP Using Weighted OBDD. In [133].

23. Mouhssine Bouzoubaa, Bertrand Neveu, and Geir Hasle. Houria: A Solver for Equational Constraints in a Hierarchical System. In Michael Jampel, Eugene Freuder, and Michael Maher, editors, *OCS'95: Workshop Notes, Workshop on Over-Constrained Systems at CP'95*, Cassis, Marseilles, 18 September 1995.

24. James Bowen, Robert Lai, and Dennis Bahler. Lexical Imprecision in Fuzzy Constraint Networks. In *AAAI'92*, San Jose, July 1992.

25. S. Breitinger and H. C. R. Lock. Using Constraint Logic Programming for Industrial Scheduling Problems. In C. Beierle and L. Plümer, editors, *Logic Programming: Formal Methods and Practical Applications*, chapter 9, pages 273–299. Elsevier, 1995.

26. Gerhard Brewka. Preferred Subtheories: An Extended Logical Framework for Default Reasoning. In *IJCAI'89: Proceedings 11th International Joint Conference on Artificial Intelligence*, pages 1043–1048, August 1989.

27. Gerhard Brewka, Hans Werner Guesgen, and Joachim Hertzberg. Constraint Relaxation and Nonmonotonic Reasoning. Technical Report TR-92-002, ICSI, Berkeley, 1992.

28. E. Bryant. Graph-based Algorithhms for Boolean Function Manipulation. *IEEE Transactions on computers*, C-35:677–691, 1986.

29. E. Bryant. Symbolic Boolean Manipulation with Ordered BDD. *ACM Computing Surveys, Vol 24 No.3*, September 1992.

30. Alan Bundy. Incidence Calculus: A Mechanism for Probabilistic Reasoning. *Journal of Automated Reasoning*, 1:263–283, 1985.

31. C. A. Carter and W. R. LaLonde. The Design of a Program Editor Based on Constraints. Technical Report CS TR 50, Carleton University, May 1984.

32. C. Cayrol, M.C. Lagasquie-Schiex, and T. Schiex. Non-monotonic Reasoning: From Complexity to Algorithms. *4th International Symposium on Artificial Intelligence and Mathematics, Fort Lauderdale, USA*, 1996.

33. T.Y. Chen, J-L. Lassez, and G. Port. Maximal Unifiable Subsets and Minimal Non-unifiable Subsets. *New Generation Computing*, 4:133–152, 1986.

34. John Chinneck. Finding Minimal Infeasible Sets of Constraints in Infeasible Mathematical Programs. Technical Report SCE-93-01, Department of Systems and Computer Engineering, Carleton University, Ottawa, 1993.

35. John Chinneck. Finding the Most Useful Subset of Constraints for Analysis in an Infeasible Linear Program. Technical Report SCE-93-07, Department of Systems and Computer Engineering, Carleton University, Ottawa, 1993.

36. John Chinneck. Computer Codes for the Analysis of Infeasible Linear Programs. Technical Report SCE-94-21, Department of Systems and Computer Engineering, Carleton University, Ottawa, 1994.

37. John Chinneck. MINOS(IIS): Infeasibility Analysis Using MINOS. *Computers and Operations Research*, 21(1):1–9, 1994.

38. John Chinneck. Analyzing Infeasible Nonlinear Programs. *Computational Optimization and Applications*, 1995.

39. John Chinneck and Erik Dravnieks. Locating Minimal Infeasible Constraint Sets in Linear Programs. *ORSA Journal on Computing*, 3(2):157–168, Spring 1991.

40. John Chinneck and M. A. Saunders. MINOS(IIS) Version 4.2: Analyzing Infeasibilities in Linear Programs. *European Journal of Operational Research*, 1994. (Technical Note).

41. Philippe Codognet and Daniel Diaz. Boolean Constraint Solving Using clp(FD). In *ILPS'93: Proceedings 3rd International Logic Programming Symposium*, pages 525–539, Vancouver, 1993.

42. Ellis S. Cohen, Edward T. Smith, and Lee A. Iverson. Constraint-Based Tiled Windows. *IEEE Computer Graphics and Applications*, pages 35–45, May 1986.

43. Jacques Cohen. Constraint Logic Programming Languages. *Communications of the ACM*, 33(7):52–68, July 1990.

44. Alain Colmerauer. An Introduction to Prolog III. *Communications of the ACM*, 33(7):69–90, July 1990.

45. Alain Colmerauer. Prolog II Reference Manual and Theoretical Model. Technical report, Groupe Intelligence Artificielle, Université Aix – Marseille II, October 1982.

46. M. Cooper. *Visual Occlusion and the Interpretation of Ambiguous Pictures*. Ellis Horwood, Chicester, 1992.

47. H. Cormen, C. Leiserson, and R. Rivest. *Introduction to Algorithms*. MIT Press-McGraw-Hill, 1991.

48. M. Corsini and A. Rauzy. Toupie User's Manual. Technical report, LABRI, Université de Bordeaux I, France, 1993.

49. O. Coudert and J.C. Madre. A Logically Complete Reasonning Maintenance System Based on a Logical Constraint Solver. *IJCAI91*, I:294–299, 1991.

50. Xavier Cousin. Meilleures Solutions en Programmation Logique. In *Proceedings Avignon'91*, 1991. In French.

51. B. A. Davey and H. A. Priestley. *Introduction to Lattices and Order*. Cambridge University Press, 1990.

52. E. Davis. Constraint Propagation with Interval Labels. *Artificial Intelligence*, 32:281–331, 1987.

53. Bruno De Backer and Henri Beringer. An Efficient Solver for HCLP($\mathcal{R}$). Technical report, IBM France Scientific Center, 1991. (ICLP'91, Pre-Conference Workshop on CLP).

54. Bruno De Backer and Henri Beringer. Intelligent Backtracking for CLP Languages, An Application to CLP($\mathcal{R}$). In *ILPS'91: International Logic Programming Symposium*, pages 405–419, 1991.

55. T.L. Dean and M. Boddy. An Analysis of Time-dependent Planning. In *Proc. AAAI-88*, pages 49–54, St. Paul, Minnesota, 1988.

56. Romuald Debruyne. DnAc6. Research Report 94-054, Laboratoire d'Informatique, de Robotique et de Micro-électronique de Montpellier, 1994. In French.

57. Rina Dechter. Enhancement schemes for constraint processing: backjumping, learning, and cutset decomposition. *Artif. Intell.*, 41:273–312, 1990.

58. Rina Dechter and Avi Dechter. Belief Maintenance in Dynamic Constraint Networks. In *AAAI'88*, Saint Paul, USA, pages 37–42, 1988.

59. Rina Dechter, A. Dechter, and Judea Pearl. Optimization in Constraint Networks. In R.M Oliver and J.Q. Smith, editors, *Influence Diagrams, Belief Nets and Decision Analysis*, chapter 18, pages 411–425. John Wiley & Sons Ltd., 1990.

60. Rina Dechter and Judea Pearl. Tree Clustering for Constraint Networks. *Artificial Intelligence*, 38:353–366, 1989.

61. N. Dendris, L. Kirousis, Y. Stamatiou, and D. Thilikos. Partiality and Approximation Schemes for Local Consistency in Networks of Constraints. In *FSTTCS'95: Proceedings of the 15th Conf. on Foundations of Software Technology and Theoretical Computer Science*, LNCS. Springer, December 1995.

62. N. Dendris, L. Kirousis, Y. Stamatiou, and D. Thilikos. Partial Arc Consistency. In [133].

63. Daniel Dennett. Cognitive Wheels: The Frame Problem of AI. In C. Hookway, editor, *Minds, Machines, and Evolution: Philosophical Studies*, pages 129–151. Cambridge U.P., 1984.

64. Yannick Descotte and Jean-Claude Latombe. Making Compromises Among Antagonist Constraints in a Planner. *Artificial Intelligence*, 27:183–217, 1985.

65. Didier Dubois, Hélène Fargier, and Henri Prade. The Calculus of Fuzzy Restrictions as a Basis for Flexible Constraint Satisfaction. In *Proceedings 2nd IEEE Conference on Fuzzy Sets*, San Francisco, March 1993.

66. Didier Dubois, Hélène Fargier, and Henri Prade. Propagation and Satisfaction of Flexible Constraints. In R. R. Yager and L. A. Zadeh, editors, *Fuzzy Sets, Neural Networks and Soft Computing*. Kluwer Academic Press, 1993. (Also available as Rapport IRIT/92-59-R, IRIT).

67. Didier Dubois and Henri Prade. *Fuzzy Sets and Systems: Theory and Applications*. Academic Press, New York, 1980.

68. Didier Dubois and Henri Prade. A Class of Fuzzy Measures Based on Triangular Norms: A General Framework for the Combination of Uncertain Information. *Int. Journal of Intelligent Systems*, 8(1):43–61, 1982.

69. M. Anton Ertl and Andreas Krall. Optimal Instruction Scheduling Using Constraint Logic Programming. In J. Maluszyński and M. Wirsing, editors, *PLILP'91: Proceedings 3rd International Symposium on Programming Language Implementation and Logic Programming*, LNCS 528, Passau, Germany, August 1991. Springer.

70. David W. Etherington. Formalizing Nonmonotonic Reasoning Systems. *Artificial Intelligence*, 31(1):41–85, January 1987.

71. O. Evans. Factory Scheduling Using Finite Domains. In G. Comyn, N. E. Fuchs, and M. J. Ratcliffe, editors, *Logic Programming in Action*, LNCS 636, pages 45–53. Springer, 1992.

72. François Fages, Julian Fowler, and Thierry Sola. Handling Preferences in Constraint Logic Programming with Relational Optimization. In *PLILP'94*, Madrid, September 1994.

73. François Fages, Julian Fowler, and Thierry Sola. A Reactive Constraint Logic Programming Scheme. In *ICLP'95: International Conference on Logic Programming*, Tokyo, 1995.

74: François Fages. On the Semantics of Optimization Predicates in CLP Languages. In *FSTTCS'93: Proceedings 13th Conf. on Foundations of Software Technology and Theoretical Computer Science*, 1993.

75. Boi Faltings. Arc-consistency for Continuous Variables. *Artificial Intelligence*, 65:363–376, 1994.

76. Hélène Fargier and Jérôme Lang. Uncertainty in Constraint Satisfaction Problems: A Probabilistic Approach. In *ESQARU'93*, Grenada. LNCS 747. Springer, November 1993.

77. Hélène Fargier, Jérôme Lang, and Thomas Schiex. Selecting Preferred Solutions in Fuzzy Constraint Satisfaction Problems. In *EUFIT'93: Proceedings of the First European Congress on Fuzzy and Intelligent Technologies*, Germany, 1993.

78. R. Feldman and M. Golumbic. Optimization algorithms for student scheduling via constraint satisfiability. *Comput. J.*, 33:356–364, 1990.

79. Mark Fox. *Constraint-Directed Search: A Case Study of Job-Shop Scheduling.* Morgan Kaufmann, Los Altos, California, 1987.

80. Bjorn Freeman-Benson. Multiple Solutions from Constraint Hierarchies. Technical Report 88-04-02, University of Washington, Seattle, WA, April 1988.

81. Bjorn Freeman-Benson. Constraint Imperative Programming. Technical Report 91-07-02, University of Washington, Seattle, July 1991. (PhD Dissertation).

82. Bjorn Freeman-Benson and Alan Borning. The Design and Implementation of Kaleidoscope'90: A Constraint Imperative Programming Language. In *Proceedings IEEE Computer Society International Conference on Computer Languages*, pages 174–180, Oakland, April 1992.

83. Bjorn Freeman-Benson and John Maloney. The DeltaBlue Algorithm: An Incremental Constraint Hierarchy Solver. In *Proceedings of the Eighth Annual IEEE Phoenix Conference on Computers and Communications*, Scottsdale, Arizona, March 1989. IEEE.

84. Bjorn Freeman-Benson, John Maloney, and Alan Borning. An Incremental Constraint Solver. *Communications of the ACM*, 33(1):54–63, January 1990.

85. Bjorn Freeman-Benson, Molly Wilson, and Alan Borning. DeltaStar: A General Algorithm for Incremental Satisfaction of Constraint Hierarchies. In *Proceedings 11th IEEE Phoenix Conference on Computers and Communications*, pages 561–568, Scottsdale, Arizona, April 1992.

86. Eugene Freuder. Synthesizing Constraint Expressions. *CACM*, 21(11), 1978.

87. Eugene Freuder. A Sufficient Condition for Backtrack-Free Search. *Journal of the ACM*, 29(1):24–32, January 1982.

88. Eugene Freuder. Backtrack-free and Backtrack-bounded Search. In Kanal and Kumar, editors, *Search in Artificial Intelligence*. Springer-Verlag, 1988.

89. Eugene Freuder. Partial Constraint Satisfaction. In *IJCAI'89: Proceedings 11th International Joint Conference on Artificial Intelligence*, pages 278–283, August 1989.

90. Eugene Freuder. Exploiting Structure in Constraint Satisfaction Problems. In B. Mayoh, E. Tyugu, and J.Penjaam, editors, *Constraint Programming: Proceedings 1993 NATO ASI Parnu, Estonia*, NATO Advanced Science Institute Series, pages 54–79. Springer, 1994.

91. Eugene Freuder and Richard Wallace. Partial Constraint Satisfaction. *Artificial Intelligence*, 58:21–70, 1992. (Reprinted in Eugene Freuder and Alan Mackworth,

editors, *Constraint-Based Reasoning*, MIT Press, 1994. Also reprinted in [133].)

92. D. Frost and R. Dechter. Dead-End Driven Learning. In *Proceedings AAAI-94*, pages 294–300, 1994.

93. Thom Frühwirth, Alexander Herold, Volker Küchenhoff, Thierry Le Provost, Pierre Lim, Eric Monfroy, and Mark Wallace. Constraint Logic Programming: An Informal Introduction. In G. Comyn, N. E. Fuchs, and M. J. Ratcliffe, editors, *Logic Programming in Action*, LNCS 636, pages 3–35. Springer, 1992. (Also available as Technical Report ECRC-93-5).

94. Michel Gangnet and Burton Rosenberg. Constraint Programming and Graph Algorithms. In *2nd International Symposium on Artificial Intelligence and Mathematics*, January 1992.

95. M. Garey, D. Johnson, and L. Stockmeyer. Some Simplified NP-complete Graph Problems. *Theoretical Computer Science*, 1:237–267, 1976.

96. Esther Gelle and Ian Smith. Dynamic Constraint Satisfaction with Conflict Management in Design. In [133].

97. Khaled Ghedira. Dynamic Partial Constraint Satisfaction by Multi-Agent and Simulated Annealing. In Thomas Schiex and Christian Bessiére, editors, *Proceedings ECAI'94 Workshop on Constraint Satisfaction Issues raised by Practical Applications*, Amsterdam, August 1994.

98. Matthew Ginsberg, editor. *Readings in Nonmonotonic Reasoning*. Morgan Kaufmann, Los Altos, California, 1987.

99. F. Glover. Tabu Search: A Tutorial. *Interfaces*, 20:74–94, 1990.

100. Lluis Godo and Lluis Vila. Possibilistic Temporal Reasoning Based on Fuzzy Temporal Constraints. In Chris Mellish, editor, *IJCAI'95: Proceedings International Joint Conference on Artificial Intelligence*, Montreal, August 1995.

101. H.-J. Goltz. Reducing Domains for Search in CLP(FD) and its Application to Job-shop Scheduling. In Ugo Montanari and Francesca Rossi, editors, *CP'95: Proceedings of the First International Conference on Principles and Practice of Constraint Programming 1995*, LNCS 976, pages 549–562. Springer, 1995.

102. James A. Gosling. *Algebraic Constraints*. PhD thesis, Carnegie-Mellon University, May 1983. Published as CMU Computer Science Department Technical Report CMU-CS-83-132.

103. Harvey J. Greenberg. Diagnosing Infeasibility in Min-Cost Network Flow Problems Part I: Dual Infeasibility. *IMA Journal of Mathematics Applied in Business and Industry*, 1:99–109, 1987.

104. Harvey J. Greenberg. Diagnosing Infeasibility in Min-Cost Network Flow Problems Part II: Primal Infeasibility. *IMA Journal of Mathematics Applied in Business and Industry*, 2:39–50, 1988/89.

105. Harvey J. Greenberg. An Empirical Analysis of Infeasibility Diagnosis for Instances of Linear Programming Blending Models. *IMA Journal of Mathematics Applied in Business and Industry*, 4:163–210, 1992.

106. Harvey J. Greenberg. How to Analyze the Results of Linear Programs Part 3: Infeasibility Diagnosis. Technical report, Mathematics Department, University of Colorado at Denver, February 1993. (Revised from December 1992 and January 1993).

107. Harvey J. Greenberg and Frederic H. Murphy. Approaches to Diagnosing Infeasible Linear Programs. *ORSA Journal on Computing*, 3(3):253–261, Summer 1991.

108. H.W. Guesgen. A Formal Framework for Weak Constraint Satisfaction Based on Fuzzy Sets. In *Proc. ANZIIS-94*, pages 199–203, Brisbane, Australia, 1994.

109. H.W. Guesgen and J. Hertzberg. *A Perspective of Constraint-Based Reasoning.* LNAI 597. Springer, Berlin, Germany, 1992.
110. H.W. Guesgen and J. Hertzberg. A Constraint-based Approach to Spatiotemporal Reasoning. *Applied Intelligence (Special Issue on Applications of Temporal Models)*, 3:71–90, 1993.
111. Paul Halmos. *Naive Set Theory.* Springer, 1974.
112. Djamila Haroud, S. Boulanger, E. Gelle, and I. Smith. Strategies for Conflict Management in Preliminary Engineering Design. *AIEDAM Special Issue on Conflict Management in Design*, 1995, pages 313–323.
113. Djamila Haroud and Boi Faltings. Global Consistency for Continuous Constraints. *Proc. of the ECAI-94*, 1994.
114. Alois Haselböck, Thomas Havelka, and Markus Stumptner. Revising Inconsistent Variable Assignments in Constraint Satisfaction Problems. In Manfred Meyer, editor, *Constraint Processing: Proceedings of the International Workshop at CSAM'93, St. Petersburg, July 1993*, Research Report RR-93-39, pages 113–122, DFKI Kaiserslautern, August 1993.
115. Nevin C. Heintze, Spiro Michaylov, Peter J. Stuckey, and Roland H.C. Yap. *The CLP(R) Programmer's Manual, Version 1.2.* IBM Thomas J. Watson Research Center, PO Box 704, Yorktown Heights, NY 10598, U.S.A., September 1992.
116. J. Hertzberg, H.W. Guesgen, A. Voß, M. Fidelak, and H. Voß. Relaxing Constraint Networks to Resolve Inconsistencies. In *GWAI'88*, pages 61–65, Eringerfeld, Germany, 1988.
117. H. Hosobe, K. Miyashita, S. Takahashi, S. Matuoka, and A. Yonezawa. Locally Simultaneous Constraint Satisfaction. In Alan Borning, editor, *Principles and Practice of Constraint Programming*, LNCS 874. Springer, May 1994. (PPCP'94: Second International Workshop, Orcas Island, Seattle, USA).
118. Walter Hower. Proper Constraint Relaxation. In João Martins and Ernesto Morgado, editors, *EPIA'89: 4th Portuguese Conference on Artificial Intelligence*, volume 2, Lisbon, September 1989.
119. Walter Hower. The Relaxation of Unsolvable CSPs: General Problem Formulation and Specific Illustration in the Scheduling Domain. In *Workshop on Constraint Processing at IJCAI'89*, page 154, Detroit, aug 1989. (Abstract).
120. Walter Hower. Sensitive Relaxation of an Overspecified Constraint Network. In *ISAI'89: 2nd International Symposium on Artificial Intelligence*, Monterrey, Mexico, October 1989.
121. Walter Hower. On Conflict Resolution in Inconsistent Constraint Networks. Forschungsbericht AIDA-90-02, Fachgebiet Intellektik, Fachbereich Informatik, Technische Hochschule Darmstadt, April 1990.
122. S. Huard and E. Freuder. A debugging assistant for incompletely specified constraint network knowledge bases. *International Journal of Expert Systems*, 6(4):419–446, JAI Press, 1993.
123. Paul Hubbe and Eugene Freuder. An efficient cross-product representation of the constraint satisfaction problem search space. In *AAAI'92*, pages 421–427, San Jose, CA, 1992.
124. Tien Huynh and Catherine Lassez. A CLP(R) Options Trading Analysis System. In Robert A. Kowalski and Kenneth A. Bowen, editors, *JICSLP'88: Proceedings 5th International Conference and Symposium on Logic Programming*, pages 59–69, Seattle, Washington, U.S.A., 1988. MIT Press.
125. James Ignizio. Generalized Goal Programming. *Computers and Operations Research*, 10(4):277–290, 1983.

126. James Ignizio. *Introduction to Linear Goal Programming.* Sage Publications, Beverly Hills, 1985. Sage University Paper Series on Qualitative Applications in the Social Sciences, 07-056.

127. Joxan Jaffar and Jean-Louis Lassez. Constraint Logic Programming. Technical Report 86/74, Monash University, Victoria, Australia, June 1986.

128. Joxan Jaffar and Jean-Louis Lassez. Constraint Logic Programming. In *POPL'87: Proceedings 14th ACM Symposium on Principles of Programming Languages*, pages 111–119, Munich, 1987. ACM.

129. Joxan Jaffar and Michael Maher. Constraint Logic Programming: A Survey. *Journal of Logic Programming*, 19/20:503–581, 1994.

130. Joxan Jaffar, Spiro Michaylov, Peter Stuckey, and Roland Yap. The CLP($\mathcal{R}$) Language and System. *TOPLAS: ACM Transactions on Programming Languages and Systems*, 14(3):339–395, July 1992.

131. Michael Jampel. A Compositional Theory of Constraint Hierarchies (Operational Semantics). In [133].

132. Michael Jampel. *Over-Constrained Systems.* PhD thesis, Department of Computer Science, City University, 1996. (To be submitted.)

133. Michael Jampel, Eugene Freuder, and Michael Maher, editors. *Over-Constrained Systems*, LNCS. Springer, August 1996.

134. Michael Jampel and David Gilbert. Fair Hierarchical Constraint Logic Programming. In Manfred Meyer, editor, *Proceedings ECAI'94 Workshop on Constraint Processing*, Amsterdam, August 1994.

135. Michael Jampel and Sebastian Hunt. Composition in Hierarchical CLP. In Chris Mellish, editor, *IJCAI'95: Proceedings International Joint Conference on Artificial Intelligence*, Montreal, August 1995.

136. Michael Jampel, Jean-Marie Jacquet, and David Gilbert. A General Framework for Integrating HCLP and PCSP. In Walter Hower and Zsofi Ruttkay, editors, *ECAI'96 Workshop on Non-Standard Constraint Processing*, Budapest, August 1996.

137. Michael Jampel, Jean-Marie Jacquet, David Gilbert, and Sebastian Hunt. Transformations between HCLP and PCSP. In Eugene Freuder, editor, *CP'96: Proceedings of the 2nd International Conference on Principles and Practice of Constraint Programming.* Springer, August 1996.

138. P. Janssen, P. Jégou, B. Nougier, M.C. Vilarem, and B. Castro. SYNTHIA : Assisted Design of Peptide Synthesis Plans. *New Journal of Chemistry*, 14-12:969–976, 1990.

139. Philippe Jégou. A Logical Approach to Solve Dynamic CSPs: Preliminary Report. In Thomas Schiex and Christian Bessiére, editors, *Proceedings ECAI'94 Workshop on Constraint Satisfaction Issues raised by Practical Applications*, Amsterdam, August 1994.

140. Narendra Jussien and Patrice Boizumault. Implementing Constraint Relaxation over Finite Domains using ATMS. In [133].

141. Johan de Kleer. An Assumption-Based TMS. *Artificial Intelligence*, 28:127–162, 1986.

142. Johan de Kleer. A Comparison of ATMS and CSP Techniques. In *IJCAI-89: Proceedings 11th International Joint Conference on Artificial Intelligence*, pages 290–296, Detroit, 1989.

143. M. Kosaka and H. Mizuno *et. al.*. Applications of Fuzzy Logic/Neural Networks to Securities Trading Decision Support Systems. *IEEE Transactions*, 1991. (Systems Development Laboratory, Hitachi Ltd, Japan).

144. Antonis Kotzamanidis. *Intelligent Backtracking in Logic Programming with Constraints over the Real Numbers.* PhD thesis, Department of Computing, Imperial College, London, October 1995.

145. M. Krentel. The complexity of optimization problems. *Journal of Computer and System Sciences,* 36:490–509, 1988.

146. V. Kumar. Algorithms for constraint satisfaction problems: a survey. *AI Magazine,* 13(1), 1992.

147. M. Lacroix and P. Lavency. Preferences: Putting more knowledge into queries. In *Proceedings 15th International Conference on Very Large Databases,* pages 217–225, Brighton, 1987.

148. D. Lehmann. Another Perspective on Default Reasoning. Technical report, Leibniz Center for Research in Computer Science. Hebrew University of Jerusalem. Israel, 1992.

149. Qiangyi Luo, Peter Hendry, and Iain Buchanan. A New Algorithm for Dynamic Distributed Constraint Satisfaction Problems. Technical Report KEG-4-92, University of Strathclyde, February 1992.

150. Alan Mackworth. Consistency in Networks of Relations. *Artificial Intelligence,* 8(1):99–118, 1977.

151. Alan Mackworth. *Encyclopedia of AI,* chapter Constraint Satisfaction, pages 205–211. Springer Verlag, 1988.

152. Alan Mackworth and Eugene Freuder. The Complexity of Some Polynomial Network Consistency Algorithms for Constraint Satisfaction Problems. *Artificial Intelligence,* 25:65–73, 1985.

153. Alan Mackworth and Eugene Freuder. The Complexity of Constraint Satisfaction Revisited. *Artificial Intelligence,* 59(1–2):57–62, February 1993. Special Volume on Artificial Intelligence in Perspective.

154. Michael Maher and Peter Stuckey. Expanding Query Power in Constraint Logic Programming Languages. In Ewing Lusk and Ross Overbeek, editors, *NACLP'89: Proceedings North American Conference on Logic Programming,* pages 20–37, Cleveland, Ohio, October 1989.

155. John Maloney. *Using Constraints for User Interface Construction.* PhD thesis, Department of Computer Science and Engineering, University of Washington, August 1991. Published as Department of Computer Science and Engineering Technical Report 91-08-12.

156. V.W. Marek and M. Truszczynski. *Nonmonotonic Logic* Springer-Verlag, 1993.

157. Kim Marriott and Peter Stuckey. Semantics of CLP Programs with Optimization. *ACM TOPLAS,* 1993.

158. R. Martin-Clouaire. CSP Techniques with Fuzzy Linear Contraints: Practical Issues. In *EUFIT'93: Proceedings of the First European Congress on Fuzzy and Intelligent Technologies,* Aachen, September 1993.

159. R. Martin-Clouaire. Dealing with Soft Constraints in a Constraint Satisfaction Problem. In *Proceedings of the International Conference on Information Processing of Uncertainty in Knowledge based Systems,* pages 37–40, Palma de Mallorca, July 1992.

160. Luděk Matyska. Logic Programming with Fuzzy Sets. Technical Report TCU/CS/1993/4, Department of Computer Science, City University, London, December 1993.

161. John McCarthy. Circumscription — A Form of Non-Monotonic Reasoning. *Artificial Intelligence,* 13(1,2):27–39, April 1980.

162. Francisco Menezes and Pedro Barahona. Defeasible Constraint Solving. In [133].

163. Francisco Menezes, Pedro Barahona, and Philippe Codognet. An Incremental Hierarchical Constraint Solver. In Paris Kanellakis, Jean-Louis Lassez, and Vijay Saraswat, editors, *PPCP'93: First Workshop on Principles and Practice of Constraint Programming*, Providence RI, 1993.

164. Harald Meyer auf'm Hofe and Bidjan Tschaitschian. PCSPs with Hierarchical Constraint Orderings in Real World Scheduling Applications. In Michael Jampel, Eugene Freuder, and Michael Maher, editors, *OCS'95: Workshop Notes, Workshop on Over-Constrained Systems at CP'95*, Cassis, Marseilles, 18 September 1995.

165. S. Minton, M. D. Johnston, A. B. Philips, and P. Laird. Minimizing Conflicts: A Heuristic Repair Method for Constraint Satisfaction and Scheduling Problems. *Artificial Intelligence*, 58:161–205, 1992.

166. Sanjay Mittal and Brian Falkenhainer. Dynamic Constraint Satisfaction Problems. In *AAAI'90*, pages 25–32, Boston, 1990.

167. R. Mohr and G. Masini. Good Old Discrete Relaxation. In *ECAI'88*, pages 651–656, Munich, 1988.

168. Ugo Montanari. Networks of Constraints: Fundamental Properties and Applications to Picture Processing. *Information Science*, 7(2):95–132, 1974.

169. Ugo Montanari and Francesca Rossi. Constraint Relaxation May Be Perfect. *AI Journal*, 48:143–170, 1991.

170. Ugo Montanari and Francesca Rossi. Constraint Satisfaction, Constraint Programming and Concurrency. In Paris Kanellakis, Jean-Louis Lassez, and Vijay Saraswat, editors, *PPCP'93: First Workshop on Principles and Practice of Constraint Programming*, Providence RI, 1993.

171. H. Moulin. *Axioms for Cooperative Decision Making*. Cambridge University Press, 1988.

172. Katta Murty. *Linear Programming*. Wiley, 1983.

173. B. A. Nadel. Constraint Satisfaction Algorithms. *Comput. Intell.*, 5(4):188–224, November 1989.

174. G. Nelson. Juno, A Constraint-Based Graphics System. *Computer Graphics, ACM SIGGRAPH*, pages 235–243, 1985.

175. Bertrand Neveu. Contraintes et Problèmes Hiérarchiques. (Draft). *In French*, December 1993.

176. Francis Newberry Paulisch and Walter Tichy. Edge: An Extendible Graph Editor. *Software — Practice and Experience*, 20(S1):63–88, June 1990.

177. Hayato Ohwada and Fumio Mizoguchi. A Constraint Logic Programming Approach for Maintaining Consistency in User-Interface Design. In *Proceedings of the 1990 North American Conference on Logic Programming*, pages 139–153. MIT Press, October 1990.

178. P. J. Pacini and B. Kosko. Adaptive Fuzzy Systems for Target Tracking. *Intelligent Systems Engineering*, pages 3–21, Autumn 1992.

179. C. M. Papadimitriou. *Computational Complexity*. Addison-Wesley Publishing Company. 1994.

180. A. Philpott. Fuzzy Constraint Satisfaction. Master's thesis, University of Auckland, Auckland, New Zealand, 1995.

181. Patrick Prosser. Reactive Factory Scheduling as a Dynamic Constraint Satisfaction Problem. Technical Report AISL-31-88, University of Strathclyde, August 1988.

182. Patrick Prosser, Chris Conway, and Claude Muller. A Constraint Maintenance System for the Distributed Resource Allocation Problem. *Intelligent Systems Engineering*, pages 76–83, Autumn 1992.

183. J. Prust and W. Vonolfen. Dokumentation PARABOL 1.0. Unpublished documentation document, GMD, 1993.

184. A. Rauzy. Cedre version 0.2: User's Guide. Technical report, LaBRI URA CNRS 1304, 1994.

185. Raymond Reiter. A Logic for Default Reasoning. *Artificial Intelligence*, 13(1,2):81–132, April 1980.

186. A. Robinson. *Non-Standard Analysis*. North-Holland Publishing Company, Amsterdam, 1966.

187. Azriel Rosenfeld, Robert Hummel, and Steven Zucker. Scene Labelling by Relaxation Operations. *IEEE Trans. Sys. Man, and Cybern.*, 6(6), 1976.

188. Zsofia Ruttkay. Fuzzy Constraint Satisfaction. In *Proceedings 3rd IEEE Conference on Fuzzy Systems*, pages 1263–1268. IEEE, June 1994.

189. Mark Ryan. Defaults and Revision in Structured Theories. In *LICS'91: Proceedings of IEEE Symposium on Logic in Computer Science*, Amsterdam, July 1991.

190. Mark Ryan. *Ordered Presentations of Theories*. PhD thesis, Department of Computing, Imperial College, 1992.

191. Mark Ryan. Belief Revision and Ordered Theory Presentations. In A. Fuhrmann and H. Rott, editors, *Logic, Language and Information*. De Gruyter, 1994. (Presented at 8th Amsterdam Colloquium on Logic, December 1991).

192. Michael Sannella. The SkyBlue Constraint Solver and Its Applications. In Paris Kanellakis, Jean-Louis Lassez, and Vijay Saraswat, editors, *PPCP'93: First Workshop on Principles and Practice of Constraint Programming*, Providence RI, 1993.

193. Michael Sannella. Analyzing and Debugging Hierarchies of Multi-way Local Propagation Constraints. In Alan Borning, editor, *Principles and Practice of Constraint Programming*, LNCS 874. Springer, May 1994. (PPCP'94: Second International Workshop, Orcas Island, Seattle, USA).

194. Michael Sannella and Alan Borning. Multi-Garnet: Integrating Multi-Way Constraints with Garnet. Technical Report 92-07-01, Department of Computer Science and Engineering, University of Washington, September 1992.

195. Michael Sannella, John Maloney, Bjorn Freeman-Benson, and Alan Borning. Multi-way versus One-way Constraints in User Interfaces: Experience with the DeltaBlue Algorithm. *Software—Practice and Experience*, 23(5):529–566, May 1993.

196. Vijay Saraswat. *Concurrent Constraint Programming Languages*. PhD thesis, Carnegie-Mellon University, Computer Science Department, January 1989.

197. Vijay Saraswat, Johan de Kleer, and Brian Williams. ATMS-based Constraint Programming. Technical report, Xerox PARC, October 1991.

198. Ken Satoh. Formalizing Nonmonotonic Reasoning by Preference Order. Technical Report TR-440, ICOT, December 1988.

199. Ken Satoh. Formalizing Soft Constraints by Interpretation Ordering. In *ECAI'90: Proceedings European Conference on Artifical Intelligence*, 1990.

200. Ken Satoh and Akira Aiba. Computing Soft Constraints by Hierarchical Constraint Logic Programming. Technical Report TR-610, Institute for New Generation Computer Technology, Tokyo, January 1991. (Also *Journal of Information Processing*, 7, 1993).

201. Ken Satoh and Akira Aiba. The Hierarchical Constraint Logic Language CHAL. Technical Report TR-592, Institute for New Generation Computer Technology, Tokyo, September 1991.

202. Abdul Sattar, Aditya Ghose, and Randy Goebel. Specifying Over-Constrained Problems in Default Logic. In [133].

203. Thomas Schiex. Possibilistic Constraint Satisfaction Problems or "How to handle soft constraints". In *8th International Conference on Uncertainty in Artificial Intelligence*, Stanford, July 1992.

204. Thomas Schiex. Private Communication, April 1996.

205. Thomas Schiex, Hélène Fargier, and Gerard Verfaillie. Valued Constraint Satisfaction Problems: Hard and Easy Problems. In Chris Mellish, editor, *IJCAI'95: Proceedings International Joint Conference on Artificial Intelligence*, Montreal. Morgan Kaufman August 1995.

206. Thomas Schiex and Gérard Verfaillie. No-Good Recording for Static and Dynamic CSP. In *ICTAI'93: Proceeding of the 5th IEEE International Conference on Tools with Artificial Intelligence*, pages 48–55, Boston, November 1993.

207. Thomas Schiex and Gérard Verfaillie. Nogood Recording for Static and Dynamic Constraint Satisfaction Problems. *International Journal of Artificial Intelligence Tools*, 3(2):187–207, 1994.

208. Thomas Schiex and Gérard Verfaillie. Stubborness: An Enhancement Scheme for Backjumping and Nogood Recording. In *ECAI'94*, pages 165–169, Amsterdam, August 1994.

209. A. Schrijver. *Theory of Linear and Integer Programming*. John Wiley & Sons, 1986.

210. B. Selman and H. A. Kautz. An Empirical Study of Greedy Local Search for Satisfiability Testing. In *Proceedings AAAI-93*, pages 46–51, 1993.

211. B. Selman, H. Levesque, and D. Mitchell. A New Method for Solving Hard Satisfiability Problems. In *AAAI'92*, pages 440–446, San Jose, July 1992.

212. G. Shafer. An Axiomatic Study of Computation in Hypertrees. Working paper 232, University of Kansas, School of Business, Lawrence, 1991.

213. Linda Shapiro and Robert Haralick. Structural Descriptions and Inexact Matching. *IEEE Transactions on Pattern Analysis and Machine Intelligence*, 3:504–519, 1981.

214. Wolfgang Slany. Fuzzy Scheduling Bibliography. Available by ftp as `mira.dbai.tuwien.ac.at:/pub/slany/fuzzy-scheduling.ps.Z`, and also by email from `listproc@vexpert.dbai.tuwien.ac.at` with the message body "`GET LISTPROC fuzzy-scheduling.bib`".

215. Paul Snow and Eugene Freuder. Improved Relaxation and Search Methods for Approximate Constraint Satisfaction with a Maximin Criterion. In *Proc. of the 8th Biennal Conf. of the Canadian Society for Comput. Studies of Intelligence*, pages 227–230, May 1990.

216. Guy Steele. *The Definition and Implementation of a Computer Programming Language Based on Constraints*. PhD thesis, MIT, August 1980. Published as MIT-AI TR 595, August 1980.

217. Ivan Sutherland. Sketchpad: A Man-machine Graphical Communication System. In *Proceedings of the Spring Joint Computer Conference*, pages 329–346. IFIPS, 1963.

218. Gilles Trombettoni. CCMA*: A Complete Constraint Maintenance Algorithm Using Constraint Programming. In Manfred Meyer, editor, *Constraint Processing: Proceedings of the International Workshop at CSAM'93, St. Petersburg, July*

1993, Research Report RR-93-39, pages 123–132, DFKI Kaiserslautern, August 1993.

219. Fujio Tsusumi. An Efficient Algorithm of Logic Programming with Constraint Hierarchy. In Jean-Pierre Jouannaud, editor, *CCL'94: Proceedings 1st International Conference on Constraints in Computational Logics*, Munich, LNCS 854, pages 170–182. Springer, September 1994.

220. Pascal Van Hentenryck. *Constraint Satisfaction in Logic Programming*. MIT Press, Cambridge, MA, 1989.

221. Pascal Van Hentenryck. Incremental Constraint Satisfaction in Logic Programming. In *ICLP'90: Proceedings 7th International Conference on Logic Programming*, pages 189–202, Jerusalem, June 1990. MIT Press.

222. Pascal Van Hentenryck, Yves Deville, and Choh-Man Teng. A Generic Arc-consistency Algorithm and its Specializations. *Artificial Intelligence*, 57(2–3):291–321, October 1992.

223. Pascal Van Hentenryck and Thierry Le Provost. Incremental Search in Contraint Logic Programming. *New Generation Computing*, 9:257–275, 1991.

224. Bradley Vander Zanden. *An Incremental Planning Algorithm for Ordering Equations in a Multilinear system of Constraints*. PhD thesis, Department of Computer Science, Cornell University, April 1988.

225. N.R. Vempaty. Solving Constraint Satisfaction Problems using Finite State Automata. *AAAI*, San Jose, USA:453–458, 1992.

226. Gerard Verfaillie and Thomas Schiex. Dynamic Backtracking for Dynamic CSPs. In Thomas Schiex and Christian Bessière, editors, *Proceedings ECAI'94 Workshop on Constraint Satisfaction Issues raised by Practical Applications*, Amsterdam, August 1994.

227. Gerard Verfaillie and Thomas Schiex. Solution Reuse in Dynamic Constraint Satisfaction Problems. In *AAAI'94: Proceedings of the National Conference on Artificial Intelligence*, pages 307–312, August 1994.

228. Richard Wallace. Directed Arc Consistency Preprocessing as a Strategy for Maximal Constraint Satisfaction. In Manfred Meyer, editor, *Proceedings ECAI'94 Workshop on Constraint Processing*, Amsterdam, August 1994.

229. Richard Wallace. Cascaded Directed Arc Consistency and No-Good Learning for the Maximal Constraint Satisfaction Problem. In [133].

230. Richard Wallace and Eugene Freuder. Conjunctive Width Heuristics for Maximal Constraint Satisfaction. In *AAAI-93: Proceedings of the 11th National Conference on Artificial Intelligence*, Washington, DC, 1993. American Association for Artificial Intelligence.

231. Richard Wallace and Eugene Freuder. Applying Algorithms for Constraint Satisfaction to Maximum Satisfiability. In Thomas Schiex and Christian Bessière, editors, *Proceedings ECAI'94 Workshop on Constraint Satisfaction Issues raised by Practical Applications*, Amsterdam, August 1994.

232. Richard Wallace and Eugene Freuder. Comparing Constraint Satisfaction and Davis-Putnam Algorithms for the Maximal Satisfiability Problem. In D. S. Johnson and M. A. Trick, editors, *Cliques, Coloring and Satisfiability: Second DIMACS Implementation Challenge*, (to appear). Amer. Math. Soc., 1996.

233. Richard Wallace and Eugene Freuder. Heuristic Methods for Over-Constrained Constraint Satisfaction Problems. In [133].

234. Rolf Weißschnur. Die Projektion von Constraint-Satisfaction-Problemen auf Boltzmann-Maschinen. Master's thesis, Universität Bonn, Institut für Informatik, May 1994.

235. Rolf Weissschnur, Joachim Hertzberg, and Hans Werner Guesgen. Experiences in Solving Constraint Relaxation Networks with Boltzmann Machines. In [133].

236. Molly Wilson. *Hierarchical Constraint Logic Programming*. PhD thesis, University of Washington, Seattle, May 1993. (Also available as University of Washington Technical Report 93-05-01).

237. Molly Wilson and Alan Borning. Extending Hierarchical Constraint Logic Programming: Nonmonotonicity and Inter-Hierarchy Comparison. In Ewing Lusk and Ross Overbeek, editors, *NACLP'89: Proceedings North American Conference on Logic Programming*, pages 3–19, Cleveland, Ohio, 1989.

238. Molly Wilson and Alan Borning. Hierachical Constraint Logic Programming. *The Journal of Logic Programming*, 16(3 & 4):277–318, July and August 1993.

239. Armin Wolf. Transforming Ordered Constraint Hierarchies into Ordinary Constraint Systems. In [133].

240. Christopher van Wyk. *A Language for Typesetting Graphics*. PhD thesis, Department of Computer Science, Stanford, June 1980.

241. Karine Yvon. A Solver for Fair Hierarchical Constraint Logic Programming. BSc Final Year Project, Department of Computer Science, City University, London, June 1995.

242. Lotfi Zadeh. Fuzzy Sets. *Information and Control*, 8:338–353, 1965.

243. Lotfi Zadeh. Calculus of Fuzzy Restrictions. In K. Tanaka L.A. Zadeh, K.S. Fu and M. Shimura, editors, *Fuzzy Sets and Their Applications to Cognitive and Decision Processes*. Academic Press, 1975.

Constraint Hierarchies

Alan Borning, Bjorn Freeman-Benson, and Molly Wilson

Department of Computer Science and Engineering, University of Washington,
Box 352350, Seattle, Washington 98195-2350, USA.
Please send correspondence to Alan Borning, borning@cs.washington.edu

Abstract. Constraints allow programmers and users to state declaratively a relation that should be maintained, rather than requiring them to write procedures to maintain the relation themselves. They are thus useful in such applications as programming languages, user interface toolkits, and simulation packages. In many situations, it is desirable to be able to state both *required* and *preferential* constraints. The required constraints must hold. Since the other constraints are merely preferences, the system should try to satisfy them if possible, but no error condition arises if it cannot. A *constraint hierarchy* consists of a set of constraints, each labeled as either required or preferred at some strength. An arbitrary number of different strengths is allowed. In the discussion of a theory of constraint hierarchies, we present alternate ways of selecting among competing possible solutions, and prove a number of propositions about the relations among these alternatives. We then outline algorithms for satisfying constraint hierarchies, and ways in which we have used constraint hierarchies in a number of programming languages and systems.

This paper was originally published in *Lisp and Symbolic Computation*, Vol. 5 No. 3, (September 1992), pages 223–270.

1 Introduction

A constraint describes a relation that should be satisfied. Examples of constraints include:

- a constraint that a resistor in a circuit simulation obey Ohm's Law
- a constraint that two views of the same data remain consistent (for example, bar graph and pie chart views)
- a default constraint that parts of an object being edited remain fixed, unless there is some stronger constraint that forces them to change.

Constraints are useful in programming languages, user interface toolkits, simulation packages, and other systems because they allow users to declare that a relation is to be maintained, rather than requiring users to write, and invoke, procedures to do the maintenance. In general constraints are *multi-directional*. For example, a constraint that $A + B = C$ might be used to find a value for any of A, B, or C. In general there may be many interrelated constraints in a given application; it is left up to the system to sort out how they interact and to keep them all satisfied.

1.1 The Refinement versus The Perturbation Model

We can roughly classify constraint-based languages and systems as using one of two approaches: the *refinement* model or the *perturbation* model. In both cases constraints restrict the values that variables may take on. In the refinement model, variables are initially unconstrained; constraints are added as the computation unfolds, progressively refining the permissible values of the variables. This approach is more or less universally adopted in the logic programming community, for example, in the Constraint Logic Programming language scheme [11, 40] and in the cc (concurrent constraint) languages [66, 65].

In contrast, in the perturbation model, at the beginning of an execution cycle variables have specific values associated with them that satisfy the constraints. The values of one or more variables are perturbed (usually by some outside influence, such as an edit request from the user), and the task of the system is to adjust the values of the variables so that the constraints are again satisfied. The perturbation model has often been used in constraint-based applications such as the interactive graphics systems Sketchpad [75], ThingLab I [3], Magritte [33], and Juno [58], and user interface construction systems such as Garnet [56, 57]. We can also view the ubiquitous spreadsheet as using the perturbation model: formulas are constraints relating the permissible values in cells. Before a user action, cells have values that satisfy the constraints (formulas). The user edits the value in a cell, or edits a formula, and the system must change the values of other cells as needed so that the constraints are again satisfied.

In the perturbation model, there will generally be many ways to update the current state so that the constraints are again satisfied. As a trivial example, suppose we have a constraint $A + B = C$, and edit the value of B. Should we change just A, change just C, change both A and C, undo the change to B, or what? At some cost in generality, we can use *read-only annotations* to limit this choice. A common special case is to use *one-way constraints*, that is, constraints in which all but one of the variables are declared to be read-only. For the $A + B = C$ constraint, if A and B are declared to be read-only, it is clear what to do when B is edited (change C), at least if there are no circularities in the constraint graph.

Except for systems that are restricted to non-circular one-way constraints, a problem with the perturbation model is that it is often unclear which variables to alter to re-satisfy the constraints. A variety of heuristics were used in earlier systems (see Section 6.1). However, none of these methods was entirely satisfactory: sometimes they gave counter-intuitive solutions. Worse, it was difficult to specify declaratively which solutions were preferred and to alter these preferences, since the heuristics were buried in the procedural code of the satisfier.

1.2 Requirements and Preferences

Constraint hierarchies were originally devised to solve the problem of specifying declaratively what to change when perturbing a constraint system [6]. In a constraint hierarchy, the programmer or user can state both *required* and *preferential* constraints (also known as *hard* and *soft* constraints). The required constraints must hold. The system should try to satisfy the preferential constraints if possible, but no error condition arises if it can't. We allow an arbitrary number of levels of preference, each successive level being more weakly preferred than the previous one.

Thus, in the $A + B = C$ example, we could also include weak constraints that A and B remain unchanged, and a weaker constraint that C remain the same. Given this hierarchy, if we edit A, the system will change C rather than B to re-satisfy the constraints. One use, therefore, of constraint hierarchies is to take a problem for which the perturbation model is more natural, and turn it into a more declarative "refinement" problem.

However, constraint hierarchies have numerous other applications as well—anywhere that we would like to state preferences as well as requirements—for example, planning, scheduling, or layout. As a simple example, consider the problem of laying out a table in a document. We would like the table to fit on a single page while still leaving adequate white space between rows. This can be represented as the interaction of two constraints: a hard constraint that the height of the blank space between lines be greater than zero, and a soft constraint that the entire table fit on one page. As another example, suppose we are moving a part of a constrained geometric figure around on the display using the mouse. While the part moves, other parts may also need to move to keep all the constraints satisfied. However, if the locations of all parts aren't determined, we would prefer that they remain where they were, rather than flailing wildly about. Further, there may be choices about which parts to move and which to leave fixed; the user may have preferences in such cases. Again, constraint hierarchies provide a convenient way of stating these desires.

In the remainder of this paper, we first present a theory of constraint hierarchies; this theory is the paper's primary focus. As part of this presentation, we discuss a number of alternate ways of selecting among competing possible solutions, and prove several propositions about the relations among these alternatives. Following this, we outline several algorithms for satisfying constraint hierarchies, and describe how we have used constraint hierarchies in a number of programming languages and systems, including HCLP (a logic programming language scheme), CIP (a hybrid constraint-imperative language scheme), and ThingLab II (a constraint-based simulation environment). Finally, we discuss some previous and related work in more detail; we describe in particular how these other systems handle problems involving defaults and preferences, and show how to classify their behavior in terms of the constraint hierarchy theory.

2 A Theory of Constraint Hierarchies

In this section we present a theory of constraint hierarchies. In later sections, we describe some extensions to this basic theory, and then how these notions have been embedded in a variety of systems and languages, including logic programming and object-oriented languages.

2.1 Definitions

A constraint is a relation over some domain $\mathcal{D}$. The domain $\mathcal{D}$ determines the constraint predicate symbols $\Pi_{\mathcal{D}}$ of the language, so that a constraint is an expression of the form $p(t_1, \ldots, t_n)$ where p is an n-ary symbol in $\Pi_{\mathcal{D}}$ and each t_i is a term.

A *labeled constraint* is a constraint labeled with a strength, written sc, where s is a strength and c is a constraint. For clarity in writing labeled constraints, we give symbolic names to the different strengths of constraints. In both the theory and in our implementations of languages and systems that include constraint hierarchies, we then map each of these names onto the integers $0 \ldots n$, where n is the number of non-required levels. Strength 0, with the symbolic name *required*, is always reserved for required constraints.

A *constraint hierarchy* is a multiset of labeled constraints. Given a constraint hierarchy H, H_0 denotes the required constraints in H, with their labels removed. In the same way, we define the sets $H_1, H_2, \ldots, H_n$ for levels $1, 2, \ldots, n$. We also define $H_k = \emptyset$ for $k > n$.

A *solution* to a constraint hierarchy H is a valuation for the free variables in H, i.e., a function that maps the free variables in H to elements in the domain $\mathcal{D}$. We wish to define the set S of all solutions to H. Clearly, each valuation in S must be such that, after it is applied, all the required constraints hold. In addition, we desire each valuation in S to be such that it satisfies the non-required constraints as well as possible, respecting their relative strengths. To formalize this desire, we first define the set S_0 of valuations such that all the H_0 constraints hold. Then, using S_0, we define the desired set S by eliminating all potential valuations that are worse than some other potential valuation using the comparator predicate *better*. (In the definition, $c\theta$ denotes the boolean result of applying the valuation θ to c, and we say that "$c\theta$ holds" if $c\theta = \textbf{true}$.)

$$S_0 = \{\theta \mid \forall c \in H_0 \; c\theta \text{ holds}\}$$
$$S = \{\theta \mid \theta \in S_0 \wedge \forall \sigma \in S_0 \; \neg better(\sigma, \theta, H)\}$$

There are many plausible candidates for comparators. We insist that *better* be irreflexive and transitive:

$$\forall \theta \forall H \; \neg better(\theta, \theta, H)$$
$$\forall \theta, \sigma, \tau \forall H \; better(\theta, \sigma, H) \wedge better(\sigma, \tau, H) \rightarrow better(\theta, \tau, H)$$

However, in general, *better* will not provide a total ordering—there may exist θ and σ such that θ is not better than σ and σ is not better than θ. We also insist that *better* respect the hierarchy—if there is some valuation in S_0 that completely satisfies all the constraints through level k, then all valuations in S must satisfy all the constraints through level k:

$$\text{if } \exists \theta \in S_0 \wedge \exists k > 0 \text{ such that}$$
$$\forall i \in 1 \ldots k \; \forall p \in H_i \; p\theta \text{ holds}$$
$$\text{then } \forall \sigma \in S \; \forall i \in 1 \ldots k \; \forall p \in H_i \; p\sigma \text{ holds}$$

We now define several different comparators. In the definitions, we will need an error function $e(c\theta)$ that returns a non-negative real number indicating how nearly constraint c is satisfied for a valuation θ. This function must have the property that $e(c\theta) = 0$ if and only if $c\theta$ holds. For any domain $\mathcal{D}$, we can use the trivial error function that returns 0 if the constraint is satisfied and 1 if it is not. A comparator that uses this error function is a *predicate* comparator. For a domain that is a metric space, we can use its metric in computing the error instead of the trivial error function. (For example, the error for $X = Y$ would be the distance between X and Y.) Such a comparator is a *metric* comparator.

The first of the comparators, *locally-better*, considers each constraint in H individually.

Definition. A valuation θ is *locally-better* than another valuation σ if, for each of the constraints through some level $k - 1$, the error after applying θ is equal to that after applying σ, and at level k the error is strictly less for at least one constraint and less than or equal for all the rest.

$$locally\text{-}better(\theta, \sigma, H) \equiv$$
$$\exists k > 0 \ \text{ such that}$$
$$\forall i \in 1 \ldots k - 1 \ \forall p \in H_i \ e(p\theta) = e(p\sigma)$$
$$\wedge \ \exists q \in H_k \ e(q\theta) < e(q\sigma)$$
$$\wedge \ \forall r \in H_k \ e(r\theta) \leq e(r\sigma)$$

Next, we define a schema *globally-better* for global comparators. The schema is parameterized by a function g that combines the errors of all the constraints H_i at a given level.

Definition. A valuation θ is *globally-better* than another valuation σ if, for each level through some level $k - 1$, the combined errors of the constraints after applying θ is equal to that after applying σ, and at level k it is strictly less.

$$globally\text{-}better(\theta, \sigma, H, g) \equiv$$
$$\exists k > 0 \ \text{ such that}$$
$$\forall i \in 1 \ldots k - 1 \ g(\theta, H_i) = g(\sigma, H_i)$$
$$\wedge \ g(\theta, H_k) < g(\sigma, H_k)$$

Using *globally-better*, we now define three global comparators, using different combining functions g. The weight for constraint p is denoted by w_p. Each weight is a positive real number.

$$weighted\text{-}sum\text{-}better(\theta, \sigma, H) \equiv globally\text{-}better(\theta, \sigma, H, g)$$
$$\text{where} \ \ g(\tau, H_i) \equiv \sum_{p \in H_i} w_p e(p\tau)$$

$$worst\text{-}case\text{-}better(\theta, \sigma, H) \equiv globally\text{-}better(\theta, \sigma, H, g)$$
$$\text{where} \ \ g(\tau, H_i) \equiv \max \{w_p e(p\tau) \mid p \in H_i\}$$

$$least\text{-}squares\text{-}better(\theta, \sigma, H) \equiv globally\text{-}better(\theta, \sigma, H, g)$$

$$\text{where} \quad g(\tau, H_i) \equiv \sum_{p \in H_i} w_p e(p\tau)^2$$

Orthogonal to the choice of *locally-better* or one of the instances of *globally-better*, we can choose an appropriate error function for the constraints. *Locally-predicate-better* is *locally-better* using the trivial error function that returns 0 if the constraint is satisfied and 1 if it is not. *Locally-metric-better* is *locally-better* using a domain metric in computing the constraint errors. *Weighted-sum-predicate-better*, *weighted-sum-metric-better*, and so forth, are all defined analogously.

Unsatisfied-count-better is a special case of *weighted-sum-predicate-better*, using weights of 1 on each constraint; it counts the number of unsatisfied constraints in making its comparisons. The predicate versions of the other two global comparators aren't particularly useful: *worst-case-predicate-better* has an all-or-nothing behavior which doesn't filter out solutions as well as one might like; and *least-squares-predicate-better* always gives the same results as *weighted-sum-predicate-better* (since $1^2 = 1$).

2.2 Illustrative Examples

The first example in this subsection illustrates that constraints in stronger levels dominate those in weaker levels, while the second illustrates the various solutions that different comparators can produce.

First, consider the following constraint hierarchy, which includes the canonical Celsius–Fahrenheit constraint:

Level	Constraints
H_0	required Celsius * 1.8 = Fahrenheit − 32.0
H_1	strong Fahrenheit = 212
H_2	weak Celsius = 0

The set S_0 consists of all valuations such that the H_0 (required) constraints hold. For this hierarchy, the set S_0 is infinite, and consists of all valuations with valid temperature pairs $\langle C, F \rangle$, i.e.,

$$S_0 = \{\ldots, \langle -60, -76 \rangle, \langle -40, -40 \rangle, \langle 0, 32 \rangle, \langle 10, 50 \rangle, \langle 100, 212 \rangle, \ldots\}$$

while the set S consists of the single pair $\theta = \langle 100, 212 \rangle$. For example, S does not contain the pair $\sigma = \langle 10, 50 \rangle$ because θ satisfies the level H_1 constraint whereas σ does not. Thus, $\exists k > 0$ (namely $k = 1$) such that $\exists c \in H_k$ for which $e(c\theta) < e(c\sigma)$. Therefore, *locally-better*(θ, σ, H). Further, S does not contain the pair $\rho = \langle 0, 32 \rangle$ because although ρ satisfies the H_2 constraint that θ does not, θ satisfies the H_1 constraint that ρ does not. Intuitively, because θ satisfies the stronger H_1 constraint better than ρ, *locally-better*(θ, ρ, H). This example produces exactly the same answer whether *locally-predicate-better*, *locally-metric-better*, or one of the *globally-better* comparators is used. However, this would not be the case in general. (Some propositions concerning the relations between the comparators are discussed in Section 2.3.)

As a second example, consider the following constraint hierarchy H for the domain $\mathcal{R}$, and its solutions under each of the useful comparators. (Note that H_0 is empty for this hierarchy.)

Level	Constraint	Weight
H_1	weak $X = 0$	1.0
H_1	weak $X \geq 2$	1.0
H_1	weak $X = 4$	0.25

Comparator	Solutions
locally-predicate-better	$X = 0.0$ or $X = 4.0$
locally-metric-better	$0.0 \leq X \leq 4.0$
weighted-sum-predicate-better	$X = 4.0$
weighted-sum-metric-better	$X = 2.0$
worst-case-metric-better	$X = 1.0$
least-squares-metric-better	$X = 1.3333$

Using the *weighted-sum-metric-better* comparator, the solution consists of exactly one valuation: $\theta = \{X \mapsto 2.0\}$. Thus, θ is *weighted-sum-metric-better* than all other valuations including, for example, $\sigma = \{X \mapsto 0.46\}$. The following table summarizes the computation of $g(\theta, H_1)$ and $g(\sigma, H_1)$, verifying that $g(\theta, H_1) < g(\sigma, H_1)$.

Constraints	$\theta = \{X \mapsto 2.0\}$ Error	Weighted	$\sigma = \{X \mapsto 0.46\}$ Error	Weighted
$X = 0$	2.0	2.0	0.46	0.46
$X \geq 2$	0.0	0.0	1.54	1.54
$X = 4$	2.0	0.5	3.54	0.89
Weighted Sum		2.5		2.89

Using the *locally-predicate-better* comparator, the solution consists of two valuations: $\theta = \{X \mapsto 0.0\}$ and $\rho = \{X \mapsto 4.0\}$. Both valuations are better than all the other valuations (including $\sigma = \{X \mapsto 0.46\}$), but neither one is better than the other. For example, the first of the following two tables illustrates that *locally-predicate-better* (θ, σ) is true and thus $\sigma \notin S$.

Constraints	$\theta = \{X \mapsto 0.0\}$ Trivial Error	Comparison	$\sigma = \{X \mapsto 0.46\}$ Trivial Error
$X = 0$	0	$<$	1
$X \geq 2$	1	$\leq$	1
$X = 4$	1	$\leq$	1
	$\exists q \in H_1\ e(q\theta) < e(q\sigma)$	$\wedge$	$\forall r \in H_1\ e(r\theta) \leq e(r\sigma)$

This second table illustrates that neither *locally-predicate-better* (θ, ρ) nor *locally-predicate-better* (ρ, θ) is true, and thus both $\rho \in S$ and $\theta \in S$.

Constraints	$\theta = \{X \mapsto 0\}$ Trivial Error	Comparison	$\rho = \{X \mapsto 4\}$ Trivial Error
$X = 0$	0	$<$	1
$X \geq 2$	1	$>$	0
$X = 4$	1	$>$	0
	$\neg \forall r \in H_1 \; e(r\theta) \leq e(r\rho)$	$\wedge$	$\neg \forall r \in H_1 \; e(r\theta) \geq e(r\rho)$

2.3 Remarks on the Comparators

The definitions of the global comparators include weights on the constraints. For the local comparators, adding weights would be futile, since the result would be the same with or without the weights.

One might argue that allowing an arbitrary number of constraint strengths is unnecessary: since soft constraints can have weights on them, one could make do with only two levels (required and one preferential level), and use appropriate weights to achieve the desired effects. There are three reasons we believe such an argument is not valid: two conceptual, and the other pragmatic. To illustrate the first reason, consider moving a line with a mouse in an interactive graphics application. The line has a strong constraint that it be horizontal, and another strong constraint that one endpoint follow the mouse. There is also a weaker constraint that the line be attached to some fixed point in the diagram. The user's expectations in this case are likely that the line will remain exactly horizontal and will precisely follow the mouse (letting the weaker attachment constraint be unsatisfied), rather than keeping the line nearly horizontal, or quite close to the mouse, but letting the weaker constraint have a bit of influence on the result. Second, since adding weights to constraints is futile for the local comparators, we would need to give up these comparators and use only global ones. Third, solutions to constraint hierarchies in which one level completely dominates the next can often be found much more efficiently than solutions to systems with only one preferential level and weights on the constraints—see Section 4.

Most of the concepts in constraint hierarchies derive from concepts in subfields of operations research such as linear programming [53], multiobjective linear programming [53], goal programming [39], and generalized goal programming [38]. The domain of the constraints in operations research is usually the real numbers, or sometimes the integers (for integer programming problems). The notion of constraint hierarchies is preceded by the approach to multiobjective problems of placing the objective functions in a priority order. The concept of a *locally-better* solution is derived from the concept of a *vector minimum* (or *pareto optimal solution*, or *nondominated solution*) to a multiobjective linear programming problem. Similarly, the concepts of *weighted-sum-better* and *worst-case-better* solutions are both derived from analogous concepts in multiobjective linear programming problems and generalized goal programming.

There are a number of relations that hold between local and global comparators.

Proposition 1. *For a given error function e,*

$$\forall\theta\forall\sigma\forall H \ locally\text{-}better(\theta,\sigma,H) \rightarrow weighted\text{-}sum\text{-}better(\theta,\sigma,H)$$

Proof. Suppose *locally-better*(θ,σ,H) holds. Then there is some level $k > 0$ in H such that the error after applying θ to each of the constraints through levels $k-1$ is equal to that after applying σ. It then follows that the sum of the weighted errors after applying θ to the constraints through levels $k-1$ is equal to that after applying σ. Furthermore, at level k the error after applying θ is strictly less for at least one constraint and less than or equal for all the rest. This implies that the weighted sum of the errors after applying θ to the constraints at level k is strictly less than that after applying σ. Therefore *weighted-sum-better*(θ,σ,H) also holds.

Corollary 2. *For a given constraint hierarchy, let* S_{LB} *denote the set of solutions found using the* locally-better *comparator, and* S_{WSB} *that for* weighted-sum-better. *Then* $S_{\mathrm{WSB}} \subseteq S_{\mathrm{LB}}$.

Proposition 3. *For a given error function e,*

$$\forall\theta\forall\sigma\forall H \ locally\text{-}better(\theta,\sigma,H) \rightarrow least\text{-}squares\text{-}better(\theta,\sigma,H)$$

The proof is similar to that for Proposition 1.

Corollary 4. *Let* S_{LSQ} *denote the set* S *of solutions found using the* least-squares-better *comparator. Then* $S_{\mathrm{LSQ}} \subseteq S_{\mathrm{LB}}$.

Propositions 1 and 3 concern particular instances of the *globally-better* schema. However, *locally-better* does not imply *globally-better* for an arbitrary combining function g. In particular, *locally-better* does not imply *worst-case-better*.

2.4 Errors for Inequalities

A problem arises in connection with metric predicates and strict inequalities. For example, what should be the error function for the constraint $X > Y$, where X and Y are reals? If X is greater than Y, then the error must be 0. If X isn't greater than Y, we'd like the error to be smaller the closer X is to Y. Thus, an obvious error function is $e(X > Y) = 0$ if $X > Y$, otherwise $Y - X$. This isn't correct, however, since it gives an error of 0 if X and Y are equal. However, if the error when X and Y are equal is some positive number d, then we get a smaller error when Y is equal to $X + d/2$ than when Y is equal to X, thus violating our desire that the error become smaller as X gets closer to Y.

To solve this problem, we introduce an infinitesimal number ϵ [61], which is greater than 0 and less than any positive standard real number. Using ϵ we can then define

$$e(X > Y) = \begin{cases} Y - X & \text{if } X < Y \\ \epsilon & \text{if } X = Y \\ 0 & \text{if } X > Y \end{cases}$$

$$e(X \neq Y) = \begin{cases} 0 & \text{if } X \neq Y \\ \epsilon & \text{if } X = Y \end{cases}$$

$$e(X < Y) = \begin{cases} 0 & \text{if } X < Y \\ \epsilon & \text{if } X = Y \\ X - Y & \text{if } X > Y \end{cases}$$

Note that ϵ is only being added to the range of the error function, not to the domain $\mathcal{D}$. If we did try to change the domain itself to be the hyperreal numbers, we would end up with the same problem as before.[1]

2.5 Existence of Solutions

If the set of solutions S_0 for the required constraints is non-empty, intuitively one might expect that the set of solutions S for the hierarchy would be non-empty as well. However, this is not always the case. Consider the hierarchy *required* $N > 0$, *strong* $N = 0$ for the domain of the real numbers, using a metric comparator. Then S_0 consists of all valuations mapping N to a positive number, but S is empty, since for any valuation $\{N \mapsto d\} \in S_0$, we can find another valuation, for example $\{N \mapsto d/2\}$, that better satisfies the soft constraint $N = 0$.

However, the following proposition, especially relevant for floating point numbers, does hold:

Proposition 5. *If S_0 is non-empty and finite, then S is non-empty.*

Proof. Suppose to the contrary that S is empty. Pick a valuation θ_1 from S_0. Since $\theta_1 \notin S$, there must be some $\theta_2 \in S_0$ such that $better(\theta_2, \theta_1, H)$. Similarly, since $\theta_2 \notin S$, there is an $\theta_3 \in S_0$ such that $better(\theta_3, \theta_2, H)$, and so forth for an infinite chain $\theta_4, \theta_5, \ldots$. Since *better* is transitive, it follows by induction that $\forall i, j > 0 \; [i > j \to better(\theta_i, \theta_j, H)]$. The irreflexivity property of *better* requires that $\forall i > 0 \; \neg better(\theta_i, \theta_i, H)$. Thus all the θ_i are distinct, and so there are an infinite number of them. But, by hypothesis S_0 is finite, a contradiction.

For most (if not all) practical applications of constraint hierarchies, H will be finite. For example, for a CIP or HCLP program, if the program terminates, the resulting set of constraints will be finite. The next proposition tells us that in many cases of practical importance, if the required constraints can be satisfied, then solutions to the hierarchy exist.

---

[1] What would be the error for the constraint $0 > \epsilon/2$? According to the definition, the error would be $\epsilon/2$. But this is less than the error for $0 > 0$, even though the $0 > 0$ constraint is more nearly satisfied.

Proposition 6. *If S_0 is non-empty, if H is finite, and if a predicate comparator is used, then S is non-empty.*

Proof. Suppose to the contrary that S is empty. Using the same argument as before, we show that there must be an infinite number of distinct valuations $\theta_i \in S_0$. However, if the comparator is predicate, one valuation cannot be better than another if both valuations satisfy exactly the same subset of constraints in H. Therefore each of the θ_i must satisfy a different subset of the constraints in H. However, this is a contradiction, since H is finite.

3 Extensions to the Constraint Hierarchy Theory

3.1 Read-only Annotations

As noted in Section 1.1, perturbation-based constraint systems often use read-only annotations to help limit the choice of which variables should be updated to re-satisfy the constraints after some change to the system. Constraint hierarchies provide an alternative method for specifying this choice, without giving up the generality of multi-way constraints. However, even in a multi-way constraint system with hierarchies, read-only annotations can still be useful. One use is in constraints that reference an external input device or other outside source of information. If we have a constraint that a point follow the mouse, the constraint should be read-only on the mouse position (unless, of course, the mouse is equipped with a small computer-controlled motor). Another use is in constraints describing a change over time, where the constraint relates an old and a new state. Here, we may wish to make the old state read-only, so that the future can't alter the past.

Intuitively, when choosing the best solutions to a constraint hierarchy, constraints should not be allowed to affect the choice of values for their read-only variables, i.e., information can flow out of the read-only variables, but not into them. (Alternatively we can say that constraints are only allowed to affect the choice of values for their unannotated variables.) However, we still want the constraints to be satisfied if possible (respecting their strengths). In particular, required constraints must be satisfied, even if they contain read-only annotations.

We now give an informal outline of the definition. One way of preventing a constraint from affecting the choice of values for a variable is to replace that occurrence of the variable by a constant. Thus, we begin the definition of the set of solutions to a constraint hierarchy H by forming a set Q of constraint hierarchies, where each element of Q is a constraint hierarchy with arbitrary domain elements substituted for the read-only variables. (Note that the same variable v may have read-only occurrences and normal occurrences. Only the read-only occurrences are replaced when forming elements of Q.) Intuitively, we guess a valuation for v, and then form a hierarchy using that guess. After making all possible guesses, we weed out solutions arising from incorrect ones. (Note that this is purely a specification of the meaning of read-only annotations, not a reasonable algorithm for actually solving such constraint hierarchies! Algorithms are discussed in Section 4.)

Here is an example:

Original H	$q \in Q$ formed by replacing $Y?$ with $d \in D$			
	$Y? \mapsto 9.83$	$Y? \mapsto 3$	$Y? \mapsto -6.2$	$\cdots$
required $X = Y?$	$X = \mathbf{9.83}$	$X = \mathbf{3}$	$X = \mathbf{-6.2}$	
strong $\quad X = 4$	$X = 4$	$X = 4$	$X = 4$	$\cdots$
weak $\quad Y = 3$	$Y = 3$	$Y = 3$	$Y = 3$	

Next we solve the constraint hierarchies in Q, discarding any valuations that map the remaining unannotated occurrences of a variable to something different from what was substituted for its read-only occurrences. (In other words, we discard all valuations in which we guessed incorrectly.) This ensures that the permissible values for a variable won't be affected by read-only occurrences of that variable, but that they will be consistent with the read-only occurrences. Continuing the example:

	$q \in Q$ formed by replacing $Y?$ with $d \in D$		
replacement ρ	$Y? \mapsto 9.83$	$Y? \mapsto 3$	$\cdots$
hierarchy q	required $X = \mathbf{9.83}$ strong $\quad X = 4$ weak $\quad Y = 3$	required $X = \mathbf{3}$ strong $\quad X = 4$ weak $\quad Y = 3$	$\cdots$
valuation θ	$\{Y \mapsto 3, X \mapsto 9.83\}$	$\{Y \mapsto 3, X \mapsto 3\}$	$\cdots$
consistency	$Y\theta \neq Y?\rho$	$Y\theta = Y?\rho$	$\cdots$
outcome	Discard	Keep	$\cdots$

The valuation $\{Y \mapsto 3, X \mapsto 3\}$ is the only consistent solution, and thus is the solution to the original hierarchy.

We now give a formal definition of the meaning of read-only annotations. In the definition, we will introduce new variables w_i, which we will want to omit in the final solution. We therefore define an operator *omitting*.

Definition. Let θ be a valuation. Let the domain of θ be the variables $v_1, \ldots, v_n$. Then

$$\theta \text{ omitting } w_1, \ldots, w_m$$

is the valuation σ such that the domain of σ is $\{v_1, \ldots, v_n\} - \{w_1, \ldots, w_m\}$, and such that $\sigma v = \theta v$ for all v in the domain of σ. Similarly, if Θ is a set of valuations,

$$\Theta \text{ omitting } w_1, \ldots, w_m = \{\theta \text{ omitting } w_1, \ldots, w_m \mid \theta \in \Theta\}$$

Definition. Let H be a constraint hierarchy containing read-only annotations, and let D be the domain of the constraints. Let $v_1, \ldots, v_m$ be the variables in H that have one or more read-only occurrences. Let $w_1, \ldots, w_m$ be new variables not occurring in H, and let J be the hierarchy that results from substituting

w_i for each read-only occurrence of the corresponding variable v_i. (The w_i are no longer annotated as read-only in J; also, occurrences of the variables v_i that aren't annotated as read-only are unaffected.) Define Q as the set of all hierarchies $J\rho$, where each ρ is formed by substituting arbitrary domain elements for the w_i:

$$Q = \{J\rho \mid d_1 \in \mathcal{D}, \ldots, d_m \in \mathcal{D}, \; \rho = \{w_1 \mapsto d_1, \ldots, w_m \mapsto d_m\}\}$$

Let $solutions(J\rho)$ be the set of solutions to $J\rho$. (Here we are using the definition of "solutions" given in the basic theory section (2), since J has no variables with read-only annotations.) Let the set of *consistent solutions* to $J\rho$ be defined as:

$$consistent(J\rho) = \{\theta \mid \theta \in solutions(J\rho) \wedge$$
$$w_1\rho = v_1\theta \wedge \ldots \wedge w_m\rho = v_m\theta\}$$

In English, to be a consistent solution, if ρ maps w_i to some domain element d_i, then θ must map the corresponding v_i to the same domain element d_i (i.e., we guessed correctly).

The desired set of solutions to H is the set of all consistent solutions, omitting the mappings for the newly introduced variables w_i:

$$solutions(H) = \left(\bigcup_{J\rho \in Q} consistent(J\rho) \right) \text{ omitting } w_1, \ldots w_m$$

Proposition 7. *For a constraint hierarchy H containing only required constraints, let H' be the same hierarchy, but with the read-only annotations removed. Then $solutions(H) = solutions(H')$.*

Proof.
  $solutions(H) \supseteq solutions(H')$
Let $v_1, \ldots, v_m, w_1, \ldots, w_m$, and J be defined as above. Let θ be a solution for H'. Define $\rho = \{w_i \mapsto v_i\theta, \ldots, w_m \mapsto v_m\theta\}$. (In other words, if θ maps v_i to d_i, then ρ maps the corresponding w_i to d_i.) Then clearly $\theta \in solutions(J\rho)$ and θ is consistent. So $\theta \in solutions(H)$.
  $solutions(H) \subseteq solutions(H')$
Now assume θ is a solution for H. By definition, θ is a consistent solution to $J\rho$ for some ρ. As H consists only of required constraints and as θ is consistent with ρ, θ also satisfies all of the constraints in H'.

Blocked Hierarchies Even with this definition, it is possible for a constraint to restrict the values that its read-only annotated variables can take on. For example, consider the following constraint hierarchy for the domain $\mathcal{R}$:

  required $V > 0$
  required $V? = 1$

The $V > 0$ constraint contains the only unannotated occurrence of V, and thus only $V > 0$ is allowed to affect the choice of values for V, and not $V? = 1$. However, the solutions to the first constraint by itself, $V > 0$, includes $V \mapsto 0.3$, $V \mapsto 1.728$, and so forth, in addition to $V \mapsto 1$, while $solutions(H) = \{V \mapsto 1\}$. Thus, the choice of values for V is being affected by the $V? = 1$ constraint. We therefore impose an additional check, $blocked(H)$, that tests for this situation.

The $blocked(H)$ predicate is true if any constraint in H limits the permissible values for one of its read-only annotated variables. In such a case, additional constraints can be added to the hierarchy so that the set of solutions can be found without any constraints limiting the permissible values for the read-only annotated variables.

The definition of $blocked(H)$ is based on the following observation: if there is a domain element d such that there are no solutions when d replaces all occurrences of a variable (both annotated and unannotated), but there are solutions when d replaces only the unannotated occurrences, then the annotated (read-only) occurrences are eliminating d from $solutions(H)$. Thus, if such a d exists, the annotated occurrences are restricting the values that the variable can take on, and $blocked(H)$ is true.

Definition.

$$blocked(H) \equiv \exists d \in \mathcal{D} \; \exists i \in [1 \ldots m] \text{ such that}$$
$$solutions(J\rho\theta\sigma) = \emptyset \; \wedge \; solutions(J\theta\sigma) \neq \emptyset$$
$$\text{where } \rho = \{w_i \mapsto d\}, \; \theta = \{v_i \mapsto d\}, \text{ and}$$
$$\sigma = \{w_1 \mapsto v_1, \ldots, w_{i-1} \mapsto v_{i-1},$$
$$w_{i+1} \mapsto v_{i+1}, \ldots, w_m \mapsto v_m\}$$

If there are no read-only annotations on the variables in H, then clearly $blocked(H)$ is false.

Within the logic programming community, read-only annotations were originally introduced in Concurrent Prolog [71] for an entirely different purpose than ours, namely for the control of communication and synchronization among networks of processes. In our work, having a blocked solution is an unusual and undesirable state, which would arise only if a design or other error had been made in specifying the constraints. In contrast, in concurrent logic programming, blocking caused by read-only annotations is ubiquitous and essential in controlling program execution.

There were problems with the original formulation of read-only annotations in Concurrent Prolog (see [64] for a discussion), and a number of alternatives have been proposed. For example, Maher [48] describes ALPS, a class of languages that incorporates constraints into a flat committed-choice logic language. The definition of $blocked$ was directly inspired by the ALPS work.

Illustrative Examples of Using Read-only Annotations Consider the hierarchy H for the domain $\mathcal{R}$:

required $C * 1.8 = F? - 32.0$
strong $C = 0.0$
weak $F = 212.0$

Without the read-only annotation on F, the solution to this hierarchy would be $\{\{C \mapsto 0.0, F \mapsto 32.0\}\}$.

However, to find the solution while accommodating the read-only annotation, the hierarchy J is formed by replacing $F?$ by a newly introduced variable W:

required $C * 1.8 = W - 32.0$
strong $C = 0.0$
weak $F = 212.0$

Q is the set of all hierarchies resulting from substituting an arbitrary real number for W. For example, the hierarchy resulting from the substitution $\rho = \{W \mapsto 14.0\}$ is:

required $C * 1.8 = 14.0 - 32.0$
strong $C = 0.0$
weak $F = 212.0$

which has the singleton set of solutions $\{\theta = \{C \mapsto -10.0, F \mapsto 212.0\}\}$, but is not consistent because $W\rho \neq F\theta$ ($14.0 \neq 212.0$).

The only hierarchy in Q with a consistent solution results from $\rho = \{W \mapsto 212.0\}$:

required $C * 1.8 = 212.0 - 32.0$
strong $C = 0.0$
weak $F = 212.0$

and so the set of solutions to the original hierarchy H is $\{\{C \mapsto 100.0, F \mapsto 212.0\}\}$. (Note that the *strong* $C = 0.0$ constraint is not satisfied because there is no consistent solution that satisfies it.)

Now consider the motivating example in Section 3.1 for which *blocked* is true:

required $V > 0$
required $V? = 1$

To illustrate the definition of *blocked*, form the new hierarchy J by replacing $V?$ with W:

required $V > 0$
required $W = 1$

There exists a $d \in \mathcal{R}$, for example $d = 6$, such that, for the substitutions $\rho = \{W \mapsto 6\}$, $\theta = \{V \mapsto 6\}$, and $\sigma = \{\}$, $J\rho\theta\sigma$ has no solutions, but $J\theta\sigma$ does have a solution:

$$\frac{Jp\theta\sigma}{\begin{array}{ll} required & 6 > 0 \\ required & 6 = 1 \end{array}} \qquad \frac{J\theta\sigma}{\begin{array}{ll} required & 6 > 0 \\ required & W = 1 \end{array}}$$

$$no\ solutions \qquad \{\{W \mapsto 1\}\}$$

Hence *blocked* is true for this hierarchy. However, if we added the additional constraint *required* $V = 1$ to the original hierarchy, then *blocked* would become false.

Practical Examples of Using Read-only Annotations A trivial but useful example is a spreadsheet-like constraint that $A? + B? + C? = Sum$. The read-only annotations prevent the user from editing *Sum* and having the change propagate back to A, B, or C, but still allow the user to edit A, B, or C.

As noted in the introduction, an important use of read-only annotations is in constraints that reference an external input device or other outside source of information. For example, if we have a constraint that a point P follow the mouse, the constraint should be read-only on the mouse position:

$$P = mouse.position?$$

As another example, suppose we have a simple scrollbar displayed on the screen. When the "thumb" is dragged up and down, we want the top and bottom of the scrollbar to remain fixed. However, we want to be able to reposition the scrollbar as a whole, so simply anchoring the top and bottom isn't the correct solution.[2] To handle this problem cleanly, we define a constraint relating the position of the thumb, the top, the bottom, and a number *percent*, in which the the top and bottom are annotated as read-only:

$$percent = \frac{thumb - bottom?}{top? - bottom?}$$

The read-only annotations on *top* and *bottom* are specific to this constraint, so the whole scrollbar can be repositioned by some other "move" constraint.

Circularities While the sets of solutions to many hierarchies are intuitively clear, this clarity often vanishes when the hierarchy contains cycles. We present two such examples here. These are pathological cases that would not arise in realistic applications—but nevertheless the theory should and does specify how they are to be handled.

The following two hierarchies both contain a cycle through variables annotated as read-only. In the first hierarchy, none of the constraints in the cycle is more restrictive than the others and so, intuitively, information can flow properly and still yield a solution.

---

[2] We could almost achieve the desired result by putting strong (but not required) anchors on the top and bottom of the mouse. However, if other constraints on the output value from the slider became too strong, then the top or bottom would move; we would prefer a more robust object.

required $X? = Y + 1$
required $\quad X = Y? + 1$

For this hierarchy, *blocked* is false and the set of solutions is the infinite set $\{\{X \mapsto d + 1, Y \mapsto d\} \mid d \in \mathcal{R}\}$.

In the second hierarchy, however, the *required* $X? = Y + 1$ constraint is more restrictive than the *required* $X \geq Y?$ one. Thus the "unequal" information flow results in *blocked* being true.

required $X? = Y + 1$
required $\quad Y = 20$
required $\quad X \geq Y?$

For this hierarchy, the set of solutions is $\{\{X \mapsto 21, Y \mapsto 20\}\}$; however, *blocked* is true.

3.2 Write-only Annotations

In addition to read-only annotations, it is also convenient if *write-only anno-tations* are available. Intuitively, if a variable is annotated as write-only in a constraint, we only want information to be able to flow from the constraint into that variable, and not back. We could define the effect of write-only annotations from first principles, in a manner analogous to the definition for read-only an-notations. However, it is simpler to define write-only annotations in terms of read-only annotations.

Definition. Let H be a constraint hierarchy containing write-only annota-tions (it may contain read-only annotations as well), and let $\mathcal{D}$ be the domain of the constraints. Let $v_1, \ldots, v_m$ be the variables in H that have one or more write-only occurrences. Let $w_1, \ldots, w_m$ be new variables not occurring in H, and let J be the hierarchy that results from substituting w_i for each write-only occurrence of the corresponding variable v_i. Let J' be the hierarchy formed by augmenting J with the additional required constraints $v_i = w_i?$ for $1 \leq i \leq m$. The desired set of solutions to H is the the set of solutions to J', with the mappings for the w_i omitted:

$$solutions(H) = solutions(J') \text{ omitting } w_1, \ldots w_m$$

The definition of the set $solutions(J')$ used above is, of course, that given in Section 3.1.

For example, let H be:

required $X! = Y$
strong $\quad X = 4$
weak $\quad Y = 3$

Intuitively, even though the constraint $X = 4$ is stronger than the constraint $Y = 3$, information will only be allowed to flow from Y to X in the $X! = Y$ constraint, since X is annotated as write-only. Tracing through the definition, the hierarchy J' is formed by replacing $X!$ by a newly introduced variable W, and adding the required constraint $X = W?$.

```
required  W = Y
required  X = W?
strong    X = 4
weak      Y = 3
```

The set of solutions to J' is $\{\{W \mapsto 3, X \mapsto 3, Y \mapsto 3\}\}$. The desired set of solutions to H is the same, but with the mapping for W omitted: $\{\{X \mapsto 3, Y \mapsto 3\}\}$.

3.3 Partially Ordered Hierarchies

In some applications, imposing a total order on the constraint strengths may be over-specifying the problem. We therefore also define the set of solutions to a *partially ordered* constraint hierarchy. A partially ordered hierarchy must still have a distinguished *required* strength. However, the other constraint strengths need only be placed in a partial order, rather than a total order.

Informally, we define the set of solutions to a partially ordered constraint hierarchy by forming the set of all totally ordered hierarchies that are *consistent* with the original one. These totally ordered hierarchies are formed by adding any additional, permissible orderings between the partially ordered strengths: less than, greater than, or equal. The desired set of solutions is then the union of the sets of solutions to the totally ordered hierarchies.

Definition. If P is a partially ordered hierarchy, a totally ordered hierarchy H is *consistent* with P if (1) both hierarchies contain the same constraints, and (2) there is a mapping m from the strengths of P to the strengths of H such that if $s_1 < s_2$ in P then $m(s_1) < m(s_2)$ in H, and (3) $\forall i, s_i c_i \in P$ iff $m(s_i) c_i \in H$.

Definition. Let P be a partially ordered hierarchy. Then

$$solutions(P) = \bigcup_{H \in \mathcal{H}} solutions(H)$$

where $\mathcal{H}$ is the set of all totally ordered hierarchies consistent with P.

As a trivial example, consider the following hierarchy:

```
wimpy      X = 3
indecisive X = 4
```

Strengths *wimpy* and *indecisive* are both non-required, but no ordering is specified between them. The total orders that are consistent with this partial order make *wimpy* stronger than *indecisive*, *wimpy* weaker than *indecisive*, and *wimpy* the same strength as *indecisive*. The *locally-predicate-better* solutions to these hierarchies are $\{\{X \mapsto 3\}\}$, $\{\{X \mapsto 4\}\}$, and $\{\{X \mapsto 3\}, \{X \mapsto 4\}\}$ respectively. Therefore, the set of *locally-predicate-better* solutions to the original partially ordered hierarchy is $\{\{X \mapsto 3\}, \{X \mapsto 4\}\}$.

The definition involves adding all possible orderings between the strengths, including equality. For the local comparators, equality is unnecessary—any solution for a totally ordered hierarchy formed using an equality relation will also be a solution for one of the other totally ordered hierarchies formed using just inequality. This is, however, not the case for the global comparators. For example, if the *least-squares-better* comparator is used, the solutions to the totally ordered hierarchies are $\{\{X \mapsto 3\}\}$, $\{\{X \mapsto 4\}\}$, and $\{\{X \mapsto 3.5\}\}$ respectively, so that the set of *least-squares-better* solutions to the original partially ordered hierarchy is $\{\{X \mapsto 3\}, \{X \mapsto 3.5\}, \{X \mapsto 4\}\}$.

We have also considered a variant definition for the solutions to partially ordered hierarchies. In the variant, not only would the constraints from two partially ordered strengths be combined into a single strength (i.e., the equality ordering), but also all possible weightings between the constraints would be used. In the above example, for *least-squares-better*, the following infinite set of totally ordered hierarchies would be considered:

strong $X = 3$
weak $X = 4$

strong $X = 4$
weak $X = 3$

$medium[w_1]\ X = 3$
$medium[w_2]\ X = 4$
  for all positive numbers (weights) w_1 and w_2.

The set of solutions in this case would map X to all numbers between 3 and 4 inclusive, i.e. $\{\{X \mapsto a\} \mid a \in [3 \ldots 4]\}$.

3.4 Objective Functions

In a standard linear programming problem [53], we wish to minimize (or maximize) the value of a linear function $z(x_1, \ldots, x_k) = a_1 x_1 + \ldots + a_k x_k$ in k real-valued variables $x_1, \ldots, x_k$, subject to the non-negativity constraints $x_1 \geq 0, \ldots, x_k \geq 0$, and also subject to m additional linear equality or inequality constraints on $x_1, \ldots, x_k$. The function to be minimized or maximized is called the *objective function*.

If the objective function is to be minimized, and if its coefficients z_i are all non-negative, then we can easily represent the linear programming problem as a constraint hierarchy. The k non-negativity constraints and the m additional linear equality and inequality constraints can be represented as required constraints, and the objective function can be represented as a soft constraint $z(x_1, \ldots, x_k) = 0$, since we know *a priori* a lower bound (namely 0) on the value of the objective function. However, if a lower bound isn't known *a priori*, then this transformation would not be appropriate. We could instead set a goal g for the objective function, and decide that we would be completely satisfied if we reach or exceed the goal. (This is the goal programming approach.) In this case, we can represent the objective function as the soft constraint $z(x_1, \ldots, x_k) \leq g$.

Another approach would be to represent the objective function as the soft constraint $z'(x_1, \ldots, x_k) = 0$ where

$$z'(x_1, \ldots, x_k) = \begin{cases} -1/z(x_1, \ldots, x_k) & \text{if } z(x_1, \ldots, x_k) < \text{-}1 \\ z(x_1, \ldots, x_k) + 2 & \text{if } z(x_1, \ldots, x_k) \geq \text{-}1 \end{cases}$$

However, this approach has the disadvantage that it has converted a linear problem into a nonlinear one, making it much harder to solve.

Similar arguments apply for the case of maximizing an objective function.

To overcome these difficulties, we can again extend the basic constraint hierarchy theory to include objective functions explicitly. A *constraint hierarchy with objective functions* is a constraint hierarchy, along with a set of objective functions, also labeled with strengths (which must all be non-required). To simplify the definition, we first replace any objective function $z(x_1, \ldots, x_k)$ to be maximized by $0 - z(x_1, \ldots, x_k)$, which should be minimized. Let Z_i be the set of objective functions at the ith level of the hierarchy. We can then extend the definition of *locally-better* as follows. (The expression $z\theta$ denotes the value of $z(x_1\theta, \ldots, x_k\theta)$, i.e. the value of z when applied to the values for $x_1, \ldots, x_k$ defined by θ.)

> *locally-better*$(\theta, \sigma, H) \equiv$
> $\exists k > 0$ such that
> $\forall i \in 1 \ldots k - 1 \ (\forall p \in H_i \ e(p\theta) = e(p\sigma) \ \wedge \ \forall z \in Z_i \ z\theta = z\sigma)$
> $\wedge \ (\exists q \in H_k \ e(q\theta) < e(q\sigma) \ \vee \ \exists z \in Z_k \ z\theta < z\sigma)$
> $\wedge \ \forall r \in H_k \ e(r\theta) \leq e(r\sigma)$
> $\wedge \ \forall z \in Z_k \ z\theta \leq z\sigma$

In other words, for θ to be *locally-better* than σ, θ must do exactly as well as σ on both the constraints and objective functions through level $k - 1$; at level k, θ must do as well or better on all the constraints and objective functions, and it must do strictly better for at least one constraint or objective function.

In keeping with its nature, *locally-better* considers constraints and objective functions individually. The *globally-better* comparators combine the errors for the constraints at a given level of the hierarchy. The constraint errors are bounded below by 0, while in general the objective function has no definite minimum value—so combining these values into one composite value seems unwise. For the global comparators, therefore, we restrict the constraint hierarchy with objective functions to have at each level either just constraints, or just a single objective function. (Multiple objective functions at a given level should be replaced by a single function that combines the values appropriately.)

The extended *globally-better* schema is:

> *globally-better*$(\theta, \sigma, H, g) \equiv$
> $\exists k > 0$ such that
> $\forall i \in 1 \ldots k - 1 \ (\ g(\theta, H_i) = g(\sigma, H_i) \ \wedge \ z_i\theta = z_i\sigma \ )$
> $\wedge \ (\ g(\theta, H_k) < g(\sigma, H_k) \ \vee \ z_k\theta < z_k\sigma \ )$

Here, if i is a level containing constraints, $g(\tau, H_i)$ is defined in the usual way and $z_i\tau$ is 0; if i is a level containing an objective function, $g(\tau, H_i)$ is defined to be 0, and $z_i\tau$ is the value of the objective function at that level.

3.5 Comparing Solutions Arising from Different Hierarchies

In some applications—in particular, in many HCLP($\mathcal{R}$) programs that we have written—to rule out unintuitive solutions, it is useful to compare not just solutions to a given constraint hierarchy, but also solutions from several different hierarchies. (In logic programming, these different hierarchies are generated by alternate choices of rules.) We have extended the theory described above to include such comparisons [86], but, for the sake of brevity, we don't discuss this extension here.

4 Constraint Satisfaction Algorithms

Searching for an efficient constraint satisfaction algorithm that works for all domains, comparators, and kinds of constraints would be a futile endeavor. Rather, we need to look for algorithms specialized by one or more attributes. In [25] we outline a number of algorithms for solving constraint hierarchies, each of which makes a different engineering trade-off between generality and efficiency. Much of our research so far has used the *locally-predicate-better* comparator over arbitrary domains. When there are no circularities in the constraint graph, we have an efficient incremental algorithm for this comparator. For arbitrary linear constraints, we also have an efficient algorithm based on linear programming techniques. In the following sections, we briefly discuss these two algorithms. For more details on the incremental acyclic algorithm, the reader is referred to [23, 24, 25, 50]; [25] and [50] include proofs of correctness and complexity results. References [26, 27, 85] discuss the linear programming algorithm.

4.1 Blue and DeltaBlue: Algorithms for Acyclic Hierarchies

Among the most common techniques for satisfying constraints is *local propagation*. In local propagation, a constraint can be used to determine the value of one of its variables whenever the values of the other $n - 1$ of its variables are known. This may then allow some other constraint to determine another variable's value, and so forth. Local propagation is similar in this respect to propagating values through a dataflow network. The difference is that while a dataflow network has a single (partially ordered) propagation path, a set of multi-way constraints typically has many potential propagation paths. Thus the constraint solver must in general decide which path to use, and in the case of a constraint hierarchy solver, ensure that this is a path that computes a "best" solution.

For local propagation, each constraint supplies one or more *methods*: procedures that, if executed, will cause the constraint to be satisfied. Each method determines a value for one or more variables (outputs) from its other variables (inputs). For example, the plus constraint $A + B = C$ has three methods: $A \leftarrow C - B$, $B \leftarrow C - A$, and $C \leftarrow A + B$. A local propagation constraint solver produces a propagation path by selecting, and perhaps executing, a method for each constraint in the hierarchy (or, if the constraint cannot be satisfied, no method).

Because local propagation solutions are based on these "all or nothing" methods rather than on some error metric, local propagation constraint solvers are restricted to the predicate comparators from Section 2.1. Similarly, because local propagation paths utilize at most one method (i.e., at most one constraint) per output variable, they are unable to solve cyclic constraints such as those produced by a set of simultaneous equations.

We christened our local propagation algorithm for constraint hierarchies "Blue". Subsequently, to improve response time for large constraint graphs, we developed an incremental version of the algorithm which we named DeltaBlue. The Blue algorithm is $O(N^2)$ in the total number of constraints, whereas the DeltaBlue algorithm is $O(cN)$ in the number of affected constraints [31].

Local propagation algorithms, such as Blue and DeltaBlue, can easily accommodate read-only and write-only annotations as well as partially ordered hierarchies. The read-only and write-only annotations are handled by not including certain methods. For example, $A? + B = C$ would have two, instead of three, methods: $B \leftarrow C - A$ and $C \leftarrow A + B$, but not $A \leftarrow C - B$. Similarly, $A + B! = C$ would have just one method: $B \leftarrow C - A$. Partially ordered hierarchies are easily handled as well by the basic Blue and DeltaBlue algorithms. The basic Blue and DeltaBlue algorithms find a single *locally-predicate-better* solution to the constraint hierarchy. However, both algorithms can be modified to return all solutions, as in the ThingLab I Multiple Solutions Browser [28].

We have implemented and used both Blue and DeltaBlue in Smalltalk, C, C++, Object Pascal, and Common Lisp. All of these implementations support read-only and write-only annotations, but only the Smalltalk implementation accommodates partially ordered hierarchies.

4.2 Algorithms for Linear Equality and Inequality Constraints

One disadvantage of local propagation algorithms is that they cannot reliably handle cycles in the constraint graph. In some cases these algorithms will find an acyclic solution to a cyclic graph, but this behavior is not guaranteed; the algorithms often halt with a "cyclic constraint graph" error message instead. Further, if the constraints are truly simultaneous, then local propagation algorithms simply cannot find a solution. Therefore, we designed another set of algorithms that can solve constraint hierarchies consisting of arbitrary collections of linear equality and inequality constraints using the *weighted-sum-metric-better*, *worst-case-metric-better*, and *locally-metric-better* comparators. These algorithms are instances of our general DeltaStar [26, 27] framework and are collectively referred to as the Orange algorithms.

The DeltaStar framework is an algorithm for incrementally solving a constraint hierarchy, based on an alternate, but provably equivalent, description of the constraint hierarchy theory [26, 29, 85]. Whereas the basic constraint hierarchy theory in Section 2 emphasizes the dichotomy between the hard and soft levels, the alternative theory emphasizes the hierarchical refinement of the set of solutions.

The Orange algorithms use the basic DeltaStar framework by transforming the constraint hierarchy into a series of linear programming problems—one problem for each level in the hierarchy. All three Orange algorithms have been implemented in Smalltalk and Common Lisp. However, none of these implementations supports partially ordered hierarchies or read-only and write-only annotations.

4.3 Other Algorithms

Although not designed for solving constraint hierarchies, many other constraint solving techniques are available, including augmented term rewriting [46], relaxation [3, 44, 75], and searching for a solution over a finite domain. Augmented term rewriting is an equation rewriting technique borrowed from functional programming languages, with added support for objects and multi-directional constraints. Relaxation is an iterative numerical technique, in which the value of each real-valued variable is repeatedly adjusted to minimize the error in satisfying its constraints. Relaxation will converge on a *least-squares-better* solution, unless it gets trapped in a local but suboptimal minimum. Mackworth [47], Van Hentenryck [78], and others describe efficient algorithms for solving sets of constraints on variables ranging over finite domains.

5 Using Constraint Hierarchies

In the following sections, we discuss four systems in which we have used constraint hierarchies: ThingLab, ThingLab II, HCLP($\mathcal{R}$) (a language that integrates constraint hierarchies with logic programming), and Kaleidoscope (a hybrid constraint-imperative programming language); we also list a number of systems built by other researchers that have applied this theory as well.

5.1 Systems for Building Simulations and User Interfaces

ThingLab [3] was a constraint-based laboratory that allowed a user to construct simulations of such things as electrical circuits, mechanical linkages, demonstrations of geometric theorems, and graphical calculators using interactive direct-manipulation techniques. ThingLab used two kinds of local propagation, as well as relaxation, to solve constraints. It would propagate known values "forward" and degrees of freedom "backward" through the graph. Later versions of ThingLab incorporated such features as explicit constraint hierarchies (as described here), incremental compilation, and a graphical facility for defining new kinds of constraints [4, 6, 19, 51]. The Animus system [5, 15] was an animation system implemented on top of ThingLab. Animus added *temporal constraints* to ThingLab where a temporal constraint is a relation that is required to hold between the existence of a stimulus event and a response in the form of a stream of new events. ThingLab II is a complete rewrite of the original ThingLab, oriented toward building user interfaces [50, 51]. ThingLab II supports constraint hierarchies, and includes an implementation of the DeltaBlue incremental constraint satisfaction algorithm. ThingLab II also includes a compiler that optimizes structured, constrained objects by discarding unnecessary structure and compiling the constraints into native code [19].

In other research on using constraint hierarchies in user interfaces, Epstein and LaLonde [17] used our constraint hierarchy theory in implementing a layout system for Smalltalk windows. They used constraints to define the relation between the canvas size, window size, and scale factors. By default, all parameters were variable. However, the user could add a stronger constraint that one or more of the parameters stayed fixed, thus creating a fixed canvas, fixed size, or fixed scale window. TRIP and TRIP II [43, 77] also use constraint hierarchies for user interfaces, with a two-level constraint hierarchy consisting of required constraints and one level of soft constraints, with weights on each soft constraint. Delta TRIP is a version of TRIP II using the DeltaBlue algorithm as its constraint satisfier. Finally, constraint hierarchies were used to simulate the physiological affects of open-heart surgery in a system for supporting anesthesiologists in the operating room [62].

5.2 Constraint Hierarchies in Logic Programming Languages

In standard logic programming, as exemplified by Prolog, rules are of the form

$$p(\mathbf{t}) \ :- \ q_1(\mathbf{t}), \ldots, q_m(\mathbf{t}).$$

where $p, q_1, \ldots, q_m$ are predicate symbols, and $\mathbf{t}$ denotes a list of terms. The Constraint Logic Programming (CLP) scheme [40] is a general scheme for extending logic programming to include constraints, and is parameterized by $\mathcal{D}$, the domain of the constraints. In a CLP language, rules are of the form

$$p(\mathbf{t}) \ :- \ q_1(\mathbf{t}), \ldots, q_m(\mathbf{t}), c_1(\mathbf{t}), \ldots, c_n(\mathbf{t}).$$

where $p, q_1, \ldots, q_m$ are as before, and $c_1, \ldots, c_n$ are constraints over the domain $\mathcal{D}$.

Operationally, in a CLP language we can think of executing the Prolog part of the program in the usual way, accumulating constraints on logic variables as we go, and either verifying that the constraints are solvable or else backtracking if they are not. The program can terminate with substitutions being found for all variables in the input, or with some constrained variables still unbound, in which case the output would include the remaining constraints on these variables.

Hierarchical Constraint Logic Programming (HCLP) [7, 85, 86] is a generalization of the CLP scheme, and is again parameterized by the domain $\mathcal{D}$ of the constraints. In HCLP rules are of the form

$$p(\mathbf{t}) \ :- \ q_1(\mathbf{t}), \ldots, q_m(\mathbf{t}), s_1 c_1(\mathbf{t}), \ldots, s_n c_n(\mathbf{t}).$$

where each s_i is a symbolic name indicating the strength of the corresponding constraint c_i.

Operationally, goals are satisfied as in CLP, temporarily ignoring the non-required constraints, except to accumulate them. After a goal has been successfully reduced, there may still be non-ground variables in the solution. In this event, the accumulated hierarchy of non-required constraints is solved, using a method appropriate for the domain and comparator, thus further refining the values of these variables. Additional answers may be produced by backtracking. As with CLP, constraints can be used multi-directionally, and the scheme can accommodate collections of constraints that cannot be solved by simple forward propagation methods.

To test our ideas, and to allow us to experiment with HCLP programs, we have written two different HCLP interpreters. Our first interpreter is written in $CLP(\mathcal{R})$, allowing it to take advantage of the underlying $CLP(\mathcal{R})$ constraint solver and backtracking facility. As a result, it is small (2 pages of code) and clean. However, it is not incremental—rather, it recomputes all the *locally-predicate-better* answers for each derivation, instead of incrementally updating its answers as constraints are added and deleted due to backtracking, and thus the interpreter is not particularly efficient. Our second HCLP interpreter is again for the domain of the real numbers, but supports the *weighted-sum-metric-better*, *worst-case-metric-better*, and *locally-metric-better* comparators instead. The comparator to be used in a given program is indicated by a declaration at the beginning of an HCLP program. The second interpreter is implemented in Common Lisp, and uses the DeltaStar algorithm mentioned in Section 4.2. The second interpreter includes some evaluable predicates for performing input and graphical output, so that we can use HCLP for interactive graphics applications. Further details regarding both implementations may be found in [85].

5.3 Constraint Hierarchies in Imperative Languages and Systems

Imperative languages, such as those in the Algol family, have the standard notions of state and destructive assignment. Pure constraint languages, on the other hand, are declarative, without state and assignment. Constraint imperative programming languages, such as Kaleidoscope'90 and '91, are an attempt to merge these two apparently incompatible paradigms.

In CIP (Constraint Imperative Programming), the two paradigms are reconciled by using imperative statements to provide control flow and constraint expressions to provide computation. Imperative assignment statements are translated into constraints between the previous and current states of the object. In other words, X:=X+1 is defined as the constraint $X_t = X_{t-1}? + 1$. (The read-only annotation is used to prevent any computations in the present from changing the past.) Objects are represented as a stream of values over time, as in Lucid [83], where time is defined by the execution of subsequent imperative statements. A weak equality constraint between each pair of values ensures that the object does not change randomly: $\forall t \ weak \ X_t = X_{t-1}?$. When a variable is assigned to, the stronger "assignment" constraint will override the weaker stay constraint, and the object's state will change. The new value will be propagated forward via the weak stay constraints until the variable is assigned to again.

Constraints do not typically refer to time, whereas time (or rather, sequencing) is crucial to an imperative language. Thus the Kaleidoscope languages use *constraint templates* to create constraints over a variety of intervals, including: just once (e.g., an assignment constraint), until some condition is false (e.g., asserting a constraint while the mouse button is held down), or always (e.g., a data invariant).

Additionally, the Kaleidoscope languages are object-oriented, supporting both user defined objects and user defined constraints over those objects. These latter constraints are defined using *constraint constructors*: side-effect-free procedures that define the meaning of complex constraints over objects in terms of more primitive constraints over the objects' component parts.

Further details regarding the semantics and implementations of both Kaleidoscope'90 and '91 can be found in [20, 21, 22, 29].

5.4 User Interface Issues

There are a number of user interface issues that arise in supporting constraint hierarchies, three of which are discussed here: how to express constraints, how to show alternate solutions to the constraint hierarchy, and how to achieve good performance in an interactive graphical constraint-based system.

Expressing Constraints Expressing constraint hierarchies in a textual language presents no particular difficulty; once we have a syntax for the constraints themselves, we can annotate them with strengths. In ThingLab, our approach has been to manipulate graphical objects that carry the constraints, rather than graphically representing the constraints themselves. For example, when constructing a graphical calculator, we insert *Plus*, *Times*, *Printer*, and other sorts of objects, each of which holds state, icon, and constraint information. This approach carries over naturally to constraint hierarchies: objects can carry both required and preferential constraints. Objects will normally have weak stay constraints on their parts to give stability to them and to any larger containing object, in addition to any other constraints they may have.

Showing Alternate Solutions A given constraint hierarchy may have several solutions (even infinitely many). The technique used in HCLP($\mathcal{R}$) to present multiple solutions is the same as in other logic programming languages such as Prolog and CLP($\mathcal{R}$). A single *answer* may represent one or more solutions. For example, the answer $X > 5$ compactly represents the infinite set of solutions mapping X to each real number greater than 5. Answers are presented, one at a time. The user can reject an answer, and backtracking will produce a new one (if one exists). As in CLP($\mathcal{R}$), a given answer can contain variables, perhaps with constraints on them. For example, consider the following short HCLP($\mathcal{R}$) program:

(a) `banana(X) :- artichoke(X), weak X>6.`
(b) `artichoke(X) :- strong X=1.`
(c) `artichoke(X) :- required X>0, required X<10, weak X<4.`

Given the goal `?- banana(A)`, the first answer would be produced using the `banana` clause (a) and the first of the `artichoke` clauses (b), yielding the hierarchy $strong\ X = 1$, $weak\ X > 6$. There is a single answer to this hierarchy, namely $X = 1$, which would then be displayed. Upon backtracking, the second `artichoke` clause (c) is selected, resulting in the hierarchy $required\ X > 0$, $required\ X < 10$, $weak\ X < 4$, $weak\ X > 6$. Using the *locally-predicate-better* comparator, this hierarchy has two answers. The first answer to this hierarchy, but the second to the goal, is $X > 0, X < 4$. Upon further backtracking the third and final answer to the goal, namely $X > 6, X < 10$, would be displayed. Thus, this program produces two constraint hierarchies and three answers:

Clauses	Hierarchies		Answers	
a, b	$strong\quad X = 1$ $weak\qquad X > 6$	$X = 1$		
a, c	$required\ X > 0$ $required\ X < 10$ $weak\qquad X < 4$ $weak\qquad X > 6$	$0 < X < 4$	$6 < X < 10$	

Both of our HCLP($\mathcal{R}$) implementations have primarily textual interfaces. In a system with a graphical interface, presenting multiple solutions raises some interesting problems. ThingLab II adopts the simple strategy of just picking one solution. In a previous version of ThingLab [28], we did allow the user to browse through multiple solutions graphically. For overconstrained problems (i.e., cases in which HCLP would return additional answers on backtracking), the multiple solution browser would pop up a menu of alternate solutions, so that the user could browse through the different alternatives. For underconstrained problems (i.e., cases where HCLP would return an answer with one or more variables not bound to a unique value), the multiple solution browser would allow the user to move interactively through the space of possible solutions. The user would select an underconstrained part, and the system would respond by displaying a control icon in a new pane and by setting up constraints relating the position of the icon to underconstrained variables in the selected part. The user could then move the control icon in either one or two dimensions, depending on how many degrees of freedom remained for the underconstrained part. (Our implementation didn't support manipulating parts with more than two degrees of freedom, although it could be so extended.) Based on the position of the icon, the system would satisfy the constraints and display the solution. Both techniques (for overconstrained and underconstrained problems) would be used simultaneously if needed.

Performance Issues In an interactive application, keeping the perceived response time low is perhaps more important than achieving the fastest speed. We use two terms in discussing response time: latency, the delay between the input event and the first time the constraints are satisfied; and repetition time, the time it takes to re-satisfy the constraints each time the screen image is updated.

In a naive implementation of constraint hierarchies, the system, in response to each new input event, would first remove any old constraints from previous input events, and then add one or more constraints to the constraint hierarchy. Thus, each new input event would result in a new constraint hierarchy, a new invocation of the constraint solver, and a new set of solutions. The latency and repetition time would be identical. For example, if the scroll bar of a window is being moved by the mouse, the mouse motion events remove and add a sequence of individual constraints: *ScrollBar* = 15, *ScrollBar* = 16, ..., *ScrollBar* = 25, etc.

The implementations of the DeltaBlue algorithm in ThingLab and ThingLab II divide the task of solving the constraints into two parts: structure-directed solving, and data-directed solving. The structure solver, or planner, finds one or more solutions to a constraint hierarchy based only on the structure of the constraints (which variables they constrain, whether or not they constrain their variables uniquely, and so forth). The data solver uses the structure solution to satisfy the constraint hierarchy for specific data values. The structure solution is known as a "plan" because it embodies the procedure for solving the hierarchy. The same plan can be used to solve for multiple data values, until the hierarchy is altered by adding or removing constraints.

Because the data solver is much faster than the structure solver, an "active" (or "edit") constraint is used to modify data values without changing the hierarchy. In the scroll bar example, this means that rather than adding and removing the sequence of constraints, the single active constraint *ScrollBar* = *Mouse* is used. The *Mouse* variable injects the current position of the mouse into the constraint hierarchy, and the data-driven solver can use the existing plan to produce a new solution. The run-time of this technique is $1S + nD$ (1 Structure solution + n Data solutions) whereas the run time if constraints are added and removed for each new value is $n(S + D)$—substantially slower.

ThingLab II has two techniques for executing the plan. The first is interpretation; the second is compilation into native code, followed by execution of that code. When the plan is interpreted, the latency is moderate and the repetition time is moderate. When the plan is compiled, the latency is very high and the repetition time is low. Thus, if the same plan will be used repeatedly, the average run-time will be decreased by compiling the plan. However, during prototyping and development, the constraint hierarchy is in a constant state of flux, causing compiled plans to become obsolete and be discarded. Thus, to decrease the variability of response time, our ThingLab II work has emphasized fast interpretation. Once the constraint hierarchy for an object has been designed, implemented, and tested, the ThingLab II compiler [19] can be used to compile the constraints into efficient native code.

6 Other Related Work

Much of the previous and related work on constraint-based languages and systems can be grouped into the following areas: geometric layout, spreadsheets and similar systems, user interface support, general-purpose programming languages, and artificial intelligence applications. In this section we discuss a number of these related efforts. Since this body of related work is very large, here we concentrate on work, in addition to that described in Section 5, involving combinations of hard and soft constraints. Other bibliographies and discussions may be found in [25], [29], and [46].

6.1 Geometric Layout

Geometric layout is a natural application for constraints, and was also their first area of application, in the venerable Sketchpad system [75, 76]. Sketchpad allowed the user to build up geometric figures using primitive graphical entities and constraints, such as point-on-line, point-on-circle, collinear, and so forth. When possible, constraints were solved using local propagation. When this technique was not applicable, Sketchpad would resort to relaxation. Although the primitive constraints were hard-coded into the system, new primitive constraints could be added by programming an error function in the underlying implementation language. In addition to its geometric applications, Sketchpad was used for simulating mechanical linkages. Sketchpad was a pioneering system in interactive graphics and object-oriented programming as well as in constraints. Its requirements for CPU cycles and display bandwidth were such that the full use of its techniques had to await cheaper hardware years later.

Juno [58] is a constraint based system for geometric layout similar to ThingLab. The major innovation of Juno was its dual presentation of the constraints: one window contained the graphical layout defined by the constraints while the other window contained the textual definition of the same constraints. Both representations were editable, and the results were reflected in both windows simultaneously. Other constraint-based geometric layout systems include IDEAL [79, 80], Magritte [33], COOL [43], Converge [73] for 3-d geometric modeling, and [2] for laying out cyclic graphs.

All of the interactive geometric layout systems had to deal in some way with the problem of default constraints. As discussed in Section 1.1, given a collection of geometric objects with constraints on them, if a part is moved, in general there are many ways to readjust the objects so that the constraints are satisfied. For example, if we move one endpoint of a horizontal line, we don't expect that it will suddenly triple in length (even though the constraint that it be horizontal would still be satisfied). In Sketchpad, the old x and y locations of points are the starting values for the iterative relaxation routine. Even when using local propagation, Sketchpad would solve for values using an individual constraint by considering the constraint error and finding a new value that would make the error go to zero. Thus, if one views the old values as "stay" constraints, and the user's input as a required constraint, Sketchpad would find a *locally-metric-better* solution to the constraints. If only relaxation were used and not local propagation, the solution would also be close to a *least-squares-better* solution. Sketchpad also supported read-only annotations on variables (Sutherland called them "reference-only variables"). Sutherland notes that misusing reference-only variables can lead to instabilities in the relaxation algorithm.

The original version of ThingLab followed Sketchpad's lead, and added local propagation methods to constraints, and constraints over arbitrary domains (not just the real numbers). All the explicit constraints were required; the user's edit requests were implicitly treated as strong preferences rather than requirements, so that if the edit conflicted with a required constraint, the user's constraint would be overridden. In addition, there were implicit weak or very weak constraints that parts of an object keep their old values as the object was being manipulated by the user, unless it was necessary for them to change to satisfy the user's edit or the explicit required constraints. Some of these implicit weak constraints needed to be stronger than others to achieve intuitive behavior. For example, suppose that we have a simple graphical calculator, which includes a constraint $A + B = C$. Now suppose the user edits the value of A. We expect that the system will re-satisfy the plus constraint by changing C, rather than by changing B. To achieve this, the local propagation methods of a constraint were ordered to indicate which ones should be used in preference to others. (For $A + B = C$, the method for updating C would be listed first.) This (usually) gave the same effect as making the stay constraint on C weaker than the ones on A and B. Also, the user's input—for example, moving something with the mouse—was considered as a preference rather than a requirement, so that an anchor or constant could cause it to be overridden. Thus ThingLab would usually also find a *locally-metric-better* solution.

Neither Sketchpad nor ThingLab used a separate, declarative theory of constraints; these choices were embedded in the procedural code of the constraint satisfier. This situation became increasingly troublesome when we tried to improve on ThingLab's constraint satisfier, since there was no declarative specification that we could use to decide whether a particular optimization would lead to a correct answer. In response, the constraint hierarchy theory described in this paper was developed, and was used in later versions of the system.

Similar considerations obtain for the other interactive geometric layout systems. In Magritte [33], the system performed a breadth-first search to change as few variables as possible. This often gives similar answers to *unsatisfied-count-better*, but without too much trouble one can come up with problems where it doesn't give a reasonable answer. For example, consider the constraint $X_1 + \ldots + X_n = Sum$, which is represented as a chain of three-argument plus constraints. If X_i is changed, the breadth-first search solution would be to update either X_{i-1} or X_{i+1}; but the user might well intend that plus have its normal directional bias, so that Sum would be updated instead. Constraint hierarchies allow either of these solutions to be preferred by suitable choice of comparator and strength of the stays. Vander Zanden's algorithm [82] uses a heuristic that attempts to minimize the number of equations that must be solved; again, this is related to *unsatisfied-count-better*, but the exact choice is embedded procedurally in the satisfier.

6.2 Spreadsheets and Related Systems

Spreadsheets, such as Lotus 1-2-3 or Microsoft EXCEL, are constraint systems in that the user specifies relations to hold between values in cells, although these constraints are usually unidirectional. Spreadsheets in effect trivially implement stay constraints on unedited cells by their update algorithm. The most recent spreadsheet implementations include built-in solver and optimization packages, and thus have much of the power of the other constraint systems. TK!Solver [44] is a commercially available system that uses constraints in a "general purpose problem solving environment" targeted at mechanical and electrical engineers. It uses local propagation and relaxation as solution techniques, but when relaxation is required, it asks the user to make initial guesses of the variable's values, thus greatly improving the chances of convergence.

6.3 User Interface Toolkits

Another frequent application of constraints is in user interface toolkits, where they are used for such tasks as maintaining consistency between underlying data and a graphical depiction of that data, maintaining consistency among multiple views, specifying formatting requirements and preferences, and specifying animation events and attributes. The constraint-based user interface system with the largest user base is Garnet [56, 57]. This system is a full-fledged user interface construction set, written in Common Lisp, which provides considerable functionality beyond just a constraint system. The standard constraint portion of Garnet supports only unidirectional constraints and not multidirectional ones, but does include support for constraints containing arbitrary pointer variables [81]. We recently extended Garnet to Multi-Garnet, which supports multi-way constraints, constraint hierarchies, and pointer variables in an integrated framework [63]. A precursor to Garnet is the Peridot system [54, 55]; an interesting feature of Peridot is its mechanism for inferring constraints from a widget's layout. Reference [9] discusses the design of a syntax-based program editor using constraints. References [10] and [17] describe using constraint hierarchies to define the inter- and intra-window relations in a window system. Other user interface toolkits that use constraints include GROW [1], MEL [34], GITS [60], the FilterBrowser user interface construction tool [16], and the Cactus statistics exploration environment [52].

6.4 General-Purpose Programming Languages

A number of researchers have investigated general-purpose languages that use constraints, in addition to those mentioned in Section 5. Steele's Ph.D. dissertation [74] is one of the first such efforts. Leler [46] describes Bertrand, a constraint language based on augmented term rewriting. Both Steele and Leler's languages use the refinement rather than the perturbation model and don't deal with the issues of soft constraints or the stability of an existing solution when editing it. (Steele's implementation maintains dependency information to decide which deductions should be invalidated when editing the constraint graph, as well as to aid in generating explanations. However, when such edits are made, the old values are simply erased, rather than being used as defaults for the new values.) Siri [36, 37] and RENDEZVOUS [35] are other recent languages that combine constraints with imperative programming. Siri uses a graph rewriting model of execution, derived from Bertrand's. Unlike Kaleidoscope, Siri requires the programmer to state explicitly which parts of an object remain the same after a change. In addition, Siri uses a single abstraction mechanism, a *constraint pattern*, for object description, modification, and evaluation, rather than separate mechanisms for these tasks. (This uniform use of patterns is derived from BETA [45].) RENDEZVOUS includes extensive support for processes and multiple users; its intended domain of use is multi-user, multi-media systems.

Much of the recent research on general-purpose languages with constraints has used logic programming as a base. Several instances of the CLP scheme (see Section 5.2) have now been implemented, including CLP($\mathcal{R}$) [41, 42], Prolog III [12], CHIP [14, 78], CAL [68], and CLP(Σ^*) [84]. The cc family of languages [66, 65] generalizes the CLP scheme to include such features as concurrency, atomic tell, and blocking ask. Work on logic programming and constraint hierarchies other than HCLP includes that of Maher and Stuckey [49], who give a definition of constraint hierarchies similar to the one in this paper. In their definition, pre-solutions for hierarchies perform the same function as the set S_0 in our formulation. Maher and Stuckey define a pre-measure that maps pre-solutions and sets of constraints to some scale, so that they can then be compared via a lexicographic ordering. Satoh [67] proposes a theory for constraint hierarchies using a meta-language to specify an ordering on the interpretations that satisfy the required constraints. The theory is quite general, and can accommodate all of the comparators described in Section 2.1. However, since it is defined by second-order formulae, it is not in general computable. In subsequent work [69, 70], Satoh and Aiba present an alternative theory that restricts the constraints to a single domain $\mathcal{D}$, so that they can be expressed in a first-order formula. This theory is similar to the one presented here, with the following differences: first, only the *locally-predicate-better* comparator is supported; second, the semantics of constraint hierarchies is described model theoretically rather than set theoretically; and third, the class of constraints is generalized from atomic constraints to disjunctions of conjunctions of atomic constraints. Satoh and Aiba embed such constraints in the CLP language CAL [68], to yield an HCLP language CHAL [69, 70].

Ohwada and Mizoguchi [59] discuss the use of logic programming for building graphical user interfaces, including the use of default constraints. Their constraint hierarchy is implemented using the negation-as-failure rule, i.e., if the negation of a constraint is not known to hold, then the constraint can be assumed to hold. A problem with this approach is that it then becomes necessary to list all possible conflicts when a rule is being written in order to avoid inconsistencies. In contrast, in HCLP the need for consistency is assumed and there is no need to enumerate specifically those constraints that might conflict with the goal.

6.5 Artificial Intelligence Applications

There is a substantial body of research in the artificial intelligence community using constraints in planning, simulation, computer vision, and other areas. Constraints can, for example, improve the performance of an inferencing system by early pruning of the search space, i.e., by using the constraint system as a faster, but less general, inferencer that runs as a subtask of the more general system. Again, since this body of related work is very large, here we concentrate on work involving combinations of required and preferential constraints.

Descotte and Latombe [13] use required and preferential constraints in a system, Gari, for generating plans for machining parts. For example, there might be a required constraint that a particular cut be made with either a surface grinding machine or a lathe, and a preference that such cuts not be made with a lathe. Production rules are used to encode Gari's knowledge: on the left hand side of the rule are conditions that must be satisfied for the rule to be used; on the right hand side are labeled constraints that are added if the rule's conditions are satisfied. Gari supports ten levels of constraints (required and nine preferential levels). The solver finds (close to) a *locally-predicate-better* solution to the collection of constraints using an iterative search.[3] Fox [18] discusses the problem of constraint-directed reasoning for job-shop scheduling, and allows the relaxation of constraints when conflicts occur, as well as context-sensitive selection and weighted interpretation of constraints. In Fox's system, ISIS, non-required constraints include a *relaxation specification* that specifies procedurally how to generate alternative, less restrictive, versions of the constraint. ISIS searches for a solution to the soft constraints that meets a minimum *weighted-sum-better* threshold. (Due to the complexity of the search space for this domain, the system doesn't attempt to find an optimal solution, just an acceptable one.) Constraints in ISIS have a number of other attributes, such as duration and context, which in the formalism described in this paper would be handled outside the constraint system (for example, in HCLP rules or a Kaleidoscope procedure).

The constraint systems in many AI applications solve systems of constraints over finite domains [47]. Three typical applications of such CSPs (constraint satisfaction problems) are scene labeling, map interpretation, and computer system

---

[3] The definition of a correct solution in Gari is actually a bit weaker than *locally-predicate-better*: in the terminology used in this paper, a Gari solution must simply respect the hierarchy (see Section 2.1). Equivalently, one can view Gari as using a two-level constraint hierarchy (required and one preferential level), with integral weights between 1 and 9 on the preferential constraints; taking this view, it finds *worst-case-better* solutions.

configuration. Freuder [30] gives a general model for partial constraint satisfaction problems (PCSPs) for variables ranging over finite domains, extending the standard CSP model. In Freuder's model, alternate CSPs are compared with the original problem using a metric on the problem space (as opposed to a metric on the solution space, as in our work). An optimal solution s to the original PCSP would be one in which the distance between the original problem and the new problem (for which s is an exact solution) is minimal. In an earlier CSP extension, Shapiro and Haralick [72] define the concepts of exact and inexact matching of two structural descriptions of objects, and show that inexact matching is a special case of the inexact consistent labeling problem.

A classic problem in AI is the *frame problem*: the need to infer that state will not change across events.[4] In response to this problem, a substantial body of research has been done on *nonmonotonic reasoning*; reference [32] is a collection of many of the classic papers in the area. Brewka [8] describes an approach to representing default information with multiple levels of preference. In this framework, there are many levels of theories, some of which are more preferred than others. A preferred subtheory is obtained by taking a maximally consistent subset of the strongest level, and then adding as many formulas as possible from the next strongest level, and so on, without introducing any inconsistencies. Reference [86] discusses some additional aspects of the relationship between constraint hierarchies and nonmonotonic logic.

7 Conclusion

The primary contribution of this paper has been a complete presentation of the theory of constraint hierarchies: both the basic form, and a number of useful extensions. We have also outlined a number of applications and algorithms to demonstrate that constraint hierarchies are useful and practical, and have shown how the operation of a number of other systems can be categorized using the constraint hierarchy theory.

We are continuing to investigate various aspects of constraint hierarchies, including fully integrated programming languages and additional solver algorithms. One of our primary goals in this work is to support user interfaces and interactive graphics, both of which require highly efficient constraint solvers. Thus, with Michael Sannella, we are extending the DeltaBlue algorithm to accommodate cycles, simultaneous equations, and other complex constraint graphs. Additionally, we believe that there are significant benefits that arise when constraint hierarchies are fully integrated into programming languages. Thus we are further refining our second HCLP($\mathcal{R}$) implementation and, with Gus Lopez, are implementing a second generation CIP language, Kaleidoscope'91.

---

[4] The use of weak stay constraints to assert that parts of a graphical object being manipulated should remain in the same place is in fact a way of addressing the frame problem in the context of interactive graphics.

Acknowledgements

Thanks for many useful discussions, and comments on drafts of this paper, to John Maloney, Michael Sannella, and Dan Weld. John Maloney has done much of the work on ThingLab II, and Amy Martindale and Michael Maher worked with us on HCLP. Thanks to the anonymous referees for useful comments, in particular for pointing out a problem in one of the definitions and certain parts of the exposition that needed to be clarified. This project was supported in part by the National Science Foundation under Grants CCR-9107395 and IRI-9102938, by the Canadian National Science and Engineering Research Council under Grant OGP0121431, by the University of Victoria, and by graduate fellowships from the National Science Foundation and Apple Computer for Bjorn Freeman-Benson and Molly Wilson respectively.

References

1. Paul Barth. An Object-Oriented Approach to Graphical Interfaces. *ACM Transactions on Graphics*, 5(2):142–172, April 1986.
2. Karl-Friedrich Böhringer. Using Constraints to Achieve Stability in Automatic Graph Layout Algorithms. In *CHI'90 Conference Proceedings*, pages 43–52, Seattle, Washington, April 1990. ACM SIGCHI.
3. Alan Borning. The Programming Language Aspects of ThingLab, A Constraint-Oriented Simulation Laboratory. *ACM Transactions on Programming Languages and Systems*, 3(4):353–387, October 1981.
4. Alan Borning. Graphically Defining New Building Blocks in ThingLab. *Human-Computer Interaction*, 2(4):269–295, 1986.
5. Alan Borning and Robert Duisberg. Constraint-Based Tools for Building User Interfaces. *ACM Transactions on Graphics*, 5(4), October 1986.
6. Alan Borning, Robert Duisberg, Bjorn Freeman-Benson, Axel Kramer, and Michael Woolf. Constraint Hierarchies. In *Proceedings of the 1987 ACM Conference on Object-Oriented Programming Systems, Languages, and Applications*, pages 48–60. ACM, October 1987.
7. Alan Borning, Michael Maher, Amy Martindale, and Molly Wilson. Constraint Hierarchies and Logic Programming. In *Proceedings of the Sixth International Conference on Logic Programming*, pages 149–164, Lisbon, June 1989.
8. Gerhard Brewka. Preferred Subtheories: An Extended Logical Framework for Default Reasoning. In *Proceedings of the Eleventh International Joint Conference on Artificial Intelligence*, pages 1043–1048, August 1989.
9. C. A. Carter and W. R. LaLonde. The Design of a Program Editor Based on Constraints. Technical Report CS TR 50, Carleton University, May 1984.
10. Ellis S. Cohen, Edward T. Smith, and Lee A. Iverson. Constraint-Based Tiled Windows. *IEEE Computer Graphics and Applications*, pages 35–45, May 1986.
11. Jacques Cohen. Constraint Logic Programming Languages. *Communications of the ACM*, 33(7):52–68, July 1990.
12. Alain Colmerauer. An Introduction to Prolog III. *Communications of the ACM*, pages 69–90, July 1990.

13. Yannick Descotte and Jean-Claude Latombe. Making Compromises among Antagonist Constraints in a Planner. *Artificial Intelligence*, 27(2):183–217, November 1985.

14. M. Dincbas, P. Van Hentenryck, H. Simonis, A. Aggoun, T. Graf, and F. Bertheir. The Constraint Logic Programming Language CHIP. In *Proceedings Fifth Generation Computer Systems-88*, 1988.

15. Robert A. Duisberg. *Constraint-Based Animation: The Implementation of Temporal Constraints in the Animus System.* PhD thesis, University of Washington, 1986. Published as UW Computer Science Department Technical Report No. 86-09-01.

16. Raimund Ege, David Maier, and Alan Borning. The Filter Browser—Defining Interfaces Graphically. In *Proceedings of the European Conference on Object-Oriented Programming*, pages 155–165, Paris, June 1987. Association Française pour la Cybernétique Économique et Technique.

17. Danny Epstein and Wilf LaLonde. A Smalltalk Window System Based on Constraints. In *Proceedings of the 1988 ACM Conference on Object-Oriented Programming Systems, Languages and Applications*, pages 83–94, San Diego, September 1988. ACM.

18. Mark S. Fox. *Constraint-Directed Search: A Case Study of Job-Shop Scheduling.* Morgan Kaufmann, Los Altos, California, 1987.

19. Bjorn Freeman-Benson. A Module Compiler for ThingLab II. In *Proceedings of the 1989 ACM Conference on Object-Oriented Programming Systems, Languages and Applications*, pages 389–396, New Orleans, October 1989. ACM.

20. Bjorn Freeman-Benson. Kaleidoscope: Mixing Objects, Constraints, and Imperative Programming. In *Proceedings of the 1990 Conference on Object-Oriented Programming Systems, Languages, and Applications, and European Conference on Object-Oriented Programming*, pages 77–88, Ottawa, Canada, October 1990. ACM.

21. Bjorn Freeman-Benson and Alan Borning. Integrating Constraints with an Object-Oriented Language. In *Proceedings of the 1992 European Conference on Object-Oriented Programming*, pages 268–286, June 1992.

22. Bjorn Freeman-Benson and Alan Borning. The Design and Implementation of Kaleidoscope'90, A Constraint Imperative Programming Language. In *Proceedings of the IEEE Computer Society International Conference on Computer Languages*, pages 174–180, April 1992.

23. Bjorn Freeman-Benson and John Maloney. The DeltaBlue Algorithm: An Incremental Constraint Hierarchy Solver. In *Proceedings of the Eighth Annual IEEE Phoenix Conference on Computers and Communications*, Scottsdale, Arizona, March 1989. IEEE.

24. Bjorn Freeman-Benson, John Maloney, and Alan Borning. The DeltaBlue Algorithm: An Incremental Constraint Hierarchy Solver. Technical Report 89-08-06, Department of Computer Science and Engineering, University of Washington, August 1989.

25. Bjorn Freeman-Benson, John Maloney, and Alan Borning. An Incremental Constraint Solver. *Communications of the ACM*, 33(1):54–63, January 1990.

26. Bjorn Freeman-Benson and Molly Wilson. DeltaStar, How I Wonder What You Are: A General Algorithm for Incremental Satisfaction of Constraint Hierarchies. Technical Report 90-05-02, Department of Computer Science and Engineering, University of Washington, May 1990.

27. Bjorn Freeman-Benson, Molly Wilson, and Alan Borning. DeltaStar: A General Algorithm for Incremental Satisfaction of Constraint Hierarchies. In *Proceedings*

of the Eleventh Annual IEEE Phoenix Conference on Computers and Communications, pages 561–568, Scottsdale, Arizona, March 1992. IEEE.

28. Bjorn N. Freeman-Benson. Multiple Solutions from Constraint Hierarchies. Technical Report 88-04-02, University of Washington, Seattle, WA, April 1988.

29. Bjorn N. Freeman-Benson. *Constraint Imperative Programming*. PhD thesis, University of Washington, Department of Computer Science and Engineering, July 1991. Published as Department of Computer Science and Engineering Technical Report 91-07-02.

30. Eugene Freuder. Partial Constraint Satisfaction. In *Proceedings of the Eleventh International Joint Conference on Artificial Intelligence*, pages 278–283, August 1989.

31. Michel Gangnet and Burton Rosenberg. Constraint Programming and Graph Algorithms. In *Second International Symposium on Artificial Intelligence and Mathematics*, January 1992.

32. Matthew L. Ginsberg, editor. *Readings in Nonmonotonic Reasoning*. Morgan Kaufmann, Los Altos, California, 1987.

33. James A. Gosling. *Algebraic Constraints*. PhD thesis, Carnegie-Mellon University, May 1983. Published as CMU Computer Science Department Technical Report CMU-CS-83-132.

34. Ralph D. Hill. A 2-D Graphics System for Multi-User Interactive Graphics Based on Objects and Constraints. In E. H. Blake and P. Wisskirchen, editors, *Advances in Object Oriented Graphics I*, pages 67–91. Springer-Verlag, Berlin, 1990.

35. Ralph D. Hill. Languages for the Construction of Multi-User Multi-Media Synchronous (MUMMS) Applications. In Brad Myers, editor, *Languages for Developing User Interfaces*, pages 125–143. Jones and Bartlett, Boston, 1992.

36. Bruce Horn. Constraint Patterns as a Basis for Object-Oriented Constraint Programming. In *Proceedings of the 1992 ACM Conference on Object-Oriented Programming Systems, Languages, and Applications*, Vancouver, British Columbia, October 1992.

37. Bruce Horn. Properties of User Interface Systems and the Siri Programming Language. In Brad Myers, editor, *Languages for Developing User Interfaces*, pages 211–236. Jones and Bartlett, Boston, 1992.

38. James P. Ignizio. Generalized Goal Programming. *Computers and Operations Research*, 10(4):277–290, 1983.

39. James P. Ignizio. *Introduction to Linear Goal Programming*. Sage Publications, Beverly Hills, 1985. Sage University Paper Series on Qualitative Applications in the Social Sciences, 07-056.

40. Joxan Jaffar and Jean-Louis Lassez. Constraint Logic Programming. In *Proceedings of the Fourteenth ACM Principles of Programming Languages Conference*, Munich, January 1987.

41. Joxan Jaffar and Spiro Michaylov. Methodology and Implementation of a CLP System. In *Proceedings of the Fourth International Conference on Logic Programming*, pages 196–218, Melbourne, May 1987.

42. Joxan Jaffar, Spiro Michaylov, Peter Stuckey, and Roland Yap. The CLP($\mathcal{R}$) Language and System. *ACM Transactions on Programming Languages and Systems*, 14(3):339–395, July 1992.

43. Tomihisa Kamada and Satoru Kawai. A General Framework for Visualizing Abstract Objects and Relations. *ACM Transactions on Graphics*, 10(1):1–39, January 1991.

44. M. Konopasek and S. Jayaraman. *The TK!Solver Book.* Osborne/McGraw-Hill, Berkeley, CA, 1984.

45. Bent Bruun Kristensen, Ole Lehrmann Madsen, Birger Møller Pederson, and Kirsten Nygaard. Abstraction Mechanisms in the BETA Programming Language. In *Proceedings of the Tenth Annual Principles of Programming Languages Symposium*, Austin, Texas, January 1983. ACM.

46. William Leler. *Constraint Programming Languages.* Addison-Wesley, 1987.

47. Alan K. Mackworth. Consistency in Networks of Relations. *Artificial Intelligence,* 8(1):99–118, 1977.

48. Michael J. Maher. Logic Semantics for a Class of Committed-choice Programs. In *Proceedings of the Fourth International Conference on Logic Programming,* pages 858–876, Melbourne, May 1987.

49. Michael J. Maher and Peter J. Stuckey. Expanding Query Power in Constraint Logic Programming. In *Proceedings of the North American Conference on Logic Programming,* Cleveland, October 1989.

50. John Maloney. *Using Constraints for User Interface Construction.* PhD thesis, Department of Computer Science and Engineering, University of Washington, August 1991. Published as Department of Computer Science and Engineering Technical Report 91-08-12.

51. John Maloney, Alan Borning, and Bjorn Freeman-Benson. Constraint Technology for User-Interface Construction in ThingLab II. In *Proceedings of the 1989 ACM Conference on Object-Oriented Programming Systems, Languages and Applications,* pages 381–388, New Orleans, October 1989. ACM.

52. John Alan McDonald, Werner Stuetzle, and Andreas Buja. Painting Multiple Views of Complex Objects. In *Proceedings of the 1990 ACM Conference on Object-Oriented Programming: Systems, Languages, and Applications and the European Conference on Object-Oriented Programming,* pages 245–257, Ottawa, Canada, October 1990.

53. Katta G. Murty. *Linear Programming.* Wiley, 1983.

54. Brad Myers. *Creating User Interfaces by Demonstration.* PhD thesis, University of Toronto, 1987.

55. Brad A. Myers. Creating Dynamic Interaction Techniques by Demonstration. In *CHI+GI 1987 Conference Proceedings,* pages 271–278, April 1987.

56. Brad A. Myers, Dario Guise, Roger B. Dannenberg, Brad Vander Zanden, David Kosbie, Philippe Marchal, Ed Pervin, Andrew Mickish, and John A. Kolojejchick. The Garnet Toolkit Reference Manuals: Support for Highly-Interactive Graphical User Interfaces in Lisp. Technical Report CMU-CS-90-117, Computer Science Dept, Carnegie Mellon University, March 1990.

57. Brad A. Myers, Dario Guise, Roger B. Dannenberg, Brad Vander Zanden, David Kosbie, Philippe Marchal, and Ed Pervin. Comprehensive Support for Graphical, Highly-Interactive User Interfaces: The Garnet User Interface Development Environment. *IEEE Computer,* 23(11):71–85, November 1990.

58. Greg Nelson. Juno, A Constraint-Based Graphics System. In *SIGGRAPH '85 Conference Proceedings,* pages 235–243, San Francisco, July 1985. ACM.

59. Hayato Ohwada and Fumio Mizoguchi. A Constraint Logic Programming Approach for Maintaining Consistency in User-Interface Design. In *Proceedings of the 1990 North American Conference on Logic Programming,* pages 139–153. MIT Press, October 1990.

60. Dan R. Olsen, Jr. Creating Interactive Techniques by Symbolically Solving Geometric Constraints. In *Proceedings of the ACM SIGGRAPH Symposium on User*

Interface Software and Technology, pages 102–107, Snowbird, Utah, October 1990. ACM SIGGRAPH and SIGCHI.

61. A. Robinson. *Non-Standard Analysis*. North-Holland Publishing Company, Amsterdam, 1966.

62. Ernst Rotterdam. Physiological Modeling and Simulation with Constraints. Technical Report R89001, Medical Information Science, Department of Anesthesiology, Oostersingel 59, 9713 E2 Groningen, June 1989.

63. Michael Sannella and Alan Borning. Multi-Garnet: Integrating Multi-Way Constraints with Garnet. Technical Report 92-07-01, Department of Computer Science and Engineering, University of Washington, September 1992.

64. Vijay A. Saraswat. Problems with Concurrent Prolog. Technical Report CS-86-100, Carnegie-Mellon University, May 1985. Revised January 1986.

65. Vijay A. Saraswat. *Concurrent Constraint Programming Languages*. PhD thesis, Carnegie-Mellon University, Computer Science Department, January 1989.

66. Vijay A. Saraswat, Martin Rinard, and Prakash Panangaden. Semantic Foundations of Concurrent Constraint Programming. In *Proceedings of the Eighteenth Annual Principles of Programming Languages Symposium*. ACM, 1991.

67. Ken Satoh. Formalizing Soft Constraints by Interpretation Ordering. In *Proceedings of the European Conference on Artificial Intelligence*, 1990.

68. Ken Satoh and Akira Aiba. CAL: A Theoretical Background of Constraint Logic Programming and its Applications (Revised). Technical Report TR-537, Institute for New Generation Computer Technology, Tokyo, February 1990.

69. Ken Satoh and Akira Aiba. Computing Soft Constraints by Hierarchical Constraint Logic Programming. Technical Report TR-610, Institute for New Generation Computer Technology, Tokyo, January 1991.

70. Ken Satoh and Akira Aiba. The Hierarchical Constraint Logic Language CHAL. Technical Report TR-592, Institute for New Generation Computer Technology, Tokyo, September 1991.

71. Ehud Shapiro. Concurrent Prolog: A Progress Report. *IEEE Computer*, 19(8):44–58, August 1986.

72. Linda Shapiro and Robert Haralick. Structural Descriptions and Inexact Matching. *IEEE Transactions on Pattern Analysis and Machine Intelligence*, PAMI-3(5):504–519, September 1981.

73. Steven Sistare. *A Graphical Editor for Constraint-Based Geometric Modeling*. PhD thesis, Department of Computer Science, Harvard, December 1990. Published as Technical Report TR-06-9.

74. Guy L. Steele. *The Definition and Implementation of a Computer Programming Language Based on Constraints*. PhD thesis, MIT, August 1980. Published as MIT-AI TR 595, August 1980.

75. Ivan Sutherland. Sketchpad: A Man-Machine Graphical Communication System. In *Proceedings of the Spring Joint Computer Conference*. IFIPS, 1963.

76. Ivan Sutherland. *Sketchpad: A Man-Machine Graphical Communication System*. PhD thesis, Department of Electrical Engineering, MIT, January 1963.

77. Shin Takahashi, Satoshi Matsuoka, and Akinori Yonezawa. A General Framework for Bi-Directional Translation between Abstract and Pictorial Data. In *Proceedings of the ACM SIGGRAPH Symposium on User Interface Software and Technology*, pages 165–174, Hilton Head, South Carolina, November 1991.

78. Pascal Van Hentenryck. *Constraint Satisfaction in Logic Programming*. MIT Press, Cambridge, MA, 1989.

79. Christopher J. van Wyk. *A Language for Typesetting Graphics.* PhD thesis, Department of Computer Science, Stanford, June 1980.
80. Christopher J. van Wyk. A High-level Language for Specifying Pictures. *ACM Transactions on Graphics*, 1(2), April 1982.
81. Brad Vander Zanden, Brad Myers, Dario Guise, and Pedro Szekely. The Importance of Pointer Variables in Constraint Models. In *Proceedings of the ACM SIGGRAPH Symposium on User Interface Software and Technology*, pages 155–164, Hilton Head, South Carolina, November 1991.
82. Bradley T. Vander Zanden. *An Incremental Planning Algorithm for Ordering Equations in a Multilinear system of Constraints.* PhD thesis, Department of Computer Science, Cornell University, April 1988.
83. William W. Wadge and Edward A. Ashcroft. *Lucid, the Dataflow Programming Language.* Academic Press, London, 1985.
84. Clifford Walinsky. CLP(Σ^*): Constraint Logic Programming with Regular Sets. In *Proceedings of the Sixth International Conference on Logic Programming*, pages 181–196, Lisbon, June 1989.
85. Molly Wilson. *Hierarchical Constraint Logic Programming.* PhD thesis, Department of Computer Science and Engineering, University of Washington, 1992. Forthcoming.
86. Molly Wilson and Alan Borning. Extending Hierarchical Constraint Logic Programming: Nonmonotonicity and Inter-Hierarchy Comparison. In *Proceedings of the North American Conference on Logic Programming*, pages 3–19, Cleveland, October 1989.

Partial Constraint Satisfaction [*]

Eugene C. Freuder and Richard J. Wallace

Computer Science Department, Kingsbury Hall, University of New Hampshire,
Durham, NH 03824, USA

Abstract. A constraint satisfaction problem involves finding values for
variables subject to constraints on which combinations of values are allowed. In some cases it may be impossible or impractical to solve these
problems completely. We may seek to partially solve the problem, in particular by satisfying a maximal number of constraints. Standard backtracking and local consistency techniques for solving constraint satisfaction problems can be adapted to cope with, and take advantage of, the
differences between partial and complete constraint satisfaction. Extensive experimentation on maximal satisfaction problems illuminates the
relative and absolute effectiveness of these methods. A general model of
partial constraint satisfaction is proposed.

1 Introduction

Constraint satisfaction involves finding values for problem variables subject to
constraints on acceptable combinations of values. Constraint satisfaction has
wide application in artificial intelligence, in areas ranging from temporal reasoning to machine vision. *Partial constraint satisfaction* involves finding values for
a subset of the variables that satisfy a subset of the constraints. Viewed another
way, we are willing to "weaken" some of the constraints to permit additional
acceptable value combinations. partial constraint satisfaction problems arise in
several contexts:

- The problem is overconstrained and admits of no complete solution.
- The problem is too difficult to solve completely but we are willing to settle
  for a "good enough" solution.
- We are seeking the best solution obtainable within fixed resource bounds.
- Real time demands require an "anytime algorithm", which can report *some*
  partial solution almost immediately, improving on it if and when time allows.

The utility of some form of partial constraint satisfaction has been repeatedly
recognized. A variety of applications has motivated a variety of approaches. As

---

[*] This paper is reprinted (with minor changes) from *Artificial Intelligence*, volume 58,
numbers 1-3, E. C. Freuder and R. J. Wallace, Partial constraint satisfaction, pages
21-70, 1992 with kind permission from Elsevier Science - NL, Sara Burgerhartstraat
25, 1055 KV Amsterdam, The Netherlands. (This volume was a special issue that
was reprinted as an MIT Press book, *Constraint-Based Reasoning*.)

AI increasingly confronts real world problems, in expert systems and robotics, for example, we are increasingly likely to encounter situations where, rather than searching for a solution to a problem, we must, in a sense, search for a problem we can solve.

Conflicting constraints have arisen in a variety of domains. Descotte and Latombe made "compromises" among antagonist constraints in a planner for machining problems [10]. Borning used constraint "hierarchies" to deal with situations in which a set of requirements and preferences for the graphical display of a physical simulation cannot all be satisfied [2]; these hierarchies have been imbedded in a constraint logic programming language [3]. Hower used "sensitive relaxation" to resolve conflicts in floor planning [20].

Scheduling problems are a natural source of constraint satisfaction problems, and schedule conflicts a natural source of partial satisfaction problems. Fox added the concepts of constraint "relaxation" (the selection of constraint alternatives), "preferences" among relaxations and constraint "importance" to constraint representations, to cope with conflicting constraints in job-shop scheduling [12]. Feldman and Golumbic [11] used "priorities" in looking for optimal student schedules.

Machine vision has also provided motivation for work on partial constraint satisfaction. Shapiro and Haralick [35] treated inexact matching of structural descriptions using an extension of constraint satisfaction that they called the inexact consistent labeling problem, which sought a solution within a given error bound. Mohr and Masini [31] suggested a modification of local consistency processing to deal with errors, permitting values to fail to satisfy some constraints, in order to cope with noise in domains such as computer vision; Cooper [5] defined an alternative generalization of constraint satisfaction with errors.

The related problem of approximate constraint satisfaction, where weights are assigned to individual combinations of values [34] [36] is motivated by machine vision as well. The expression of preferences in database queries is also a related problem [24]. Note that the concept of optimization can play a role in constraint satisfaction problems even when all constraints are satisfied; there may be an additional criterion to optimize among alternative solutions [5] [7] [37].

Most of this paper will focus on methods for *maximal constraint satisfaction*, where we seek a solution that satisfies as many constraints as possible. We have systematically reviewed the basic backtracking and local consistency methods for constraint satisfaction [23] [29] [32], and developed analogous methods for maximal satisfaction. The maximal satisfaction context has provided new challenges and new opportunities. The algorithms we formulated were subjected to carefully designed experiments that shed light on both relative and absolute performance as a function of basic structural problem parameters.

Our algorithms also allow for *sufficient satisfaction*, where we terminate the search if we find a solution which is sufficiently good, in the sense that the number of constraint violations does not exceed some predetermined bound. Our methods easily extend to *resource-bounded satisfaction*, where we report the best

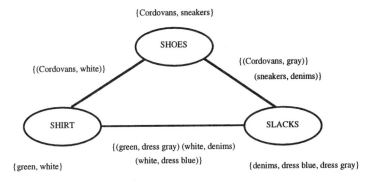

Fig. 1. Robot clothing problem.

solution in hand when a resource bound has been reached, and naturally support anytime algorithms.

Maximal satisfaction provides a form of optimization. Sufficient satisfaction incorporates a concept of acceptable error. Our methods clearly generalize to the use of more complex metrics to evaluate proposed solutions than a simple count of the number of violated constraints. In this manner, preferences can be introduced to distinguish among conflicting constraints. However, the simple metric of counting constraint violations facilitates the presentation of our algorithms, and provides a suitable context for an initial evaluation of their performance.

At the end of the paper we develop a still more general model of partial constraint satisfaction [14], in which we compare alternative problems rather than alternative solutions. We suggest viewing partial satisfaction of a problem, P, as a search through a space of alternative problems for a solvable problem "close enough" to P. We argue that a full theory of partial satisfaction should consider not merely how a partial solution requires us to violate or vitiate constraints, but how the entire solution set of the problem with these altered constraints differs from the solution set of the problem with which we started.

In this paper we will use for pedagogical purposes a simple, toy problem involving a fashion conscious robot seeking to choose matching clothes while getting dressed in the morning. (This could be regarded more seriously as a simple version of a configuration problem [30].) The problem is pictured in Fig. 1. Our robot has a minimal wardrobe: sneakers or Cordovans for footwear, a white and a dark green shirt and three pairs of slacks: denim, dress blue and dress gray. The robot has been told that: the sneakers only go with the denim slacks; the Cordovans only go with the gray slacks and the white shirt; the white shirt will go with either denim or blue slacks; the green shirt only goes with the gray slacks. These are the constraints under which it has to operate.

Section 2 discusses methods for achieving maximal constraint satisfaction. Branch and bound for maximal constraint satisfaction is the natural extension of backtracking for constraint satisfaction. Retrospective and prospective backtrack techniques for constraint satisfaction are shown to have analogues in a branch and bound setting for maximal satisfaction. Local consistency methods for constraint satisfaction have analogues in maximal satisfaction methods. Ordering techniques are if anything likely to be more important for branch and bound than for backtracking.

Section 3 describes extensive testing of maximal constraint satisfaction methods corresponding to several of the most successful constraint satisfaction methods. The results demonstrate the effectiveness of the maximal satisfaction analogues. They also illustrate the importance of taking advantage of the additional information available in the partial satisfaction domain, where the world is not just black and white (consistency or inconsistency), but shades of grey. Furthermore, the experimental design permits insights into the relationship between problem structure and the performance of the different methods.

Section 4 generalizes to other forms of partial satisfaction. Other metrics are briefly discussed. Partially ordered problem spaces are introduced. A partial constraint satisfaction problem is defined as a search through a space of alternative problems. Section 5 contains brief concluding remarks.

2   Methods

2.1   Introduction

A *constraint satisfaction problem* (*CSP*) involves a set of problem *variables*, a *domain* of potential values for each variable, and a set of *constraints*, specifying which combinations of values are acceptable. A *solution* specifies an assignment of a value to each variable that does not violate any of the constraints. We will consider here *binary, finite CSPs* where the constraints only involve two variables at a time, and the domains are finite sets of values. A constraint can therefore be represented explicitly as a set of permitted pairs of values. (If all pairs of values are allowed between two variables, then there is effectively no constraint between them; we will say that these variables do not *share* a constraint.)

For our running example: the variables are shoes, slacks and shirt; the values for shoes are Cordovans and sneakers; the constraint between shoes and shirt specifies that the only allowable combination of values is Cordovans and white shirt. Two values, like Cordovans and white shirt, that satisfy the relevant constraint, are *consistent*. A pair of values that violates a constraint is an *inconsistency*.

For now we define a *partial constraint satisfaction problem* (*PCSP*) as a CSP where we are willing to accept a solution that violates some of the constraints. A more formal approach to PCSPs is developed in Sect. 4.

Backtracking [18] is the classic algorithm for solving CSPs. A number of variations and refinements of backtracking have been developed. Several algorithms, including classical backtracking itself, utilize *retrospective* techniques, in which a new value selected to try to extend an incomplete solution is tested by "looking back" over the previously chosen values in the incomplete solution, to see if the new value is consistent with the previously chosen values. By "remembering" more about the course of the search process, some variations reduce redundant testing. Other algorithms employ *prospective* strategies. Values are tested against the domains of variables that are not yet represented in the incomplete solution, so that inconsistencies can be dealt with before values from these domains are considered for inclusion. *Ordering* techniques have been used to direct the order in which variables, values or constraints are considered during search.

Our strategy in studying PCSP algorithms was to look for analogues of successful CSP techniques, focusing on backtrack and its variations. Branch and bound [25] [33] [37] is a widely used optimization technique that may be viewed as a variation on backtracking. Thus it was a natural choice in seeking an analogue of backtracking to find optimal partial solutions for PCSPs.

We begin by applying branch and bound to constraint satisfaction problems. Then we set about finding analogues of various CSP retrospective, prospective and ordering techniques, for partial, specifically maximal, constraint satisfaction algorithms. Finding appropriate analogues presents both challenges and opportunities. In presenting these algorithms we will generally begin with a review of the CSP version, then move on to the PCSP version, highlighting the differences. We will present examples and discussions as well as the algorithms themselves.

2.2 Retrospective techniques

2.2.1. Basic branch and bound. Branch and bound for maximal constraint satisfaction is the natural analogue of backtracking for constraint satisfaction. First we will briefly review backtrack search in the context of our running example. Backtrack search will find no solutions; the problem is overconstrained. Then we will use a basic branch and bound algorithm to find a way to dress our robot while violating a minimal number of its esthetic requirements.

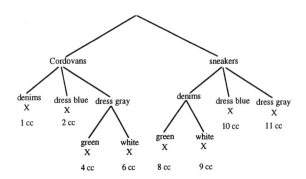

Fig. 2. Backtracking example.

A depth-first traversal of the tree in Fig. 2 traces the progress of a backtrack search on our running example. First we try Cordovans for shoes, then denims for slacks. According to the given constraints, denim slacks are not consistent with Cordovans, so we try blue slacks. These do not work so we try gray. Gray is good; we can move on to try and find a consistent shirt. The green shirt does not go with the Cordovans so we try the white shirt. It is consistent with the Cordovans but not with the gray slacks (we will assume that consistency is checked "top down" against the already chosen values). At this point we need to back up. We find that we have tried all the values for slacks, so we back up further to try another value for shoes. Ultimately all possibilities fail.

Backtrack search tries all value combinations exhaustively if necessary, but can avoid considering some combinations, by observing that a subset of values cannot be extended to a full solution, thus pruning a subtree of the search space. A standard measure of effort for CSP algorithms is the number of *constraint checks*. A constraint check occurs every time we ask a basic question of the form: is value a for variable X consistent with value b for variable Y? For example, when we ask whether Cordovans are consistent with denims, that is a constraint check. The total number of constraint checks (cc) accumulated when search has reached each leaf node of the search tree is shown in the figure. In total the search required 11 constraint checks to find that there is no solution.

We will be discussing the way different algorithms work through a search tree like this, and will need a suitable vocabulary. We will talk about the *levels* in the search tree; each level in the search tree corresponds to a problem variable. We will assume the levels are numbered from the top down; *higher levels* have smaller numbers. A shirt value will be found at a lower level or *deeper* in the search tree than a slacks value.

The branching corresponds to variable values, e.g. the choice of denim for slacks. The nodes in the search tree represent *assignments* of values to variables during the search. The green shirt value is always considered at the same level; however it is considered twice in this search. The second time the green shirt is encountered, the search has *backed up* in the interim to the shoe level. The set of assigned values along a branch of the tree, from the top down to some level, e.g (sneakers denims) is a *search path*. The search path leading down to the most recently chosen value is the *current search path*. It represents the current set of *choices* of values for variables. It represents a proposed, *incomplete* solution, unless it includes values for all the variables.

One theme that will recur in deriving our PCSP analogues is the different definition of local *failure* during CSP and PCSP search. A CSP search path fails as soon as a single inconsistency is encountered. A PCSP search path will not fail until enough inconsistencies accumulate to reach a cutoff bound. Retrospective techniques excel at determining the inconsistencies implied by past choices. Prospective techniques excel at estimating the inconsistencies implied by future choices.

Branch and bound operates in a similar fashion to backtracking in a context where we are seeking a *maximal solution*, one which satisfies as many constraints as possible. Branch and bound basically keeps track of the best solution found so far and abandons a line of search when it becomes clear that it cannot lead to a better solution. A version of backtracking that searches for all solutions, rather than the first solution, most naturally compares with the branch and bound extension to find a maximal solution.

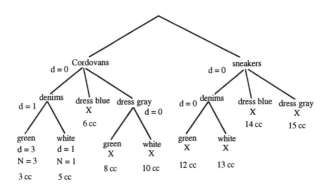

Fig. 3. Branch and bound example.

Figure 3 traces a branch and bound search for a maximally satisfying solution for our sample problem. Branch and bound is applied in this context by using as an evaluation function a count of the number of violated constraints, or inconsistencies. Where the backtrack search, looking for a *perfect solution*, that violates no constraints, said that denims were inconsistent with Cordovans and proceeded on to blue slacks, branch and bound, looking for a maximal partial solution, observes that any partial solution containing Cordovans and denims will violate at least one constraint, and proceeds to consider shirts.

Specifically it is noted that any partial solution containing Cordovans and denims will be at a distance, d, of (at least) 1 from a perfect solution. *Distance* measures the number of constraints violated by the chosen values. By the time we add a green shirt we have violated three constraints. We say that green shirt is an *extension* of the search path (Cordovans, denims). We will talk about *extending* a search path by adding one or more values, and, in particular, extending a search path to a complete solution, that contains a value for each variable. The search path leading from Cordovans, through denims, to green shirt now contains three inconsistencies, giving it an associated distance (from a perfect solution) of three. The *first inconsistent value* in the search path is denims. It is the first value in the search path to be inconsistent with another value in the search path. The *last inconsistent value* is green shirt.

Cordovans, denims and a green shirt provide a partial solution at a distance of 3 from a perfect solution. This distance is taken as the value of N. N is used during the search to store the number of inconsistencies in the best solution found up to that point in the search. N is a *necessary bound*, which we will often simply refer to as the *bound*, in the sense that to do better it is necessary to find a solution with fewer inconsistencies.

The necessary bound N can be set initially based on a priori knowledge that a solution is available that violates fewer than N constraints, or an a priori *requirement* that we are not interested in solutions that violate more than $N - 1$ constraints. As branch and bound proceeds, if a solution is found that violates $N' < N$ constraints, N is replaced by N'.

As the branch and bound search proceeds in our example, it finds a better partial solution, with a single constraint violation (Cordovans, denims, white shirt). This updates N to be 1. Now when it tries Cordovans and blue slacks (hoping that an even better, in this case perfect, solution is to be found), it recognizes that any solution involving Cordovans and blue slacks can be no better than the solution already found. Thus it does not consider matching a shirt to the Cordovans and blue slacks, but proceeds immediately to try gray slacks. As with backtrack search, the basic idea of recognizing defeat early permits pruning of the search space.

Search concludes when we find a perfect solution, not available in this case, or run out of things to try. We could also quit when we reach a preset *sufficient bound S*, which specifies that we will be satisfied if we find a partial solution that violates no more than S constraints. We may know, for example, that no exact solution is possible and thus be able to set S to 1. We may be willing to settle for a "close enough" or *sufficient solution*. Obviously the larger we set S the easier the problem is likely to be.

Circumstances may also impose resource bounds. In particular, real time processing may require immediate answers, that can be refined later if time allows. The branch and bound process is well suited to providing resource-bounded solutions. We can simply report the best solution available when, for example, a time bound is exceeded. The branch and bound process is also clearly well-suited to support an anytime algorithm, which can repeatedly provide a "best-so-far" answer when queried. It can quickly provide *some* answer, with a better one perhaps to follow as time allows.

Figure 4 provides a basic branch and bound algorithm for maximal constraint satisfaction. It also provides for a priori sufficient and necessary bounds, S and N, on an acceptable solution. If there are no such a priori bounds, S is initially 0 and N "infinity". The parameters of the P-BB procedure appear in several other algorithms in this paper. The parameter *Search-path* carries the current search path. *Distance* carries the number of constraints already violated by the values on the current search path, the number of inconsistencies in the proposed, incomplete solution. *Variables* carries a list of the variables not assigned values in the current search path, the variables at lower levels in the search tree. *Values* carries a list of the values not previously tried as extensions of the current search path; the first value in *Values* is the next value that can be tried as an instantiation of the first variable in *Variables*.

In this and several subsequent algorithms we will be employing N, S, and *Best-solution* as global variables containing the necessary and sufficient bounds

P-BB(Search-path, Distance, Variables, Values)
    if Variables = nil then {all problem variables have been assigned values
                            in Search–path}
        Best-solution ← Search–path
        N ← Distance
        if $N \leq S$ then return 'finished' {Best–solution is sufficiently close}
        else return 'keep-searching'
    else if Values = nil then {tried all values for extending search path}
        return 'keep-searching' {so will back up to see if can try another value for
                            the last variable assigned a value in Search–path}
    else if Distance = N then {already extended Search–path to assign values for
                            remaining variables without violating any additional constraints}
        return 'keep–searching' {so will see if can do better by backing up to try
                            another value for the last variable assigned a value in Search–path}
    else {try to extend Search-path}
        Current–value ← first value in Values
        New–distance ← Distance
        try choices in Search–path, from first to last, as as New–distance < N:
            if the choice is inconsistent with Current–value then
                New–distance ← New–distance + 1
        if New–distance < N and
            P-BB(Search–path plus Current–value, New–distance,
                    Variables minus the first variable,
                    values of second variable in Variables)
            = 'finished' then return 'finished' {Search–path was extended to
                                    sufficient solution}
        else {will see if can do better with another value}
            return P-BB(Search–path, Distance, Variables,
                    Values minus Current–value)

Fig. 4. Branch and bound algorithm.

and the best solution found so far. Other variables in all the algorithms are local, with the exception of some backmark arrays as indicated in Sect. 2.2.3.

The basic recursive structure of P-BB is also common to many other algorithms in this paper. P-BB works sideways in the search tree by recursing through a set of values for a variable, and deeper into the search tree by recursing through the variables. Backing up is implemented through the unwinding of the recursion.

In this algorithm as in all retrospective procedures, a value being considered for inclusion in the solution is compared with values already chosen, to determine whether constraints between the instantiated variables and the current one are satisfied. Each comparison of two values is a constraint check.

Since the total number of constraint checks is a standard measure of CSP algorithm efficiency, we wish to minimize this quantity. To this end, the new distance is compared with N after *each* constraint failure, so that if the bound is reached, the present value is not checked further. A subtle point involves the test to see if the Distance is already N before trying a new value, v for V. (Actually our implementation checked for *Distance* $\geq N$, but an equality test appears sufficient.) One might wonder, if the number of inconsistencies among the already chosen values stored in *Search-path* equals the bound, what is the algorithm doing trying to extend *Search-path* to another variable, V? However, when the algorithm began trying to extend the solution, N may have been larger. A complete assignment of values to variables, extending *Search-path* and requiring only the current N constraint violations, may have been found in the interim, using another value for V, before reaching v.

We have chosen here a depth-first implementation of the branch and bound paradigm. Other branch and bound control structures, notably a best-first approach, are possible. Depth-first is the most direct analogue of backtrack and as such facilitates the development of analogues of backtrack variations. Depth-first also supports an anytime algorithm that will almost immediately have a "best so far" solution to report. (Limited experimentation with a best-first approach was not encouraging with respect to its efficiency, but this should not preclude further study.)

In Fig. 3 the total number of constraint checks (cc) accumulated by the time each leaf node of the search tree has been processed is given at the bottom of the figure. Note that, due to our procedures for minimizing constraint checks, some checks are avoided at many points of the search tree, including at some of the lowest level nodes. In some cases subtrees are pruned; in some cases a value does not need to be checked against the entire preceeding search path.

The general worst case bound for this algorithm is of course exponential. However, it is no worse than that for a backtrack algorithm for finding a perfect solution. Both, in the worst case, will end up trying all possible combinations of values, and testing all the constraints among them. On the other hand, the exponential worst case bound is bad enough. We want to consider techniques that may help to avoid achieving that bound. As we have indicated, our strategy is to look for analogues of methods which have already proven successful for finding perfect solutions.

2.2.2. Backjumping. Backjumping [17] remembers information about previous failures to reduce the need for redundant constraint checks to rediscover them. Consider what happens when we test shirts in the branch of the backtrack search tree that begins with sneakers and denims (Fig. 2). Both shirts fail immediately upon being tested against sneakers; there is no need to see if they are consistent with denims. Yet classical, so-called "chronological", backtracking blindly backs up and tries dress blue slacks. There is no point in doing so; even if the blue slacks went with the sneakers, we would obviously fail again when we reached the shirt level. Backjumping recognizes this, and after the shirts fail to match the sneakers it immediately backtracks to the shoe level. There it tries to consider another type of shoe; since there is none, search terminates. The final two constraint checks have been avoided. Backjumping generalizes this insight.

We need to recognize that all the values tried for a given variable may not fail against the same previous value. When processing a variable, backjumping remembers the deepest level, l, in the search tree at which any of the values fails. When all the values have been discarded, backtracking can proceed directly to this level. Actually, the algorithm does not "jump" directly to this level. As the recursive calls unwind, they return the depth l; the recursion unwinds until level l is reached before trying to consider any more values. As we extend search paths we ultimately must reach levels where all value choices for extending the search path do fail to be consistent with previous choices, or else we successfully proceed all the way down to a solution. Of course, the deepest level of failure may simply be the previous level, so no real jumping back need result.

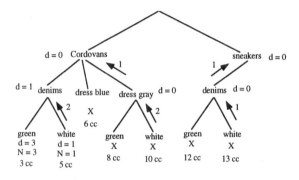

Fig. 5. Backjumping example.

In the backjumping analogue for partial constraint satisfaction, failure does not necessarily occur when an inconsistency is found. Failure occurs only when an inconsistency pushes us too far away from the perfect solution, i.e. when the accumulated number of inconsistencies reaches the necessary bound, N. For example, if search has reached the seventh variable in the sequence, and checking a value of its domain against the value for the fourth variable causes the distance to increase to the bound N, the depth of failure of that variable is 4. Figure 5 shows a trace of backjumping for a maximal solution on our running example. The numbers on the arrows show some of the values returned by P-BJ; notice how the returned depth of 1 at the bottom right of the search tree supports a jump back to the shoes level without considering additional slacks.

Figure 6 contains a backjumping algorithm for partial satisfaction. Aside from the different definition of failure, there is one major difference from the conventional backjumping algorithm. We cannot always jump back all the way to the deepest level of failure. If any values below that level were inconsistent when chosen, i.e. required an increase in *Distance* when they were chosen, we can only jump back to the level, l, of the last, deepest, one of these inconsistent values. Otherwise, minimum distance solutions can be missed that are based on other values yet to be tried at level l. This is because alternative values at level l may involve fewer inconsistencies, adding less to *Distance*, so that search can proceed from this level without encountering the bound at the same point in the search.

```
P-BJ(Search–path, Distance, Variables, Values,
       Current–depth, Return–depth, Inconsistency–depth)
    if Variables = nil then {have new best solution}
          Best-solution ← Search-path
          N ← Distance
          if N ≤ S then return 'finished' else return Current–depth − 1
    else if Values = nil or Distance = N then
          return Return–depth {may lead to backjumping}
    else {try to extend Search–path}
          Current-value ← first value in Values
          New–distance ← Distance
          try choices in Search–path, first to last, until New–distance = N or tried all:
                if the choice is inconsistent with Current–value then
                      New-distance ← New–distance + 1
                      Fail-depth1 ← level of the choice
          if New-distance < N then Fail–depth1 ← Current–depth {did not fail}
          if New–distance < N and
                the value, Fail–depth2, of
                P-BJ(Search–path plus Current–value, New–distance,
                      Variables minus the first variable,
                      values of second variable in Variables,
                      Current–depth + 1, 0,
                      Current–depth if New–distance ≠ Distance else Inconsistency–depth),
                is = 'finished' {found a sufficient solution}
                or < Current–depth {can backjump!}
                then return Fail–depth2 {backup immediately}
          else {try another value}
                return P-BJ(Search–path, Distance, Variables,
                            Values minus the first value,
                            Current–depth,
                            max(Fail–depth1, Return–depth, Inconsistency–depth),
                            Inconsistency–depth)
```

Fig. 6. Backjumping algorithm.

As an example of this phenomenon, we adapt the matching clothes problem, supposing that there are other values that were tested before, giving a bound of 2 (Fig. 7). We also change the order in which we examine the variables; the variable search order is indicated by numbers on the nodes. Each domain is checked from left to right, and vertical lines appear to the right of the values currently being considered. Current inconsistencies are indicated by the lines joining two nodes. Search has reached a dead end with the third variable (since the number of inconsistencies equals the bound), with the deepest level of failure (where the total number of inconsistencies became equal to the bound) at the level of variable 1. If this were a CSP, search would now jump back to variable 1 and the next value (gray slacks) would be selected. However, in the present problem the next value associated with variable 2 (white shirt) is compatible with the present value of variable 1 (blue slacks), so in this case no inconsistency is present between variables 1 and 2. Therefore, if search backs up to variable 2, a value can be chosen that leaves the distance at zero. In addition, the Cordovans (at variable 3) are compatible with the white shirt, but not with the blue pants, so it is possible to find a solution including the blue slacks and white shirt that gives a total distance of 1. Since N is currently 2, this is a better solution. However, it would have been overlooked if we had followed the ordinary backjumping procedure used for CSPs.

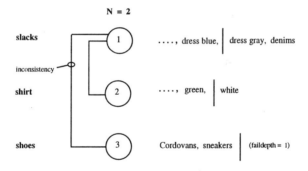

Fig. 7. Backjumping to the level of the last inconsistent choice.

To handle this situation, the backjumping analogue must keep track of the deepest level associated with an inconsistency (*Inconsistency-depth*), in addition to tracking the depth of failure. In the present implementation, *Inconsistency-depth* is passed along as search proceeds, and updated to *Current-depth*, the level at which we are currently trying to extend the search path, if there is a failure at this level of testing. Obviously, this addition to the procedure will tend to lessen the efficiency of this form of backjumping.

2.2.3. Backmarking. Backmarking [16] has the potential to avoid some redundant successful constraint checking, as well some redundant discoveries of inconsistencies. When trying to extend a search path by choosing a value for a variable V, backmarking marks the individual level, *Mark*, in the search tree at which an inconsistency is detected for each value of V. For example, if value, b, for variable, U, is consistent with the first value in the search path, but inconsistent with the second value, the *Mark* for value b is 2. (If no inconsistency is detected for a value, its *Mark* is set to the level above the level of the value.) Assuming we cannot successfully extend the search path to a complete solution, and have to back up from V, backmarking also remembers the highest level, *Backto*, to which search has backed up since the last time V was considered. When backmarking next considers a value v, for V, the *Mark* and *Backto* levels can be compared. There are two cases:

(1) *Mark* < *Backto*. If the level at which v failed before is above the level to which we have backtracked, we know, without further constraint checking, that v will fail again. The value it failed against is still there.
(2) *Backto* ≤ *Mark*. If since v last failed we have backed up to or above the level at which v encountered failure, we have to test v; however, we can start testing values against v at level *Backto*. The values above that level are unchanged since we last–successfully–tested them against v.

Figure 8 contains a backmarking algorithm for partial constraint satisfaction. Again, for partial satisfaction failure for a value, v, does not necessarily occur at the first inconsistency. It occurs at a level, *Lastmark*, where the last inconsistency is found, that which causes us to reach the bound that terminates the search path. We call this level *Lastmark*, rather than *Mark*, because, as we shall see in a moment, we also need to keep track of the level *Firstmark* where the first inconsistency with v was found. Thus we mark a range, as opposed to a single failure point. As before we store the level, *Backto*, the highest level to which search has backed up since last trying to assign a value to V. *Firstmark*, *Lastmark*, *Backto* and *Inconsistencies* are arrays, and are global, with the rows associated with the variables and the columns (except for *Backto*) with values. *Firstmark* elements are initialized to 1, *Lastmark* and *Inconsistencies* to 0, *Backto* to 1.

Again, we have two cases:

(1) *Lastmark* < *Backto*. If the level at which v failed before is above the level to which we have backtracked, we again know, without further testing, that v will fail again. All the values it failed against are still there.
(2) *Backto* ≤ *Lastmark*. If since v last failed we have backed up to or above the level at which v encountered failure, we have to test v. However, we cannot always start the new testing of values against v at level *Backto*, as in the CSP case. In the CSP case we knew that there were no inconsistencies above the level of failure. Now we only know that there are no inconsistencies above the level where the first inconsistency was found. Thus there are two further cases:
  (a) *Backto* ≤ *Firstmark*. If we have backed up to or above the level where the first inconsistency was found, we know that the unchanged values

P-BMK(Search–path, Distance, Variables, Values, Current–depth, Value–index)
    if Variables = nil then
        Best-solution ← Search-path
        N ← distance
        if $N \leq S$ then return 'finished'
        else return 'keep searching'
    else if Values = nil then
        set array Backto, beginning with current variable, to Current–depth − 1
        return 'keep searching'
    else if ((Lastmark[Current–depth, Value–index] ≥ Backto[Current–depth]
            {search has backed up to or above the level of the last inconsistency
            marked at the previous encounter with the first of Values}
            and
            adding to Distance the number of inconsistencies found between the
            first of Values and previous values in Search–path starting at level
            min(Backto[Current–depth], Firstmark[Current–depth, Value–index])
            (updating Arrays appropriately)
            produces a New–distance < N)
            {adding the first of Values to the search path
            will not push the number of inconsistencies to the bound}
        or
            (Distance + Inconsistencies[Current–depth, Value–index] < N
            {current distance + recorded inconsistencies is less than the bound}
            and
            adding to Distance + Inconsistencies[Current–depth, Value–index]
            the number of inconsistencies found between the first of Values
            and previous values in Search-path starting at level
            Lastmark[Current–depth, Value–index] + 1
            {updating Arrays appropriately}
            produces a New-distance < N))
            {adding the first of Values to the search path
            will not push the number of inconsistencies to the bound}
        and
        P-BMK(Search–path plus Current–value, New–distance,
                Variables minus the first variable,
                values of the second variable in Variables, Current–depth+1, 1)
        = 'finished'
        then return 'finished'
    else return P-BMK(Search–path, Distance, Variables,
                Values minus the first value,
                Current–depth, Value–index+1)

Fig. 8. Backmarking algorithm.

above that level are still consistent with v, and we can start the new testing of v at level *Backto*.

(b) *Firstmark* < *Backto*. Otherwise we need to start testing at the level of the first inconsistency, *Firstmark*. Above that level, the unchanged values will still be consistent with v.

Actually the situation is even a bit subtler and more distinct from CSP backjumping than we have let on. *Lastmark* may not in fact mark a failure point at all. The previous time we considered v, the last inconsistency may not have pushed us over the bound (or as in the CSP case, we may have found no inconsistency at all, in which case *Lastmark* is again set to the level above the level of v). On the other hand, even if the inconsistencies with v did not induce failure before, they may now, since the distance, the current number of other known inconsistencies, as well as the bound N, may have changed in the interim.

Accordingly we save another piece of information the number of *Inconsistencies* between v and the values in the search path down to level *Lastmark*.

Case (1) above is really:

(1′) *Lastmark* < *Backto*.

(a) If the current *Distance* plus *Inconsistencies* is not less than N, we know that v will fail again without further testing. The values which produced those inconsistencies before are still present.

(b) Otherwise, add *Inconsistencies* to the current *Distance*, and commence further testing of v at the level below *Lastmark*. The values which caused the inconsistencies before are still there. (We feel that it may be possible to commence testing at level *Backto*, but subtle bookkeeping issues need to be resolved.)

Several snapshots of the search with the backmarking analogue for the clothes matching problem are shown in Fig. 9. Each copy of the arrays shows the values following the portion of search diagramed to its left; only changed values are shown. For each array, rows associated with the first level of search are omitted from the figure because their entries never change.

For example, consider the cells associated with the value green shirt after the first portion of the search, which is represented at the far left of Fig. 9. The green shirt value is associated with row three and column one in the arrays. The value of *Firstmark*[3,1] is 1, since the green shirt does not match the Cordovans at level one. The value of *Lastmark*[3,1] is 2, due to the mismatch between the green shirt and the denims. Since there are two mismatches, or constraint failures, associated with the green shirt, *Inconsistencies*[3,1] is 2.

Note that in the last column of arrays, the cells associated with the green shirt and the white shirt (row three) all have a value of 1, since the comparisons with sneakers at level one resulted in a constraint failure and this was sufficient to attain the bound. In addition, *Backto*[3] has value 2. When search then proceeds to the next value for the variable pants, (dress blue, not shown), there is no failure, so the current distance is zero. However, before the green shirt is tested

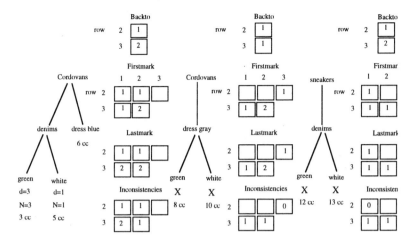

Fig. 9. Backmarking example.

again, the value of *Backto*[3] is found to be greater than *Lastmark*[3,1] –case (1')
above– and the value of *Inconsistencies*[3,1] is 1 while N is also 1, so we are in
case (1' a), and green shirt fails without any further constraint checking. The
same situation is found with the white shirt. Search returns to level 2, where the
gray pants are tested; these fail to match the sneakers and search ends with a
total of 15 constraint checks.

The version of backmarking used in the experiments described in Sect. 4
used an enhancement to minimize the number of constraint checks (though it
turned out not to make much difference). Under the second "else if" of the algorithm as described in Fig. 8, the distance was first checked against N; if N
had been reached or exceeded, there was no need for constraint checking. Unfortunately, this tactic entailed further checking after the last "else" and possible
revision of the values of *Firstmark*, *Lastmark* and *Inconsistencies* if the bound
had been reached *and* the algorithm had since backed up into the recorded range
of constraint failures. For example, if the value in *Backto* was now between the
corresponding values in *Firstmark* and *Lastmark*, *Lastmark* was set to *Firstmark*
and *Inconsistencies* set to 1.

Various refinements of PCSP backmarking are possible that have not yet
been implemented. Observe that in our running example, the second time we
encounter the green shirt, while the *Lastmark* and *Inconsistencies* machinery
does not help us, the *Firstmark* value could. Since *Firstmark* occurs at level one,
which is above the *Backto* level, and the bound is already down to one at this
point, we could infer, without any further constraint checks, that using the green
shirt will bring us to the bound. In fact, we could consider storing all levels at
which values encountered inconsistencies, and using that additional information
for further pruning without further constraint checking.

2.3 Prospective techniques

Prospective techniques "look ahead" to establish some form of local consistency before continuing the search for a global solution. Prospective techniques can prune from consideration values that do not meet local consistency criteria. Consistency techniques can be used as *preprocessing* methods prior to search (in some situations leaving little if any work for subsequent search). They can be interleaved with backtracking to form *hybrid algorithms*. There are also methods of applying local consistency techniques repeatedly to subproblems to achieve global solutions.

The most commonly used form of local consistency is *arc consistency* [26]. A problem is fully arc consistent if every value in the domain of every variable is consistent with, we also say *supported by*, at least one value in every other variable domain. Arc consistency preprocessing eliminates all unsupported values.

The most familiar hybrid algorithm is *forward checking* [19]. In forward checking, each assignment of a value, v, to a variable, V, is followed by a limited amount of arc consistency checking, in which the domains of variables that share a constraint with V are tested against v.

It is a standard branch and bound strategy (often discussed in terms of "lower bounds") to increase pruning by estimating the implications of proceeding on from the current search point. Prospective methods provide a means of implementing this strategy for partial constraint satisfaction. Until now, we have been discontinuing a search path if the number of inconsistencies on that path, the distance D, is not less than the bound N. If we had some way of determining that no matter how we sought to continue that path we would encounter at least $D\prime$ additional inconsistencies, then we could discontinue the search if $D + D\prime$ failed to be less than N. Prospective methods permit us to obtain such $D\prime$ values.

2.3.1. Pruning with arc consistency counts. In the CSP context, arc consistency permits us to eliminate values that arc consistency processing determines cannot participate in any complete solution. In our running example, sneakers can be eliminated because it is not consistent with any shirt. Furthermore the dark green shirt and the dress blue slacks can be eliminated because they are not consistent with any shoes. Notice further that although the denim slacks were originally supported by sneakers, now that sneakers have been eliminated there is no support for denims, and they can be eliminated in turn. Now there is no longer any support for the white shirt. Having eliminated both shirt possibilities arc consistency has, for this problem, discovered that there is no global solution, and no further search is necessary.

In the PCSP context, it is not possible to discard values in this way unless we have an initial value for the necessary bound, N. If we did know, for example, a priori, that there was a solution that violated only one constraint, or that any solution that did violate more than one constraint was unacceptable, we could eliminate a value for a variable that was not consistent with any value for two other variables.

However, it is possible to perform prior calculations regarding the increments in distance associated with specific values. In particular, for each value, the number of domains with no supporting values can be tallied; this number, the *arc consistency count*, is a lower bound on the increment in distance that will be incurred if this value is added to the solution. An algorithm for computing arc consistency counts is given in Fig. 10.

In the course of subsequent search, the arc consistency count for a proposed search path extension, v, can be added to the distance associated with the search path, and this sum compared with the current bound N. If the sum is not less than N, then we know that any complete solution starting with the current search path and involving v will violate too many constraints. We can fail at this point without testing v at all. Of course, v was tested once during the preprocessing; however, we may encounter v many times during subsequent search, and on a number of those occasions the preprocessing may save further testing of v. The forward checking algorithm below uses a similar strategy involving a limited, dynamic form of arc consistency count.

P-ACC(Variables, Domains, Counts)
    For all V_i belonging to Variables:
        For all $V_j \neq V_i$ such that there is a constraint between V_i and V_j
            For each value, a, in the domain of V_i:
                if there is no value, b, in the domain of V_j such that the pair (a,b)
                is allowed by the constraint between V_i and V_j
                then increment the count for value a

Fig. 10. Algorithm for computing arc consistency counts.

If arc consistency checking is used to tally the number of domains that do not support a value, then one constraint failure may be counted twice, once for each value that does not belong to an acceptable pair. This is not a problem during subsequent search because a consistency count is not incorporated into the distance except to check against the bound when first considering that value for extending the search path. (If the value is accepted, then adjustment of the distance through consistency checking is done in the same way as with basic branch and bound.) Suppose a particular value is included, and another value is considered afterwards whose arc consistency count depends in part on a constraint that also affected the former value's arc consistency count. Since the current distance is based on retrospective checking, it does not yet include this constraint, so the second count can be used without modification in computing a projected distance to compare with the upper bound.

Arc consistency counts have the advantage that they need only be computed once, before search, and can be done in $O(cd^2)$ time, where c is the number of constraints and d the maximum domain size for the variables. They can then be stored in an appropriate form for testing against the current distance. In contrast, both the retrospective and hybrid techniques require more extensive calculation to retain and update information related to distance for specific values.

Because no values are actually discarded, there is no propagation of failure in the manner of arc consistency algorithms for CSPs (e.g. where the removal of sneakers led in turn to the removal of denim). It may be possible to propagate counts in a manner analogous to ordinary constraint propagation. (Hybrid analogues based on this idea may also be possible and were suggested by Shapiro and Haralick [35] for inexact matching.) This might involve retaining information about the conditions of failure, employing conditional counts that can only be used if the supporting values are *not* used in the solution. In the arc consistency count algorithm, in contrast, which takes one pass through the variables, we are assured that the consistency counts are all unconditional.

2.3.2. Forward checking. Forward checking is a hybrid algorithm that uses a very limited amount of arc consistency checking. Each time a value, v, is assigned to a variable, V, the algorithm looks ahead to all the variables that currently have not been assigned a value, and that share a constraint with V, and removes from the domains of these variables any values inconsistent with v. For example, when Cordovans are proposed for shoes, the denim and dress blue slacks will be removed, and the green shirt.

If later we change our mind about v, the pruned values have to be restored. E.g. when we move on to consider sneakers, the denim and dress blue slacks and the green shirt must reappear. (In an implementation, recursion can handle the bookkeeping for the variable domains during backup.) Of course, since sneakers are not consistent with any shirt, forward checking with sneakers will reduce the shirt domain to the empty set, signaling a failure point.

Notice that despite the fact that it can be viewed as an integration of consistency processing and backtracking, forward checking really is almost the complement of standard backtracking. Standard backtracking checks a value for consistency against previously chosen values. With forward checking, when we propose a value we already know it is consistent with the previously chosen values (or else the consistency processing would already have pruned it away). Now we test it against the domains of the remaining, uninstantiated variables.

When used for partial satisfaction, forward checking is based on the same type of looking ahead as ordinary forward checking. However, again, the differing definition of failure comes into play. If there is an inconsistency, a value is not rejected unless the total number of currently chosen values with which it is inconsistent is at least as large as the difference between the current distance and the bound N. This means that the algorithm must dynamically keep track of the number of times a value has been found to be inconsistent with currently chosen values. This number, a form of dynamic arc consistency count, we will call the *inconsistency count* for a value.

An example of forward checking beginning to operate on the matching clothes problem in shown in Fig. 11. In this figure, counts associated with the values of variables which share a constraint with a variable, V, are shown at the point in the search at which they are calculated. For example, when Cordovans are chosen, the inconsistency count for green shirt becomes one; it increases to two when denims is chosen.

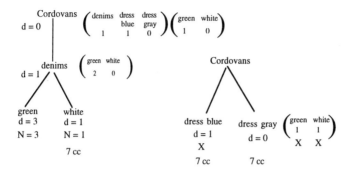

Fig. 11. Forward checking example.

Although we see more constraint checks performed in the early stages of this example than we saw with straightforward branch and bound, the counts derived from these early checks are used to avoid further constraint checks, putting forward checking ahead at a later stage.

When a value, v, is proposed, its own inconsistency count can be added to the current distance, and the total used in a manner similar to that proposed for arc consistency counts in the previous section: if the total is equal to or greater than the bound, v fails immediately. For example, when Cordovans are tried with blue slacks, the bound is already one (set by the Cordovans, denim slacks and white shirt combination), and the inconsistency count of blue slacks is one (set when Cordovans was chosen). This tells us that we cannot hope to choose a shirt that will permit us to do any better than our current best solution. The blue slacks fail, without any further testing.

Notice also that when gray slacks are tried with Cordovans, although they are consistent, together they eliminate all the shirt values. Cordovans raises the inconsistency count of green shirt to one, eliminating it, since the bound is one. Gray slacks raises the inconsistency count of white shirt to one, eliminating it, and leaving us with an empty domain for shirts. Reducing the domain of an uninstantiated variable to empty also, of course, signals a failure point.

Shapiro and Haralick [35] generalized the CSP look ahead technique in their study of the "inexact matching problem". (The algorithm they call "forward checking" does more looking ahead than ours; it employs the "extended forward checking" discussed in Sect. 2.5) They defined the "inexact consistent labeling problem", which involves searching for all solutions within a given error bound. The count of violated constraints that we use could be viewed as the "error". However, Shapiro and Haralick's algorithms did not seek optimal solutions and were not full branch and bound algorithms in the sense that they did not store and compare with the "best so far" solutions; all comparisons were with the error bound.

```
P-FC(Search-path, Distance, Variables, Domains, Inconsistency–counts)
    {Domains holds values for variables not yet assigned a value in Search–path
    that are consistent with the values in Search–path}
    if Variables = nil then
        Best-solution ← Search–path
        N ← Distance
        if N ≤ S then return 'finished'
        else return 'keep searching'
    else if first of Domains = nil then
        return 'keep searching'
    else if Distance + first count in Inconsistency–counts < N
        {first count in Inconsistency–counts is the number of inconsistencies
        between first value in first of Domains and values in Search–path}
        and
    .   while computing New–inconsistency-counts and New–domains,
        {check consistency, as needed, with first value in first of Domains;
        details determined by which version of P-FC used}
        New-domains retain at least one value in each domain
        and
        P-FC(Search–path plus first value in first of Domains,
            Distance + first count in Inconsistency–counts,
            Variables minus first variable,
            New-domains, New–inconsistency-counts) = 'finished'
        then return 'finished'
    else
        return P-FC(Search–path, Distance, Variables,
                    Domains after removing first value from first of Domains,
                    Inconsistency–counts minus count for first value
                    from first of Domains)
```

Fig. 12. Forward checking algorithm.

Different variants of forward checking can be devised, depending on the manner in which the counts for each value are used to minimize constraint checks. In contrast to Shapiro and Haralick, who stored counts in tables and did not discard values, corresponding lists can be used for domain values and associated counts, from which values are discarded if their counts are high enough to raise the distance to the bound.

In the most straightforward version, referred to as P-FC1, the values of a domain are checked against the latest value proposed, v, and the inconsistency counts associated with values inconsistent with v are incremented by one. During this revision of the domain, when a count for a value u is incremented, the incremented count is added to the distance of the search path down to v and the sum is tested against the bound. If this sum equals or exceeds the bound, it means that the search path down to v cannot be extended to a solution including u without reaching the bound. Thus u can be removed; u will not checked again at this or at lower levels of recursion.

In the second version (P-FC2), all counts are tested in this manner against the bound during revision, not just the incremented counts. This eliminates values whose counts might now be too large, not because the count has increased, but because the bound has been lowered since the values were last considered for elimination.

In the third version (P-FC3), we take this a step further. All counts are tested before doing any constraint checking to see if counts can be incremented. Thus values may be deleted simply because the bound has been lowered, without any consistency checking to determine if their counts need to be incremented. Here, of course, any incremented counts of values that survived the first test but fail the consistency test have to be tested again during revision of the list of values.

To summarize, in P-FC1, constraint checks are done on all values and count checks on the failures; in P-FC2, constraint checks and then count checks are done on all values; in P-FC3, count checks are done on all values, constraint checks are done on viable values and then count checks are done on values that failed the consistency test. A general P-FC algorithm that does not specify the details of constraint and count checking is shown in Fig. 12.

2.3.3. Tree-structured problems.

Local consistency methods have been used to support polynomial algorithms for CSPs with tree or tree-like structure [8] [15] [27]. When problems do not have such structure it may still be useful to view them as containing or being contained in such structures [6] [8] [9] [13] [15] [28]. A problem is *tree-structured* if the graph that results from viewing variables as vertices and constraints between variables as edges between vertices (the *constraint graph*) is tree-structured.

The idea of utilizing a tree which represents a subproblem is particularly attractive in the PCSP domain where the constraints that need to be removed to reduce a problem to the desired structure do not eventually have to be satisfied, but may be written off as unsatisfied constraints in the partial constraint satisfaction process. The algorithm P-T in Fig. 13 obtains an efficient solution for tree-structured maximal constraint satisfaction problems.

Tree-structured CSPs can be solved in time linear in the number of variables and quadratic in the maximum domain size, but at first blush one might suppose that tree-structured maximal constraint satisfaction problems would not admit such a small bound. In fact, however, algorithm P-T does achieve this bound.

Algorithm P-T:

For each variable, L, which is a leaf node of the constraint tree:

    For each value, e, of L:

        Set Cost(e)=0

For each level in the tree starting at the level above the leaves, and working upwards:

    For each variable, V, at that level:

        For each value, v, of that variable:

            For each child, U, of that variable:

                For each value, u, of that child:

                    If v is consistent with u

                        then set Cost(v,u) = Cost(u)

                        else set Cost(v,u) = Cost(u)+1

                Link v to a u such that Cost(v,u) is minimum

                Set Cost(v) to that minimum Cost(v,u)

        Delete all v except those with minimal cost at V

Return as a minimal solution a value for the root with minimal cost, along with the tree of values linked to it at the other variables.

Fig. 13. A linear algorithm for a tree-structured maximal constraint satisfaction problem.

Theorem. *Algorithm P-T finds a maximal solution for a tree-structured maximal constraint satisfaction problem and has an $O(nd^2)$ complexity bound, where n is the number of variables and d the maximum number of values in a variable domain.*

Proof. As we process the tree of variables we associate costs with values. The cost represents the total number of constraints violated if we choose that value and all values we have linked to it at descendant variables. We retain only minimal cost values at each variable. We claim that the cost of a value at a variable represents in fact the minimal number of constraints that we need to violate in order to instantiate that variable and its descendants in the variable tree, while that value and its descendant values represent in fact an optimal solution for the subtree. Thus at the root variable we will have found a minimal cost instantiation, an optimal solution for the complete tree-structured problem.

The claim is trivially true at the leaves. We work our way inductively up the tree to the root. Assume that the claim is true for all the children of a variable V in the constraint tree. The algorithm only keeps values at V, that minimize the additional cost vis à vis previously retained values for the children, i.e. that minimize the number of constraints violated. Changing those previously retained values could not improve matters: changing a value at a child to avoid an inconsistency with the parent value means replacing the child value with one whose additional cost at the very least offsets the additional consistency. Note also that the only constraints between V and its descendants are those between V and its children. Furthermore there are no constraints between variables in different subtrees of V. Thus the cost of a value at V represents the minimal number of constraints that we need to violate in order to instantiate that variable and its descendants in the variable tree, while that value and its descendant values represents an optimal solution for the subtree rooted at V.

Working up from the leaves, the algorithm builds up optimal solutions for subtrees, all the way to the root. Essentially all $n - 1$ edges in the tree have to be processed once, with each processing requiring at most d^2 consistency checks. □

It should be emphasized that these results, indeed more powerful results, have already been obtained in a closely related context [7]. The context is a CSP with multiple solutions, where the objective is to choose a solution which maximizes the value of a criterion function. Though superficially this context appears quite distinct from the PCSP context, where there may not even be a single solution, one could presumably use the criterion function to simulate a maximal satisfaction problem.

2.4 Ordering

As usual in branch and bound it is advantageous to order the search to heuristically increase the likelihood that a good, or ideally optimal, solution will be found early. The counts produced by the arc consistency method described in Sect. 2.3.1 can be used to order search. This can be done either by ordering the values in each domain according to their individual counts or by ordering the variables on the basis of some statistic derived from the counts for each domain.

In the tests carried out for this paper, values are ordered by increasing counts. This allows the values most likely to produce a good solution to be tested first, so that a minimum or near-minimum distance solution should be found more quickly on average, yielding a better bound early in search.

In the present work the statistic used for variable ordering is the mean for the counts associated with the values of each domain. (The minimum count was also considered, but, for most problems, there are too many zero counts to make this statistic sufficiently discriminating.) In addition, variables are ordered by *decreasing* mean count. This is based on the premise that, once a good bound is found, checking domains with less support early in the search will increase the likelihood that the bound will be reached at higher levels of the search tree. This argument is supported by tests in which the ordering was in the opposite direction; the results were appreciably worse in most cases (and sometimes worse than the basic branch and bound), especially for sets of harder problems.

A variety of variable ordering techniques have been studied for CSPs and could be considered in the PCSP context. More sophisticated cost estimates could be associated with the variables to support additional PCSP–specific techniques. The two forms of ordering based on arc consistency counts, as well as the pruning based on arc consistency counts described in Sect. 2.3.1, can, of course, be combined in different ways.

2.5 Extensions

The basic techniques discussed above can be extended in a variety of directions, and other techniques considered. Obviously in this paper we can only begin a research program that requires, at the least, a recapitulation of the entire history of progress on CSPs.

One obvious line of inquiry involves combining basic techniques. We emphasize in this paper development and analysis of basic "atomic" techniques. However, we conducted some experiments with an algorithm which combines a retrospective technique–backmarking–a prospective technique–arc consistency count pruning–and an ordering technique–value ordering. We call this algorithm P-RPO.

The branch and bound context suggests looking for tighter lower bounds on the distance of the minimal distance solution that includes a given set of value choices. These can be used to obtain quicker pruning of choices that have no chance of doing better than a solution already found. The arc consistency count pruning described above does this sort of thing in a simple but efficient manner; the arc consistency counts are only computed once in a preprocessing step. The forward checking analogues utilize a kind of dynamic arc consistency count.

Shapiro and Haralick [35] suggest more elaborate lower bound computations, up to dynamically utilizing a complete arc or even path consistency [26] analogue after each value choice, for their inexact matching problems. Of course, there are tradeoffs between consistency check savings and bound computation costs. We tested an analogue, which we call *extended forward checking* for partial constraint satisfaction, of the most successful algorithm that they implemented.

Forward checking for maximal constraint satisfaction, as described above, assigns an inconsistency count to a value, v, based on the inconsistencies that would be incurred by adding v to the choices, C, which have already been made. Extended forward checking goes further by forming a lower bound estimate of the number of further inconsistencies that would accrue in the course of choosing values for each of the remaining variables. For each of these variables it finds the minimum inconsistency count assigned to the values for that variable. It then adds all of these counts to the inconsistencies incurred in choosing C and v. This sum will be a lower bound estimate on the number of constraints violated by a maximal solution that contains C and v.

Thus we could turn the P-FC algorithm into an extended forward checking algorithm, P-EFC, by changing the test:

$$Distance + \text{first count in } Inconsistency\text{-}counts < N,$$

into the test:

$$Distance + \text{first count in } Inconsistency\text{-}counts$$

+ sum of minimum counts for each variable
in *Remaining-variables* $< N$.

The version of extended forward checking we tested included the forward checking refinements implemented in our P-FC3 algorithm, and is thus called P-EFC3.

Essentially we are playing a "what if" game. If we added v to the already chosen values, what would be the best we could hope for? If that is not as good as the best we have already done, we can prune v from consideration (until such time as we may change our choices, C).

3 Experiments

3.1 Overview

The algorithms described in Sect. 2 were tested in a series of experiments with random problems. In the first six experiments, we examined the relative efficiency of these algorithms and the relation between efficiency and problem structure. In these experiments each algorithm was run to completion, to find a maximal solution. Three preliminary studies compared the efficiency of related algorithms. The first included the retrospective algorithms, P-BJ and P-BMK, together with the basic branch and bound (P-BB). The second compared the three versions of P-FC, which differed in the number and placement of tests of distance plus counts against the bound. The third compared procedures based on the arc consistency algorithm, P-ACC. In the fourth and main experiment we tested the most promising algorithms from the first three experiments on a more extensive set of problems, in which basic problem parameters such as domain size were varied systematically. In the fifth experiment, the best algorithms were tested on problems of varying size (number of variables); this experiment also included the algorithms P-RPO and P-EFC3. In the sixth experiment we obtained data on overall efficiency, using time as a measure, with problems selected from the main experiment that presented different levels of difficulty for the algorithms.

For many problems finding a maximal solution may not be feasible because the problem is too hard to solve completely. In these cases a good submaximal solution may still be acceptable. Branch and bound techniques are useful in this situation because of their anytime feature: after an initial solution is found, the algorithm can stop at any time before completion with the best solution found so far. In Experiments 7 and 8, we examined the efficiency with which the different algorithms can find submaximal solutions with distances increasingly close to that of a maximal solution. The problems used in these experiments were larger versions of the random problems used in previous experiments and, in addition, a set of large coloring problems believed to be very hard to solve [4]. In these experiments resource bounds were established by placing a limit on the number of constraint checks that could be performed; the program terminated if this limit was reached.

3.2 Random problem generation

In generating random problems, there are four features to consider:

(1) number of variables, n
(2) number of constraints, c
(3) domain size, d
(4) number of value pairs included in a constraint, p.

In the present work, n was fixed for each set of problems. Then the values of the other three features were determined with a (constant) probability of inclusion method, that is best explained by example. Consider the choice of number of constraints. For problems of ten variables with a connected constraint graph, up to 36 constraints can be added. In generating a random problem, the probability of inclusion is fixed at, say, 0.3, and each of the possible constraints is considered for inclusion using random methods to simulate this probability. With a probability of 0.3, a set of problems is obtained with an expected value for number of constraints equal to $9 + (36 * 0.3) \approx 20$. Similar procedures were used to determine d and p for each domain and constraint, respectively. The only limitation in these cases was that the value of d or p was at least one. If no element was included (zero value), the procedure was repeated beginning with the first element until a non-zero value was obtained. This method has the advantages that:

(1) each parameter value can be varied in a way that is easily characterized, i.e., by a single probability value
(2) each element has the same probability of inclusion, which makes the sampling properties of each possible set of elements relatively easy to characterize.

For these experiments, problems without solutions were required. This limited the range of probability values that could be considered, because as d or p increases, it is more likely that problems with solutions will be produced. Values of 0.2, 0.4 and 0.6 were used to determine p, while values of 0.1, 0.2 and 0.3 were used for d, based on a maximum domain size equal to $2n$ or twice the number of variables. Values of 0.3, 0.6 and 0.9 were used to determine the number of constraints to be added to a spanning tree (which was itself derived by choosing pairs of variables at random). In the remainder of the paper, these probability values will be designated as p_c, p_d, p_p, for probability of constraint, domain, or constraint pair inclusion, respectively. (p_c is sometimes called the *density* of the problem, and p_p the (relative) *satisfiability*, while the complement of p_p is sometimes referred to as the *tightness* of a constraint.) The values chosen covered most of the range of possible values, while allowing a similar degree of variation in each case. After generation, a problem was tested for solutions. If a solution was found, the following strategy was used to obtain an insoluble problem with identical parameter values: a constraint pair that included two values in the solution was chosen at random and discarded, and another pair of values from the same domains was chosen at random as a new constraint pair; this procedure was repeated until a problem with no solutions was found.

3.3 Experimental design

Experiments 1–3. Experiments 1–3 were based on a set of ten-variable problems for which the probabilities of domain and value pair inclusion took on all the values mentioned in the last section, while the density was always 0.3. This gave nine categories of problems. Ten problems were generated for each category, for a total of 90 problems.

In Experiment 1, the algorithms tested were P-BB and two other retrospective algorithms, the backjumping and backmarking analogues, P-BJ and P-BMK. Experiment 2 compared the three variants of P-FC described in Sect. 2.

Experiment 3 compared several variants of P-BB that incorporated different forms of information derived from arc consistency counts, either singly or in combination. These were:

(1) pruning based on the count for a given value,
(2) ordering of values in each domain by increasing count,
(3) ordering of variables by (decreasing) mean count of the values in their domains,
(4) a combination of the value and variable ordering strategies,
(5) a combination of pruning and value ordering,
(6) a combination of pruning, value and variable ordering.

Experiment 4. Experiment 4 compared the most promising algorithms from each of the first three experiments. These were P-BMK, P-FC3, and two varieties of branch and bound that incorporated information from the counts obtained by P-ACC ((5) and (6) above). P-BB was also included for reference. For this experiment the problem set was expanded to include the other two probabilities of constraint inclusion (0.6 and 0.9). It was, therefore, a fully crossed design, with each of the three probabilities of inclusion associated with each parameter, as described in Sect. 3.2. This gave 27 categories; ten problems were generated for each category for a total of 270 problems. (These included the 90 problems used in Experiments 1–3).

Experiment 5. In this experiment the best retrospective and prospective algorithms from Experiment 4, P-BMK and P-FC3, along with P-BB, were compared on problems in which the number of variables, n, was varied. In addition, the extensions P-RPO and P-EFC3 were included. The number of variables ranged between eight and 12, with ten problems for each problem size. Based on preliminary tests of feasibility for higher n, the values of 0.3, 0.2 and 0.4 were chosen for p_c, p_d, and p_p, respectively. (In Experiments 1–4, with $n = 10$, problems in this category were relatively easy to solve.).

Experiment 6. In Experiment 6, the following algorithms were compared with respect to run time: P-BB, P-BJ, P-BMK, P-FC2 and P-FC3. Problems were selected from Experiment 4 in which the order of magnitude for constraint checks was three, four or five for branch and bound. (Similar ranges in terms of order of magnitude were obtained with the other algorithms tested.) Six problems were chosen at each level of difficulty, three for which the density, p_c, was 0.3 and three for which it was 0.9, for a total of 18 problems. These problems were also chosen so that other parameter values (probabilities of inclusion) also varied. Run times were obtained with the Lisp time function.

Experiments 7–8. In Experiment 7, problems had 12, 16, or 20 variables, with ten problems per group. The values of p_c, p_d, and p_p were the same as those in Experiment 5; in fact, the same 12-variable problems were used in both experiments. As a consequence, the distance associated with the best solution was known. For the 16-variable problems, an optimal solution was obtained using P-EFC3. This allowed a more complete evaluation of the suboptimal solutions obtained when the number of constraint checks was limited to two million. It was also used as a reference for the results with 20-variable problems, which were too large for an optimal solution to be found when the number of constraint checks was limited to five million. Five algorithms were tested with the 12-variable problems: P-BB, P-BMK, P-FC3, P-RPO and P-EFC3, and all but P-FC3 were tested with the 16- and 20-variable problems. (Since P-FC3 was always bettered by P-EFC3, it will not be discussed further.)

In Experiment 8 a similar procedure was used with nine large, "really hard" coloring problems [4]. These were classic graph coloring problems where the objective is to color every vertex of a graph with a color, chosen from a fixed number of colors, such that no vertices joined by an edge have the same color. Vertices correspond to CSP variables, edges to CSP constraints. These problems involved four colors, 130–144 variables and 620–646 constraints, giving densities of 0.05–0.06. Three algorithms were tested: P-BB, P-BMK and P-EFC3. (P-RPO was not included since arc consistency counts are zero for all values in coloring problems of this sort.)

In all experiments except the sixth, the basic measure was the number of constraint checks, although we also recorded the number of nodes searched. In Experiment 6, the measures were execution time and time per 1000 constraint checks. Garbage collection time and basic I/O and set-up time were subtracted from the total time in calculating these measures.

In these experiments, the analysis of variance (ANOVA) was used to test the statistical significance of differences due to the algorithm used and to variation in each problem parameter, as well as interactions between these factors. In these tests, one to three of the statistical factors were based on problem parameters. If more than one such factor appeared in an experiment, they were fully crossed, with each combination of factors forming a single experimental group. The algorithm used to solve the problem was a separate factor, with all problem groups "repeated" on it. (A simple fixed effects design, in which different problems are chosen from the same category for each algorithm would introduce more variation and is unecessary, since independence of different treatments (algorithms) is not an issue in this domain.) As an example consider the design of Experiment

1. Here, the factors based on p_d and p_p are crossed to form separate categories of problems, all of which are repeated on the factor related to the three algorithms tested. Each of these factors, as well as the first and second order ($p_d \times p_p \times$ algorithm) interactions was tested for statistical significance, using the standard null hypothesis of no differences between groups related to that effect. All analyses were done with log-transformed data, to reduce differences in variance among groups. If the effect of algorithms in the ANOVA was statistically significant, algorithms were compared on their mean performance using Tukey's q test for nonorthogonal pairwise comparisons [22].

In Experiment 4, we performed several further analyses to better understand performance characteristics in relation to problem parameters. Standard deviations were obtained for each algorithm on each problem set, as well as measures of skew, or asymmetry in the distributions of performance scores. Pearson product–moment correlations between P-BB and the other algorithms were also derived, using the original scores. And, for each algorithm, multiple regression analysis with respect to the problem parameters was carried out using the log-transformed scores and a zero y-intercept.

In Experiments 7–8, statistical analysis consisted of paired comparison t tests between algorithms, beginning with 100 constraint checks and including successive powers of ten up to the highest power within the response bound. (The value of 5 million constraint checks was also tested for 20-variable problems.) For P-RPO the constraint checks required for arc consistency checking were added to the total in each case.

3.4 Results

3.4.1. Experiments 1–3: Preliminary Comparisons of Similar Algorithms.
In each of the first three experiments, some of the algorithms were clearly superior to others tested on the same set of problems. In Experiment 1, P-BMK was markedly superior to both P-BB and P-BJ. In Experiment 2, P-FC3 was the most efficient in terms of constraint checks, although all three variants of P-FC were much better than P-BB. In Experiment 3, versions of P-BB that used variable and value ordering together, or pruning and value ordering, or a combination of all three strategies were generally superior to the other three variations as well as to P-BB.

These results were borne out by the statistical analysis. In all three experiments, the effect due to algorithms was statistically significant (Experiment 1, $F[2,243] = 12.89$, $p < 0.001$; Experiment 2, $F[2,243] = 4.58$, $p = 0.01$; Experiment 3, $F[5,486] = 5.85$, $p < 0.001$). In addition, the factors related to problem parameters, specifically to differences in p_d and p_p, were highly significant statistically ($F > 50$ always, $p \ll 0.001$), as was the interaction between these factors ($F > 10$ always, $p < 0.001$). With one exception in Experiment 3, none of the interactions between algorithms and the other factors was statistically significant ($F \leq 1$).

The meaning of the significant effects related to problem parameters can be understood from Fig. 14, which gives the main results for Experiment 1. As

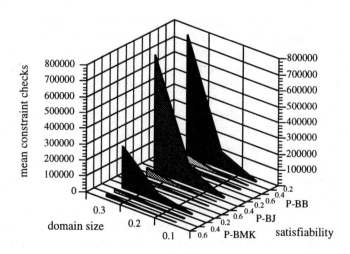

Fig. 14. Mean constraint checks for branch and bound (P-BB) and for backjumping (P-BJ) and backmarking (P-BMK) in Experiment 1, as a function of parameters for domain size (p_d) and satisfiability (p_p)

this figure shows, problems become harder as average domain size increases, and this effect is greater with increasing tightness of constraints (decreasing p). This is intuitively plausible. Of greater importance is the relatively rapid change in difficulty with more extreme values, so that really difficult problems are found in only a small part of the parameter space. These results are, therefore, consistent with Dechter and Pearl's demonstration in [8] that, on average, random problems such as these are fairly easy to solve.

In Experiment 1, P-BJ was only slightly superior to P-BB, while, as stated, P-BMK was markedly superior (Fig. 14). Most importantly, P-BMK showed substantial improvement over basic branch and bound in the most difficult parts of the parameter space, i.e. the points associated with larger domain size and greater tightness of constraints. On the other hand, it appeared that for P-BMK as well as P-BJ average *relative* improvement was greatest for larger values of satisfiability and domain size, where problems were easier to solve, although this interaction was not detected by the ANOVA, probably because it was over-shadowed by the corresponding changes that occurred with all three algorithms. Observed differences in performance were supported by individual comparisons of means. In this analysis, the overall difference between P-BMK and each of the other two algorithms was statistically significant ($q[2,89] \geq 5$, $p < 0.01$ for both comparisons), while the difference between P-BJ and P-BB was not.

In Experiment 2, both methods of forward checking that employed more count checks (P-FC2 and P-FC3) were superior to the algorithm that depended more directly on constraint checks (P-FC1). However, in the individual compar-isons, the differences that were statistically significant were related to P-FC3

$(q[2,89] = 2.82, p < 0.05$ and $q[2,89] = 4.20, p < 0.01$, for comparisons with P-FC2 and P-FC1, respectively). All versions were markedly superior to P-BB.

In Experiment 3, improvement in performance due to strategies based on arc consistency counts depended in part on domain size, which was reflected in the statistically significant interaction between algorithms and the factor based on p_d ($F[10,486] = 2.95, p = 0.001$). Consider, first, the individual strategies of pruning based on counts, or ordering values, or ordering variables on the basis of counts. For $p_d = 0.1$, pruning and value ordering each reduced the number of constraint checks relative to the basic P-BB algorithm, while variable ordering resulted in increases, which were sometimes marked. For $p_d = 0.2$ or 0.3, all three strategies improved the mean performance, although the first two did so most consistently (Fig. 15). Perhaps because of this interaction, none of the comparisons of individual strategies was statistically significant, although the compaison between variable ordering and pruning approached significance.

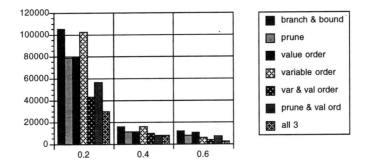

Fig. 15. Mean constraint checks required by algorithm P-BB when combined with strategies based on arc consistency counts obtained with P-ACC. Constraint checks shown as a function of constraint satisfiability, p_p; in this slice of the parameter space, $p_d = 0.2$. Results for the basic branch and bound are included for reference.

Combining strategies based on arc consistency counts resulted in greater improvement in performance (Fig. 15), and this was reflected in the individual comparisons. When pruning and value ordering were combined, performance of this algorithm was better than for either strategy alone ($q[2,89] \geq 2.8, p < 0.05$ for both comparisons). Value and variable ordering combined was also better than either strategy alone, although only the latter comparison was statistically significant ($q[2,89] = 3.43, p < 0.05$), in part because the superiority of the combination was only found for $p_d = 0.2$ or 0.3. Finally, the combination of all three strategies showed the best performance overall, although in comparisons with the combinations of two strategies, only the difference with the "double" ordering of variables and values was statistically significant ($q[2,89] = 3.30, p < 0.05$). Again, this superiority was found only for larger domain sizes.

3.4.2. Experiment 4: Comparing the best algorithms from Experiments 1–3. As indicated above, the algorithms tested in this experiment in addition to the basic branch and bound (P-BB) were P-BMK, P-FC3, and two branch and bound algorithms that used arc consistency counts: one that used pruning and value ordering and one that used all three strategies. Two algorithms based on arc consistency counts were included because the results from Experiment 3 were not altogether conclusive concerning the best algorithm in this group. Although the procedure that combined the three arc consistency count strategies was best overall, it did not perform well on problems with small domain sizes in comparison with other algorithms based on arc consistency counts, and there was considerable variability in performance even when the mean for this algorithm was better than the other means. The algorithm based on pruning and value ordering did not differ statistically from the full combination algorithm overall; moreover, unlike the combination algorithms that incorporate a variable ordering based on the counts, these strategies are less likely to result in inferior performance with respect to P-BB on individual problems.

In this experiment the factor in the ANOVA for constraint checks associated with the algorithms was highly significant statistically (Table 1; the same pattern of statistically significant results was found in the ANOVA for number of nodes checked). There was, in fact, a fairly consistent ordering over the portions of the parameter space that were tested (Fig. 16). P-BMK and P-BB incorporating arc consistency strategies generally reduced the number of consistency checks made by the basic P-BB algorithm by a factor of two to three. P-FC3 reduced this number by another factor of two. In terms of overall means the ranking of performance from worst to best was: basic branch and bound (P-BB), P-BB with the three arc consistency strategies, P-BB with pruning based on consistency counts and value ordering, P-BMK, and P-FC3. Comparisons of mean performance following the ANOVA showed statistically significant differences between each pair of algorithms, including the two that used arc consistency counts.

The ANOVA for constraint checks also showed statistically significant effects related to each problem parameter (Table 1). In addition, all interactions between these factors were statistically significant. The effects of domain size and satisfiability and the interaction between them can be observed in Fig. 16 (and Fig. 14). For all algorithms, problem difficulty increased with increasing domain size and with diminishing satisfiability. The hardest problems were those with the largest domains and the smallest number of acceptable pairs per constraint. These effects were enhanced when the density of the constraint graph increased, which accounts for the interactions that involve this factor.

The interactions between algorithms and p_c and between algorithms and p_d were also statistically significant (Table 1). In contrast, the interaction between algorithms and p_p was not significant ($F \leq 1$); nor were any of the higher-order interactions that involved the algorithms factor. Perusal of performance means suggests that the two statistically significant interactions were due in large part to the P-ACC combination algorithm. For this algorithm, the average number of consistency checks increased more dramatically with increases in p_c or p_d

Table 1. Statistically Significant Effects in ANOVA for Experiment 4

Factor	df	F	p
algorithm	4	43.53	0.0001
p_d	2	2070.00	0.0001
p_p	2	470.91	0.0001
p_c	2	423.81	0.0001
$p_d \times p_p$	4	81.58	0.0001
$p_d \times p_c$	4	8.56	0.0001
$p_p \times p_c$	4	21.19	0.0001
$p_d \times p_p \times p_c$	8	5.77	0.0001
alg $\times p_d$	8	4.68	0.0001
alg $\times p_c$	8	4.39	0.0001
error	1215		

Note: df is degrees of freedom associated with each factor, F is the value of the test statistic, p is an upper limit on the probability of obtaining an F greater or equal to this value if there are no differences associated with this factor.

than for any other algorithm. Also of interest is an apparent relation between p_p and the average difference between P-FC3 and P-BMK: for problems with higher satisfiabilities, the performance of these two algorithms was almost equal and was superior to the other algorithms; but as relative satisfiability decreased, P-FC3 became markedly superior (cf. Fig. 16). This was not reflected as an interaction in the ANOVA because for both algorithms amount of work increased dramatically with a decrease in p_p.

Differences in variability of performance within problem sets followed patterns similar to the differences in means. In 17 of the 27 sets of problems, the standard deviation for P-BB was greater than for any other algorithm; in the remaining problem sets the full combination P-ACC algorithm was the most variable. P-FC3 had the smallest standard deviation in performance in all but two sets. Almost all distributions showed a strong positive skew, i.e., the tail of the distribution on the right side of the median value was much longer than the tail on the left side.

Correlations between number of constraint checks performed by each algorithm were very high (≥ 0.97), with the exception of the P-ACC algorithm that incorporated the variable ordering; here the correlations were about 0.65. Since this correlation is a measure of the linear relation between two variables, it suggests that all of these algorithms have similar performance characteristics with respect to the problem parameters. Multiple regression analyses were very successful, in terms of accounting for most of the variance. The adjusted R^2 value was 97% in each case; in contrast, R^2 was about 25% when the original (untransformed) scores were used. Examination of residuals with normalized plots and other measures of influence of individual scores such as DFFITS [1]

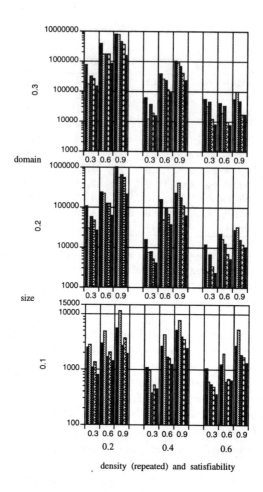

Fig. 16. Mean constraint checks per problem set for all algorithms in Experiment 4. For each combination of domain size, satisfiability and density values, algorithms are represented in the order: branch and bound (P-BB), arc consistency counts used for value and variable ordering and pruning, arc consistency counts used for value ordering and pruning, backmarking (P-BMK) and forward checking (P-FC3).

indicated that the residual values were approximately normally distributed and there were no outliers. In the multiple regression model for each algorithm, the coefficients for p_d ranged from 12.31 to 14.95 for different algorithms; for p_p the range was -0.67 to -1.00, and for p_c the range was 2.43 to 3.30. The size of these coefficients indicates the size of the effect of each parameter under the assumptions of the regression model. Clearly, domain size had the greatest influence, followed by density and then constraint satisfiability. The fundamental similarity in performance of different algorithms was also borne out.

3.4.3. Experiment 5: Effect of number of variables.

As expected, the differences in algorithm performance found in previous experiments were maintained over the range of problem sizes tested (Fig. 17). From this perspective, it appeared that P-BMK and P-FC3 have similar performance characteristics, reducing the number of constraint checks done by branch and bound by a factor of 2–3 through the range. Combining the best retrospective technique (P-BMK) with prospective and ordering techniques (pruning and value ordering based on arc consistency counts) did not materially affect this result: P-RPO reduced the effort required by P-BMK by about 1/3 throughout the range. It may be noted, however, that with few exceptions this combination outperformed P-FC3 and in a few cases (the easiest) outperformed the fastest algorithm.

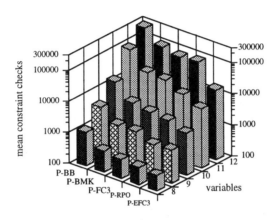

Fig. 17. Effect of number of variables on mean performance.

The best algorithm overall was clearly P-EFC3, which computed a lower bound based on the current counts of inconsistencies. Consideration of mean performance on successively greater problem sizes indicated that it, too, showed exponential growth with increasing problem size for problems with these parameters, but, as Fig. 17 shows, the rate of growth was appreciably smaller. For 12-variable problems it reduced the number of constraint checks by an order of magnitude in comparison with the other algorithms.

3.4.4. Experiment 6: Evaluation of overhead using time.

For total (corrected) time to solve a problem, the ranking of the algorithms resembled that for number of constraint checks. From longest to shortest time, the ranking was: P-BB, P-BJ, P-BMK, P-FC2 and P-FC3. To give an idea of the actual differences in time required, the means for the six hardest problems tested in this experiment (order of 10^5 constraint checks) are given in the same order: 382, 341, 214, 157 and 123 seconds. In the ANOVA, the effect due to algorithms was statistically significant ($F[4,60] = 16.46, p < 0.001$). In the individual comparisons of means, the differences between P-BB and either P-BMK or the two forward checking algorithms were statistically significant ($p < 0.01$ in each case). The difference between P-BMK and P-FC3 was also statistically significant ($p < 0.01$), but not the difference between P-BMK and P-FC2. The difference between the two forward checking variants was also not statistically significant ($q[1,17] = 1.58$).

For the average time per 1000 constraint checks, the ranking was considerably different. The following are mean values:

P-BB	1.61 sec
P-BJ	1.67
P-BMK	2.52
P-FC2	1.87
P-FC3	2.28

In the ANOVA the effect due to algorithms was statistically significant ($F[4,60] = 9.25, p < 0.001$), as well as the effect of number of constraints ($F[1,60] = 27.10$, $p < 0.001$), reflecting an increased efficiency for problems with denser graphs. On this measure, P-BB and P-BJ were both clearly superior to P-BMK and to P-FC3 ($p < 0.01$ for individual comparisons). P-FC2 was also superior to P-BMK ($p < 0.01$) and to P-FC3 ($p < 0.05$).

These results indicate that P-BMK and P-FC3 do incur a relatively large overhead in comparison with the other algorithms. In the case of backmarking, there is similar evidence for the CSP version [19]. But it is obvious that the decrease in number of constraint checks with these more elaborate algorithms yields a greater overall efficiency than the basic P-BB algorithm, reflected in the total time required to solve the problems.

3.4.5. Experiments 7–8: Finding submaximal solutions for hard problems.
The results of increasing the amount of effort applied to a problem were analyzed in two ways, by considering either, (i) the number of constraint checks required to reduce the distance by a given proportion of the total change possible (the difference between the number of constraints and the best distance), or (ii) the minimum distance found after k constraint checks. As indicated, the former measure could only be derived for 12-variable problems.

For all algorithms, after an initial drop due to the difference between the number of constraints (the initial bound) and the distance of the first solution, the relation between effort and goodness of solution approximated a simple logarithmic function (Figs. 18 and 19). Since P-EFC3 and especially P-RPO required more constraint checks to find an initial solution than P-BB or P-BMK, the latter algorithms were more effective initially ($p < 0.001$ for 100 constraint checks). (In the search phase per se, P-RPO actually found solutions with a given suboptimal distance faster than any other algorithm, but, in terms of constraint checks, this was overwhelmed by the cost of preprocessing.) However, for

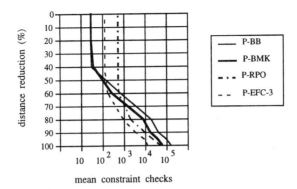

Fig. 18. Mean constraint checks required to reduce the distance as a proportion of the difference between the number of constraints and the best distance (12-variable problems).

1000 or 10,000 constraint checks, the best solutions were found by P-RPO or P-EFC3 ($p < 0.05$ for all comparisons with P-BB, for comparisons with P-BMK at 1000 constraint checks, and for the comparison between P-EFC3 and P-BMK at 10,000 constraint checks). Differences between P-RPO and P-EFC3 were never statistically significant. By 100,000 checks all algorithms except P-BB had found maximal solutions for most of the problems.

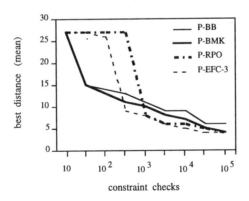

Fig. 19. Best distance found as a function of number of constraint checks for problems with 12 variables.

For the 20-variable problems (Fig. 20) the effects were similar, although here the superiority of P-EFC3 was more noticeable (and was statistically significant for each order of magnitude beginning with 1000 checks). In addition, the difference between P-RPO and P-BMK, while present, was never statistically significant. However, in general the trends found for smaller problems appear to hold as problem size is scaled up. (The results for 16-variable problems were also consistent with these results.)

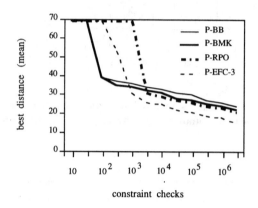

Fig. 20. Best distance found as a function of number of constraint checks for problems with 20 variables.

The results for the hard coloring problems were also similar (Fig. 21). Here, the retrospective algorithm, P-BMK, achieved a better solution after a small number of constraint checks than P-EFC3, although the latter eventually surpassed it. However, the difference between P-BMK and P-EFC3 was never great, although it eventually attained statistical significance at ten million constraint checks ($p < 0.05$). In this case, the degree of local consistency in these problems may have been responsible for the less impressive performance of P-EFC3 relative to P-BMK.

3.5 Summary of experimental results

In general, analogues of algorithms that perform well on CSPs performed well on maximal constraint satisfaction problems. Among a basic set of strategies, over the range of parameter values tested, using the measures of constraint checks and total time to obtain an optimal solution, one analogue of forward checking was generally superior. Moreover, this superiority was most evident for parts of the parameter space in which the problems were most difficult. In less extensive testing, more elaborate algorithms reduced the work required even further; here also an extension of forward checking performed better than any other algorithm. On a set of 12–variable problems, for example, the best algorithm reduced the average number of constraint checks from approximately 300,000 to approximately 20,000.

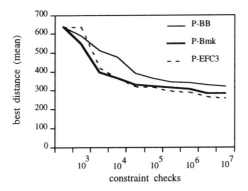

Fig. 21. Best distance found as a function of number of constraint checks for hard coloring problems.

Despite the general superiority of one type of prospective strategy (forward checking), other techniques based on local consistency (arc consistency count preprocessing) or sophisticated retrospective algorithms (P-BMK) also improved on the basic branch and bound algorithm, sometimes by a factor of 4–5. In addition, for some parts of the parameter space, P-BMK was comparable in performance to P-FC3. We conclude that these techniques merit further study, especially if they can be efficiently combined with forward checking methods.

The data from Experiment 3 suggest that for sparse problems, combining arc consistency counts with a variable ordering that puts high counts at the beginning of the search sequence was efficient on average. A key factor appears to be the size of the counts in comparison to the best distance; when the distance is small, putting higher counts at the beginning is very effective in pruning the search tree. This strategy can be incorporated into the P-RPO algorithm, and preliminary data suggest that this may be an extremely effective technique for many, but not all, problems with low densities. ($p_c = 0.1$ for the problems tested.)

The main result from the anytime experiments was that solutions within ten percent of the optimum, in terms of the total distance reduction, could be obtained with a reduction in effort equal to or greater than one order of magnitude. As in earlier experiments, P-EFC3 was the best algorithm through most of the range of effort. However, P-BB and P-BMK were more efficient if the criterion of goodness was relaxed sufficiently.

These experiments also show that random problem generation can be parameterized in a straightforward manner to produce subpopulations of problems that vary statistically in their basic features: number of constraints, domain size and satisfiability. This allowed us to examine algorithm performance in terms of the space of problems defined by these parameters. Regions of this space were delineated in which the typical problem was either easy or difficult to solve, and data on the population characteristics of algorithm performance in these regions was obtained. Such data should aid the potential user in deciding which algorithms to consider, as well as indicating the costs of such decisions.

4 Partial constraint satisfaction

We focused on maximal satisfaction to facilitate the presentation and testing of the algorithms. However, there are other forms of partial satisfaction, and to a large extent the algorithms generalize in obvious ways. In this section we work toward a general model of partial satisfaction.

4.1 Metrics

Our branch and bound algorithms have sought a solution that violates a minimal number of constraints. However, the difference between a perfect solution, and a partial solution can be measured in many different and more subtle ways. As our use of the term "distance" suggests, all the branch and bound technique requires is a metric that can compare the values being considered at a given stage of search with the "best" solution found so far.

Preferences have been expressed by ordering constraints [10], by representing their importance [12], by organizing constraints into hierarchies [2], by introducing priorities [11]. Preferences can be reflected in the branch and bound metric by assigning weights to constraints. Preferences could be associated with subsets of domains and constraints, individual values or pairs of values, as opposed to entire constraints. (Wearing a striped tie and a polka dot shirt might not be as bad as wearing a bow tie with a sweatshirt.)

Constraint deviations have been combined in both local and global fashion [2]. The branch and bound metric can sum the weights of the violated constraints, use the maximum weight, compute the average, as appropriate. The initial constraints may be viewed as ideal points which we seek to approximate by some measure.

There may be some "hidden agenda" embodied in the metric. For example, we may wish to drive a problem toward a weaker version that is easily solvable, e.g. by removing constraints to yield a tree-structured problem.

4.2 Problem spaces

We have been measuring the success of a partial solution by evaluating the number, or importance, of the violated constraints. There is another criterion that we believe should be directly considered. Weakening constraints in effect creates a different problem. Alternatives for weakening constraints provide alternative problems. We may well wish to solve a problem that is close to the original in the sense of having a solution set similar to the original. Removing one set of constraints might trivialize the problem, allowing thousands of new solutions, while removing another set might allow only a single new solution.

This consideration arises, for example, when viewing constraint knowledge base debugging as a partial constraint satisfaction problem [21]. If the knowledge base is erroneously overconstrained, a change that allows a small number of new solutions is more in keeping with Occam's Razor than one that allows many.

More formally, we can consider a space of problems with an ordering based on solution sets. A *problem space* is a partially ordered set, $(PS, \leq)$. PS is a set of constraint satisfaction problems with a partial order, $\leq$, on PS defined as follows: $P_1 \leq P_2$ iff the set of solutions to P_2 is a subset of the set of solutions to P_1. If the set of solutions to P_2 is a subset of the set of solutions to P_1, but the two sets are not equal, we will write $P_1 < P_2$ and say that P_1 is *weaker than P_2*.

One natural problem space for a partial constraint satisfaction problem with initial problem P consists of all problems Q such that $Q \leq P$ This set can be obtained by considering all the ways of weakening the constraints by allowing additional consistent combinations of values. In general, PS could contain problems, Q, which are stronger than P, $P < Q$, or problems, Q, such that neither $Q \leq P$ nor $P \leq Q$; $\leq$ is only a partial order. However, if we collect all the constraints in all the problems in PS into a single problem M, then all the problems in PS can be regarded as weakenings of M.

It may be natural to consider a space which does not include all $Q \leq P$. We may wish to specify how the problem can be weakened. Some weaker problems may make more "semantic sense". It may not be possible or desirable to violate constraints by arbitrarily allowing individual pairs of values to violate the associated constraint, as we have done until now. For example, in our dressing domain, we might decide that in a fashion emergency, or a burst of avant garde creativity, we would eschew the prohibition against mixing stripes with checks, as opposed to making an individual exception for one striped tie and one checked shirt. We may establish levels of informality in our fashion constraints corresponding to the informality of the occasion for which we are dressing.

It may make more semantic sense to have a preset hierarchy of constraints, where weakening a constraint requires moving upward in the hierarchy, as opposed to making arbitrary individual exceptions. This constraint hierarchy is reminiscent of the concept hierarchies that provide initial bias in machine learning settings, and indeed it is intriguing to think of the constraint satisfaction process as a form of concept learning, synthesizing a relationship from positive and negative information.

The specification of the problem space PS can clearly affect the efficiency of the search process. One way to specify the problem space is to specify generators, or operators, that take us from one problem P to a permitted set of problems Q_i, $Q_i \leq P$. There may be "global" restrictions on these generators, e.g. choose one constraint from column A, one from column B. As we try to move through a problem space in search of a solvable problem, it may prove desirable to take into consideration how many opportunities are opened up by altering a problem in a given way, e.g. removing one fashion restriction may be more liberating than removing another.

The process of weakening CSP's can be naturally viewed as involving four options: enlarging a variable domain (buying a new shirt), enlarging a constraint domain (deciding that an old shirt and an old tie can be worn together after all), removing a variable (deciding not to wear any tie), removing a constraint (deciding not to worry if our socks match). However, all of these can in turn be expressed in terms of the basic process of enlarging constraint domains. We can view variable domains as unary constraints. Enlarging a binary constraint until it contains all pairs of values in the specified domains for the two variables is tantamount to removing the constraint (at least until such time as the variable domains may be enlarged). Removing all the constraints on a variable in this way is tantamount to removing the variable.

4.3 Partial constraint satisfaction problems

A *partial constraint satisfaction problem* can now be specified more formally by supplying an initial constraint satisfaction problem P, a problem space, PS, containing P, a metric on that space, and necessary and sufficient solution distances, N and S. A *solution to a partial constraint satisfaction problem* can be defined as a problem P' from the problem space PS along with a solution to that problem where the metric distance of P' from P is less than N. A solution is *sufficient* if the distance is less than or equal to S. An *optimal solution* is one where the metric distance of P' from P is minimal over the problem space. The optimal solution is *dominant* if there is no optimal solution which involves a problem Q such that $P' < Q$.

The metric that evaluates our solution now compares problems rather than counting or otherwise evaluating violated constraints. Of course, one way to compare problems is to compare their constraints. Ideally, however, we might like to define the metric in terms of the partial order, defining the distance between P and P' to be the number of solutions not shared by P and P'. When $P' \leq P$, this metric, M, measures the number of solutions we have added by weakening P. This is a natural measure of how "good" our partial solution is.

Of course, computing such a metric is not likely to be easy. However, after finding a set of optimal solutions with another metric we might wish to distinguish among these by considering the different problems induced by these solutions. The problem *induced* by a partial solution S for a problem P is the problem obtained by adding to the constraints of P the inconsistent pairs of values in the partial solution S. We could compute the full solution sets for the induced problems, and look for dominant, optimal solutions using the metric M, which operates on solution sets.

We also may wish to consider how well an alternative metric does tend to reflect this natural metric M. Another natural metric is a count of the number of permitted value combinations not shared by the constraints of P and P'. To some extent, this metric does reflect the metric M based on the partial order. If P' is obtained from P by weakening the constraints then $P' \leq P$, because of the monotonic nature of constraint satisfaction problems. In other words, if for each constraint C_{ij} associated with P and constraint C'_{ij} associated with P', C_{ij} is a subset of C'_{ij}, then $P' \leq P$. In particular, the simple metric used in our maximality studies, which counts the number of violated constraints, is a metric of this form.

Viewing partial constraint satisfaction as a search through a problem space facilitates consideration of integrating branch and bound at different levels of the search process to produce different algorithms. The natural points at which to perform this integration are the failure points in a standard backtracking algorithm.

Our basic branch and bound partial constraint satisfaction algorithm integrates branch and bound at the lowest level of backtrack failure. Whenever a value c for a variable V would be rejected by normal backtracking, we can view the algorithm as determining which constraints were violated, and weakening the problem by adding precisely those constraint elements needed to permit c.

In searching through the problem space for a solvable problem, it would be desirable to avoid changing the problem in ways that do not facilitate progress. For example, if two problems are equivalent, they both do not need to be considered. A nice feature of the basic branch and bound algorithm is that for the type of partial constraint satisfaction problem for which it is designed it is able to use only the minimally different problem P required to proceed at each problem choice point.

An even simpler choice for integrating branch and bound would be to add it to backtracking upon top-level failure of backtracking, when no solution is found. A branch and bound loop can be added on the outside of the backtracking algorithm. This loop will run through the problems in the problem space, keeping track of the closest problem P' to P solved so far. Problems no closer to P than P' will be rejected immediately.

Whenever alternative problems are generated in an order which reflects their distance from the original problem, closest to furthest, generation can stop at that point when the necessary bound N is reached. If we have a top-level integration of branch and bound, this point marks the termination of the PCSP algorithm.

Branch and bound can be integrated at failure points in between these extremes. The natural compromise would occur at the points where all the values for a given variable have been exhausted in standard backtrack search. At these times options for alternative problems may be explored.

There is a tradeoff involved in the choice of how to integrate branch and bound. By integrating at a lower level we take greater advantage of backtrack pruning to avoid unnecessary effort. On the other hand by integrating at a higher level we allow greater flexibility in heuristically guiding the search through the space of alternative problems. Preliminary experiments comparing a high level and a low level approach to partial satisfaction in the domain of debugging constraint knowledge bases reflect this tradeoff [21].

In summary, the generalized view we have reached of partial constraint satisfaction as a search through a space of alternative problems has a number of potential advantages for further work in this area. It facilitates consideration of more global concerns than the suitability of a single solution. Specifically, it encourages us to consider whether our partial solution has been devalued by weakening the problem in a manner that permits too many solutions. It naturally incorporates practical concerns about available constraint modifications. It also facilitates generation of alternate problem solving strategies that take a more global view of effective means of modifying the problem.

5 Conclusion

Standard constraint satisfaction problem (CSP) solution techniques have analogues for solving partial constraint satisfaction problems (PCSPs), which both cope with and take advantage of the differences between CSP and PCSP. Branch and bound is the natural analogue of backtrack search. Local consistency count is an analogue of local consistency. We have found PCSP analogues of retrospective, prospective and ordering techniques for CSPs.

Extensive experimentation over random problems with different structural parameters revealed the effectiveness of a set of PCSP techniques as a function of these parameters. A general model of PCSPs was developed involving a standard CSP together with a partially ordered space of alternative problems and a metric to measure the distances between these problems and the original CSP.

In summary, a firm algorithmic, experimental and theoretical foundation has been laid for the study of problems for which it is impractical or impossible to satisfy fully a set of constraints.

Acknowledgement

This material is based upon work supported by the National Science Foundation under Grant No. IRI-8913040 and Grant No. IRI-9207633. Part of this work was done while the first author was a Visiting Scientist at the MIT Artificial Intelligence Laboratory. Portions of this paper were taken from [14], which is copyright (c) 1989 International Joint Conferences on Artificial Intelligence, Inc., and used by permission. We thank Peter Cheeseman and Bob Kanefsky for supplying the "really hard" coloring problems.

References

1. D. A. Belsley, E. Kuh, and R. E. Welsch. *Regression Diagnostics*. Wiley, New York, NY, 1980.
2. A. Borning, R. Duisberg, B. Freeman-Benson, A. Kramer, and M. Woolf. Constraint hierarchies. In *Proceedings 1987 ACM Conference on Object-Oriented Programming Systems, Languages and Applications*, pages 48–60, 1987.
3. A. Borning, M. Maher, A. Martindale, and M. Wilson. Constraint hierarchies and logic programming. In *Proceedings Sixth International Conference on Logic Programming*, pages 149–164, 1989.

4. P. Cheesemen, B. Kanefsky, and W. M. Taylor. Where the *really* hard problems are. In *Proceedings IJCAI-91*, pages 331–337, 1991.

5. M. Cooper. *Visual Occlusion and the Interpretation of Ambiguous Pictures*. Ellis ·Horwood, Chicester, 1992.

6. R. Dechter. Enhancement schemes for constraint processing: backjumping, learning, and cutset decomposition. *Artif. Intell.*, 41:273–312, 1990.

7. R. Dechter, A. Dechter, and J. Pearl. Optimization in constraint networks. In R. M. Oliver and J. Q. Smith, editors, *Influence Diagrams, Belief Nets and Decision Analysis*, pages 411–425. Wiley, 1990.

8. R. Dechter and J. Pearl. Network-based heuristics for constraint satisfaction problems. *Artif. Intell.*, 34:1–38, 1988.

9. R. Dechter and J. Pearl. Tree-clusstering schemes for constraint processing. *Artif. Intell.*, 38:353–366, 1989.

10. Y. Descotte and J. C. Latombe. Making compromises among antagonistic constraints in a planner. *Artif. Intell.*, 27:183–217, 1985.

11. R. Feldman and M. C. Golumbic. Optimization algorithms for student scheduling via constraint satisfiability. *Comput. J.*, 33:356–364, 1990.

12. M. Fox. *Constraint-directed Search: A Case Study of Job-Shop Scheduling*. Morgan Kaufmann, Los Altos, CA, 1987.

13. E. C. Freuder. Backtrack-free and backtrack-bounded search. In L. Kanal and V. Kumar, editors, *Search in Artificial Intelligence*, pages 343–369. Springer, New York, NY, 1988.

14. E. Freuder. Partial constraint satisfaction. In *Proceedings IJCAI-89*, pages 278–283, 1989.

15. E. Freuder. Complexity of k-tree structured constraint satisfaction problems. In *Proceedings AAAI-90*, pages 4–9, 1990.

16. J. Gaschnig. A general backtrack algorithm that elimates most redundant checks. In *Proceedings IJCAI-77*, page 457, 1977.

17. J. Gaschnig. Experimental case studies of backtrack vs. Waltz-type vs. new algorithms for satisficing assignment problems. In *Proceedings Second National Conference of the Canadian Society for Computational Studies of Intelligence*, pages 268–277, 1978.

18. S. W. Golumb and L. D. Baumert. Backtrack programming. *J. ACM*, 12:516–524, 1965.

19. R.M. Haralick and G.L. Elliott. Increasing tree search efficiency for constraint satisfaction problems. *Artificial Intelligence*, 14:263–313, 1980.

20. W. Hower. Sensitive relaxation of an overspecified constraint network. In *Proceedings Second International Symposium on Artificial Intelligence*, 1989.

21. S. Huard and E. Freuder. A debugging assistant for incompletely specified constraint network knowledge bases. *Int. J. Expert Syst.: Res. Appl.*, pages 419–446, 1993.

22. R. E. Kirk. *Experimental Design*. Brooks/Cole, Pacific Grove, CA, 2nd edition, 1982.

23. V. Kumar. Algorithms for constraint–satisfaction problems: a survey. *AI Mag.*, 13:32–44, 1992.

24. M. Lacroix and P. Lavency. Preferences: putting more knowledge into queries. In *Proceedings 15th International Conference on Very Large Data Bases*, pages 217–225, 1987.

25. E. L. Lawler and D. E. Wood. Branch–and–bound methods: a survey. *Oper. Res.*, 14:699–719, 1966.

26. A. Mackworth. Consistency in networks of relations. *Artif. Intell.*, 8:99–118, 1977.
27. A. Mackworth and E. Freuder. The complexity of some polynomial network consistency algorithms of constraint satisfaction problems. *Artif. Intell.*, 25:65–74, 1985.
28. I. Meiri, R. Dechter, and J. Pearl. Tree decomposition with applications to constraint processing. In *Proceedings AAAI-90*, pages 10–16, 1990.
29. P. Meseguer. Constraint satisfaction problems: an overview. *AI Commun.*, 2:3–17, 1989.
30. S. Mittal and B. Falkenhainer. Dynamic constraint satisfaction problems. In *Proceedings AAAI-90*, pages 25–32, 1990.
31. R. Mohr and G. Masini. Good old discrete relaxation. In *Proceedings European Conference on Artificial Intelligence*, pages 651–656, 1988.
32. B. Nadel. Constraint satisfaction algorithms. *Comput. Intell.*, 5:188–224, 1989.
33. E. M. Reingold, J. Nievergelt, and N. Deo. *Combinatorial Algorithms: Theory and Practice*. Prentice-Hall, Englewood Cliffs, NJ, 1977.
34. A. Rosenfeld, R. Hummel, and S. Zucker. Scene labeling by relaxation operations. *IEEE Trans. Syst. Man, Cybern.*, 6:420–433, 1976.
35. L. Shapiro and R. Haralick. Structural descriptions and inexact matching. *IEEE Trans. Patt. Anal. Mach. Intell.*, 3:504–519, 1981.
36. P. Snow and E. Freuder. Improved relaxation and search methods for approximate constraint satisfaction with a maximin criterion. In *Proceedings Eighth Biennial Conference of the Canadian Society for Computational Studies of Intelligence*, pages 227–230, 1990.
37. P. Van Hentenryck. *Constraint Satisfaction in Logic Programming*. MIT, Cambridge, MA, 1989.

Semiring-Based CSPs and Valued CSPs:
Basic Properties and Comparison

Stefano Bistarelli[1], Hélène Fargier[2], Ugo Montanari[1], Francesca Rossi[1],
Thomas Schiex[3], Gérard Verfaillie[4]

[1] University of Pisa, Dipartimento di Informatica, Corso Italia 40, 56125 Pisa, Italy.
E-mail: {bista,ugo,rossi}@di.unipi.it
[2] IRIT, 118 route de Narbonne, 31062, Toulouse Cedex, France.
E-mail: fargier@irit.fr
[3] INRA, Chemin de Borde Rouge, BP 27, 31326 Castanet-Tolosan Cedex, France.
E-mail: Thomas.Schiex@toulouse.inra.fr
[4] CERT/ONERA, 2 Av. E. Belin, BP 4025, 31055 Toulouse Cedex, France.
E-mail: Gerard.Verfaillie@cert.fr

Abstract. In this paper we describe two frameworks for constraint solving where classical CSPs, fuzzy CSPs, weighted CSPs, partial constraint satisfaction, and others can be easily cast. One is based on a semiring, and the other one on a totally ordered commutative monoid. We then compare the two approaches and we discuss the relationship between them. The two frameworks have been independently introduced in [2] and [28].

1   Introduction

Classical constraint satisfaction problems (CSPs) [18, 20] are a very expressive and natural formalism to specify many kinds of real-life problems. In fact, problems ranging from map coloring, vision, robotics, job-shop scheduling, VLSI design, etc., can easily be cast as CSPs and solved using one of the many techniques that have been developed for such problems or subclasses of them [9, 10, 17, 19, 20].

However, they also have evident limitations, mainly due to the fact that they are not very flexible when trying to represent real-life scenarios where the knowledge is not completely available nor crisp. In fact, in such situations, the ability of stating whether an instantiation of values to variables is allowed or not is not enough or sometimes not even possible. For these reasons, it is natural to try to extend the CSP formalism in this direction.

For example, in [6, 25, 26, 27] CSPs have been extended with the ability to associate with each tuple, or with each constraint, a level of preference, and with the possibility of combining constraints using min-max operations. This extended formalism has been called Fuzzy CSPs (FCSPs). Other extensions concern the ability to model incomplete knowledge of the real problem [7], to solve over-constrained problems [11], and to represent cost optimization problems.

In this paper we present and compare two frameworks where all such extensions, as well as classical CSPs, can be cast. However, we do not relax the assumption of a finite domain for the variables of the constraint problems.

The first framework, that we call SCSP (for Semiring-based CSP), is based on the observation that a semiring (that is, a domain plus two operations satisfying certain properties) is all that is needed to describe many constraint satisfaction schemes. In fact, the domain of the semiring provides the levels of consistency (which can be interpreted as cost, or degrees of preference, or probabilities, or others), and the two operations define a way to combine constraints together. Specific choices of the semiring will then give rise to different instances of the framework.

In classical CSPs, so-called local consistency techniques [9, 10, 17, 18, 20, 21] have been proved to be very effective when approximating the solution of a problem. In this paper we study how to generalize this notion to this framework, and we provide some sufficient conditions over the semiring operations which guarantee that such algorithms can also be fruitfully applied to our scheme. Here for being "fruitfully applicable" we mean that 1) the algorithm terminates and 2) the resulting problem is equivalent to the given one and it does not depend on the nondeterministic choices made during the algorithm.

The second framework, that we call VCSP (for Valued CSP), relies on a simpler structure, an ordered monoid (that is, an ordered domain plus one operation satisfying some properties). The values of the domain are interpreted as levels of violation (which can be interpreted as cost, or degrees of preference, or probabilities, or others) and can be combined using the monoid operator. Specific choices of the monoid will then give rise to different instances of the framework.

In this framework, we first generalize some of the usual branch and bound algorithms for finding optimal solutions to this framework. In this process, we study how to generalize the arc-consistency *property* using the notion of "relaxation" and provide sufficient conditions over the monoid operation which guarantee that the problem of checking this property on a valued CSP is either polynomial or NP-complete. Interestingly, the results are consistent with the results obtained in the SCSP framework in the sense that the conditions which guarantee the polynomiality in the VCSP framework are exactly the conditions which guarantee that k-consistency algorithms actually "work" in the SCSP framework.

The advantage of these two frameworks is that one can just see any constraint solving paradigm as an instance of either of these frameworks. Then, one can immediately inherit the results obtained for the general frameworks. This also allows one to justify many informally taken choices in existing constraint solving schemes. In this paper we study several known and new constraint solving frameworks, casting them as instances of SCSP and VCSP.

The two frameworks are not however completely equivalent. In fact, only if one assumes a total order on the semiring set, is ts possible to define appropriate mappings to pass from one of them to the other.

The paper is organized as follows. Section 2 describes the framework based on semirings and its properties related to local consistency. Then, Sect. 3 describes the other framework and its applications to search algorithms, including those relying on arc-consistency. Then, Sect. 4 compares the two approaches, and finally Sect. 5 summarizes the main results of the paper and hints at possible future developments.

2 Constraint Solving over Semirings

The framework we will describe in this section is based on a semiring structure, where the set of the semiring specifies the values to be associated with each tuple of values of the variable domain, and the two semiring operations (+ and ×) model constraint projection and combination respectively. Local consistency algorithms, as usually used for classical CSPs, can be exploited in this general framework as well, provided that some conditions on the semiring operations are satisfied. We then show how this framework can be used to model both old and new constraint solving schemes, thus allowing one both to formally justify many informally taken choices in existing schemes, and to prove that local consistency techniques can be used also in newly defined schemes. The content of this section is based on [2].

2.1 C-semirings and their properties

We associate a semiring with the standard definition of constraint problem, so that different choices of the semiring represent different concrete constraint satisfaction schemes. Such semiring will give us both the domain for the non-crisp statements and also the allowed operations on them. More precisely, in the following we will consider *c-semirings*, that is, semirings with additional properties of the two operations.

Definition 1 (c-semiring). A **semiring** is a tuple $(A, +, \times, 0, 1)$ such that

- A is a set and $0, 1 \in A$;
- $+$, called the additive operation, is a closed (i.e., $a, b \in A$ implies $a + b \in A$), commutative (i.e., $a + b = b + a$) and associative (i.e., $a + (b + c) = (a + b) + c$) operation such that $a + 0 = a = 0 + a$ (i.e., 0 is its unit element);
- $\times$, called the multiplicative operation, is a closed and associative operation such that 1 is unit element and $a \times 0 = 0 = 0 \times a$ (i.e., 0 is its absorbing element);
- $\times$ distributes over $+$ (i.e., $a \times (b + c) = (a \times b) + (a \times c)$).

A **c-semiring** is a semiring such that $+$ is idempotent (i.e., $a \in A$ implies $a + a = a$), $\times$ is commutative, and 1 is the absorbing element of $+$. $\square$

The idempotency of the $+$ operation is needed in order to define a partial ordering $\leq_S$ over the set A, which will enable us to compare different elements of the semiring. Such partial order is defined as follows: $a \leq_S b$ iff $a + b = a$.

Intuitively, $a \leq_S b$ means that a is "better" than b. This will be used later to choose the "best" solution in our constraint problems. It is important to notice that both $+$ and $\times$ are monotone on such ordering.

The commutativity of the $\times$ operation is desirable when such operation is used to combine several constraints. In fact, were it not commutative, it would mean that different orders of the constraints give different results.

If 1 is also the absorbing element of the additive operation, then we have that $1 \leq_S a$ for all a. Thus 1 is the minimum (i.e., the best) element of the partial ordering. This implies that the $\times$ operation is *extensive*, that is, that $a \leq a \times b$. This is important since it means that combining more constraints leads to a worse (w.r.t. the $\leq_S$ ordering) result. The fact that 0 is the unit element of the additive operation implies that 0 is the maximum element of the ordering. Thus, for any $a \in A$, we have $1 \leq_S a \leq_S 0$.

In the following we will sometimes need the $\times$ operation to be closed on a certain finite subset of the c-semiring. More precisely, given any c-semiring $S = (A, +, \times, 0, 1)$, consider a finite set $I \subseteq A$. Then, $\times$ is *I-closed* if, for any $a, b \in I$, $(a \times b) \in I$.

2.2 Constraint systems and problems

A constraint system provides the c-semiring to be used, the set of all variables, and their domain D. Then, a constraint over a given constraint system specifies the involved variables and the "allowed" values for them. More precisely, for each tuple of values of D for the involved variables, a corresponding element of the semiring is given. This element can be interpreted as the tuple weight, or cost, or level of confidence, or else. Finally, a constraint problem is then just a set of constraints over a given constraint system, plus a selected set of variables. These are the variables of interest in the problem, i.e., the variables of which we want to know the possible assignments compatibly with all the constraints.

Definition 2 (constraint problem). A **constraint system** is a tuple $CS = \langle S, D, V \rangle$, where S is a c-semiring, D is a finite set, and V is an ordered set of variables. Given a constraint system $CS = \langle S, D, V \rangle$, where $S = (A, +, \times, 0, 1)$, a **constraint** over CS is a pair $\langle def, con \rangle$, where $con \subseteq V$ and it is called the *type* of the constraint, and $def : D^k \to A$ (where k is the size of con, that is, the number of variables in it), and it is called the *value* of the constraint. Moreover, a **constraint problem** P over CS is a pair $P = \langle C, con \rangle$, where C is a set[5] of constraints over CS and $con \subseteq V$. □

In the following we will consider a fixed constraint system $CS = \langle S, D, V \rangle$, where $S = (A, +, \times, 0, 1)$. Note that when all variables are of interest, like in many approaches to classical CSP, con contains all the variables involved in any of the constraints of the problem. This set will be denoted by $V(P)$ and can

---

[5] Note that, if $\times$ is not idempotent, it is necessary to see C as a multiset, and not as a set.

be recovered by looking at the variables involved in each constraint: $V(P) = \bigcup_{\langle def, con' \rangle \in C} con'$.

In the $SCSP$ framework, the values specified for the tuples of each constraint are used to compute corresponding values for the tuples of values of the variables in con, according to the semiring operations: the multiplicative operation is used to combine the values of the tuples of each constraint to get the value of a tuple for all the variables, and the additive operation is used to obtain the value of the tuples of the variables of interest. More precisely, we can define the operations of *combination* ($\otimes$) and *projection* ($\Downarrow$) over constraints. Analogous operations have been originally defined for fuzzy relations in [32], and have then been used for fuzzy CSPs in [6]. Our definition is however more general since we do not consider a specific c-semiring but a general one.

Definition 3 (combination and projection). Consider two constraints $c_1 = \langle def_1, con_1 \rangle$ and $c_2 = \langle def_2, con_2 \rangle$ over CS. Then, their **combination**, $c_1 \otimes c_2$, is the constraint $c = \langle def, con \rangle$ with $con = con_1 \cup con_2$ and $def(t) = def_1(t \downarrow^{con}_{con_1}) \times def_2(t \downarrow^{con}_{con_2})$, where, for any tuple of values t for the variables in a set I, $t \downarrow^{I}_{I'}$ denotes the projection of t over the variables in the set I'. Moreover, given a constraint $c = \langle def, con \rangle$ over CS, and a subset w of con, its **projection** over w, written $c \Downarrow_w$, is the constraint $\langle def', con' \rangle$ over CS with $con' = w$ and $def'(t') = \Sigma_{\{t \mid t \downarrow^{con}_w = t'\}} def(t)$. $\square$

Using such operations, we can now define the notion of solution of a SCSP.

Definition 4 (solution). Given a constraint problem $P = \langle C, con \rangle$ over a constraint system CS, the **solution** of P is a constraint defined as $Sol(P) = (\bigotimes C) \Downarrow_{con}$, where $\bigotimes C$ is the obvious extension of the combination operation to a set of constraints C. $\square$

In words, the solution of a SCSP is the constraint induced on the variables in con by the whole problem. Such constraint provides, for each tuple of values of D for the variables in con, an associated value of A. Sometimes, it is enough to know just the best value associated to such tuples. In our framework, this is still a constraint (over an empty set of variables), and will be called the best level of consistency of the whole problem, where the meaning of "best" depends on the ordering $\leq_S$ defined by the additive operation.

Definition 5 (best level of consistency). Given a $SCSP$ problem $P = \langle C, con \rangle$, we define the **best level of consistency** of P as $blevel(P) = (\bigotimes C) \Downarrow_\varnothing$. If $blevel(P) = \langle blev, \varnothing \rangle$, then we say that P is consistent if $blev <_S \mathbf{0}$. Instead, we say that P is α-consistent if $blev = \alpha$. $\square$

Informally, the best level of consistency gives us an idea of how much we can satisfy the constraints of the given problem. Note that $blevel(P)$ does not depend on the choice of the distinguished variables, due to the associative property of the additive operation. Thus, since a constraint problem is just a set of constraints plus a set of distinguished variables, we can also apply function $blevel$ to a set of constraints only. Also, since the type of constraint $blevel(P)$ is always an empty set of variables, in the following we will just write the value of $blevel$.

Another interesting notion of solution, more abstract than the one defined above, but sufficient for many purposes, is the one that provides only the tuples that have an associated value which coincides with (the def of) $blevel(P)$. However, this notion makes sense only when $\leq_S$ is a total order. In fact, were it not so, we could have an incomparable set of tuples, whose sum (via $+$) does not coincide with any of the summed tuples. Thus it could be that none of the tuples has an associated value equal to $blevel(P)$.

By using the ordering $\leq_S$ over the semiring, we can also define a corresponding partial ordering on constraints with the same type, as well as a preorder and a notion of equivalence on problems.

Definition 6 (orderings). Consider two constraints c_1, c_2 over CS, and assume that $con_1 = con_2$. Then we define the **constraint ordering** $\sqsubseteq_S$ as the following partial ordering: $c_1 \sqsubseteq_S c_2$ if and only if, for all tuples t of values from D, $def_1(t) \leq_S def_2(t)$. Notice that, if $c_1 \sqsubseteq_S c_2$ and $c_2 \sqsubseteq_S c_1$, then $c_1 = c_2$. Consider now two $SCSP$ problems P_1 and P_2 such that $P_1 = \langle C_1, con \rangle$ and $P_2 = \langle C_2, con \rangle$. Then we define the **problem preorder** $\sqsubseteq_P$ as: $P_1 \sqsubseteq_P P_2$ if $Sol(P_1) \sqsubseteq_S Sol(P_2)$. If $P_1 \sqsubseteq_P P_2$ and $P_2 \sqsubseteq_P P_1$, then they have the same solution. Thus we say that P_1 and P_2 are **equivalent** and we write $P_1 \equiv P_2$. $\square$

The notion of problem preorder can also be useful to show that, as in the classical CSP case, also the SCSP framework is monotone: $(C, con) \sqsubseteq_P (C \cup C', con)$. That is, if some constraints are added, the solution (as well as the $blevel$) of the new problem is worse or equal than that of the old one.

2.3 Local Consistency

Computing any one of the previously defined notions (like the best level of consistency and the solution) is an NP-hard problem. Thus it can be convenient in many cases to approximate such notions. In classical CSP, this is done using the so-called local consistency techniques. Such techniques can be extended also to constraint solving over any semiring, provided that some properties are satisfied. Here we define what k-consistency [9, 10, 14] means for SCSP problems. Informally, an SCSP problem is k-consistent when, taken any set W of $k - 1$ variables and any k-th variable, the constraint obtained by combining all constraints among the k variables and projecting it onto W is better or equal (in the ordering $\sqsubseteq_S$) than that obtained by combining the constraints among the variables in W only.

Definition 7 (k-consistency). Given a $SCSP$ problem $P = \langle C, con \rangle$ we say that P is k-consistent if, for all $W \subseteq V(P)$ such that size$(W) = k - 1$, and for all $x \in (V(P) - W)$, $((\otimes\{c_i \mid c_i \in C \wedge con_i \subseteq (W \cup \{x\})\}) \Downarrow_W) \sqsubseteq_S (\otimes\{c_i \mid c_i \in C \wedge con_i \subseteq W)\}$, where $c_i = \langle def_i, con_i \rangle$ for all $c_i \in C$. $\square$

Note that, since $\times$ is extensive, in the above formula for k-consistency we could also replace $\sqsubseteq_S$ by $\equiv_S$. In fact, the extensivity of $\times$ assures that the formula always holds when $\sqsupseteq_S$ is used instead of $\sqsubseteq_S$.

Making a problem k-consistent means expliciting some implicit constraints, thus possibly discovering inconsistency at a local level. In classical CSP, this is crucial, since local inconsistency implies global inconsistency. This is true also in SCSPs. But here we can be even more precise, and relate the best level of consistency of the whole problem to that of its subproblems.

Theorem 8 (local and global α-consistency). *Consider a set of constraints C over CS, and any subset C' of C. If C' is α-consistent, then C is β-consistent, with $\alpha \leq_S \beta$.*

Proof. If C' is α-consistent, it means that $\otimes C' \Downarrow_\varnothing = \langle \alpha, \varnothing \rangle$. Now, C can be seen as $C' \otimes C''$ for some C''. By extensivity of $\times$, and the monotonicity of $+$, we have that $\beta = \otimes (C' \otimes C'') \Downarrow \varnothing \sqsupseteq_S \otimes (C) \Downarrow \varnothing = \alpha$. $\qquad\square$

If a subset of constraints of P is inconsistent (that is, its *blevel* is 0), then the above theorem implies that the whole problem is inconsistent as well.

We now define a generic k-consistency algorithm, by extending the usual one for classical CSPs [9, 14]. We assume to start from a $SCSP$ problem where all constraints of arity $k-1$ are present. If some are not present, we just add them with a non-restricting definition. That is, for any added constraint $c = \langle def, con \rangle$, we set $def(t) = 1$ for all con-tuples t. This does not change the solution of the problem, since 1 is the unit element for the $\times$ operation.

The idea of the (naive) algorithm is to combine any constraint c of arity $k-1$ with the projection over such $k-1$ variables of the combination of all the constraints connecting the same $k-1$ variables plus another one, and to repeat such operation until no more changes can be made to any $(k-1)$-arity constraint.

In doing that, we will use the additional notion of *typed locations*. Informally, a typed location is just a location (as in ordinary imperative programming) which can be assigned to a constraint of the same type. This is needed since the constraints defined in Definition 2 are just pairs $\langle def, con \rangle$, where def is a *fixed* function and thus not modifiable. In this way, we can also assign the value of a constraint to a typed location (only if the type of the location and that of the constraint coincide), and thus achieve the effect of modifying the value of a constraint.

Definition 9 (typed locations). A **typed location** is an object $l : con$ whose type is con. The assignment operation $l := c$, where c is a constraint $\langle def, con \rangle$, has the meaning of associating, in the present store, the value def to l. Whenever a typed location appears in a formula, it will denote its value. $\qquad\square$

Definition 10 (k-consistency algorithm). Consider an $SCSP$ problem $P = \langle C, con \rangle$ and take any subset $W \subseteq V(P)$ such that $\text{size}(W) = k-1$ and any variable $x \in (V(P) - W)$. Let us now consider a typed location l_i for each constraint $c_i = \langle def_i, con_i \rangle \in C$ such that $l_i : con_i$. Then a k-**consistency algorithm** works as follows.

1. Initialize all locations by performing $l_i := c_i$ for each $c_i \in C$.
2. Consider
   - $l_j : W$,
   - $A(W, x) = \otimes\{l_i \mid con_i \subseteq (W \cup \{x\})\} \Downarrow W$, and
   - $B(W) = \otimes\{l_i \mid con_i \subseteq W\}$.

   Then, if $A(W, x) \not\sqsubseteq_S B(W)$, perform $l_j := l_j \otimes A(W, x)$.
3. Repeat step 2 on all W and x until $A(W, x) \sqsubseteq B(W)$ for all W and all x.

Upon stability, assume that each typed location $l_i : con_i$ has $eval(l_i) = def'_i$. Then the result of the algorithm is a new SCSP problem $P' = k\text{-cons}(P) = \langle C', con \rangle$ such that $C' = \bigcup_i \langle def'_i, con_i \rangle$. □

Assuming the termination of such algorithm (we will discuss such issue later), it is obvious to show that the problem obtained at the end is k-consistent. This is a very naive algorithm, whose efficiency can be improved easily by using the methods which have been adopted for classical k-consistency.

In classical CSP, any k-consistency algorithm enjoys some important properties. We now will study these same properties in our SCSP framework, and point out the corresponding properties of the semiring operations which are necessary for them to hold. The desired properties are as follows: that any k-consistency algorithm returns a problem which is equivalent to the given one; that it terminates in a finite number of steps; and that the order in which the $(k-1)$-arity subproblems are selected does not influence the resulting problem.

Theorem 11 (equivalence). *Consider a SCSP problem P and a SCSP problem $P' = k\text{-cons}(P)$. Then, $P \equiv P'$ (that is, P and P' are equivalent) if $\times$ is idempotent.*

Proof. Assume $P = \langle C, con \rangle$ and $P' = \langle C', con \rangle$. Now, C' is obtained by C by changing the definition of some of the constraints (via the typed location mechanism). For each of such constraints, the change consists of combining the old constraint with the combination of other constraints. Since the multiplicative operation is commutative and associative (and thus also $\otimes$), $\otimes C'$ can also be written as $(\otimes C) \otimes C''$, where $\otimes C'' \sqsubseteq_S \otimes C$. If $\times$ is idempotent, then $((\otimes C) \otimes C'') = (\otimes C)$. Thus $(\otimes C) = (\otimes C')$. Therefore $P \equiv P'$. □

Theorem 12 (termination). *Consider any SCSP problem P where $CS = \langle S, D, V \rangle$ and the set $AD = \bigcup_{\langle def, con \rangle \in C} R(def)$, where $R(def) = \{a \mid \exists t \text{ with } def(t) = a\}$. Then the application of the k-consistency algorithm to P terminates in a finite number of steps if AD is contained in a set I which is finite and such that $+$ and $\times$ are I-closed.*

Proof. Each step of the k-consistency algorithm may change the definition of one constraint by assigning a different value to some of its tuples. Such value is strictly worse (in terms of $\leq_S$) since $\times$ is extensive. Moreover, it can be a value which is not in AD but in $I - AD$. If the state of the computation consists of the definitions of all constraints, then at each step we get a strictly worse state (in terms of $\sqsubseteq_S$). The sequence of such computation states, until stability, has finite length, since by assumption I is finite and thus the value associated with each tuple of each constraint may be changed at most size(I) times. $\quad\square$

An interesting special case of the above theorem occurs when the chosen semiring has a finite domain A. In fact, in that case the hypotheses of the theorem hold with $I = A$. Another useful result occurs when $+$ and $\times$ are AD-closed. In fact, in this case one can also compute the time complexity of the k-consistency algorithm by just looking at the given problem. More precisely, if this same algorithm is $O(n^k)$ in the classical CSP case [9, 10, 17, 19], then here it is $O(\text{size}(AD) \times n^k)$ (in [6] they reach the same conclusion for the fuzzy CSP case).

No matter in which order the subsets W of $k - 1$ variables, as well as the additional variables x, are chosen during the k-consistency algorithm, the result is always the same problem. However, this holds in general only if $\times$ is idempotent.

Theorem 13 (order-independence). *Consider a SCSP problem P and two different applications of the k-consistency algorithm to P, producing respectively P' and P''. Then $P' = P''$ if $\times$ is idempotent.* $\quad\square$

In some cases, where the given problem has a tree-like structure (where each node of the tree may be any subproblem), a variant of the above defined k-consistency algorithm can be applied: just apply the main algorithm step to each node of the tree, in any bottom-up order [21]. For such algorithm, which is linear in the size of the problem and exponential in the size of the larger node in the tree structure, the idempotency of $\times$ is not needed to satisfy the above properties (except the order independence, which does not make sense here).

Notice that the definitions and results of this section would hold also in the more general case of a local consistency algorithm which is not required to achieve consistency on every subset of k variables, but may in general make only some subsets of variables consistent (not necessarily a partition of the problem). These more general kinds of algorithms have been considered in [21], and similar properties have been shown there for the special case of classical CSPs.

2.4 Instances of the framework

We will now show how several known, and also new, frameworks for constraint solving may be seen as instances of the SCSP framework. More precisely, each of such frameworks corresponds to the choice of a specific constraint system (and thus of a semiring). This means that we can immediately know whether one can inherit the properties of the general framework by just looking at the properties of the operations of the chosen semiring, and by referring to the theorems in the previous subsection. This is interesting for known constraint solving schemes, because it puts them into a single unifying framework and it justifies in a formal way many informally taken choices, but it is especially significant for new schemes, for which one does not need to prove all the properties that it enjoys (or not) from scratch. Since we consider only finite domain constraint solving, in the following we will only specify the semiring that has to be chosen to obtain a particular instance of the SCSP framework.

Classical CSPs. A classical CSP problem [18, 20] is just a set of variables and constraints, where each constraint specifies the tuples that are allowed for the involved variables. Assuming the presence of a subset of distinguished variables, the solution of a CSP consists of a set of tuples which represent the assignments of the distinguished variables which can be extended to total assignments which satisfy all the constraints.

Since constraints in CSPs are crisp, we can model them via a semiring with only two values, say 1 and 0: allowed tuples will have the value 1, and not allowed ones the value 0. Moreover, in CSPs, constraint combination is achieved via a join operation among allowed tuple sets. This can be modeled here by taking as the multiplicative operation the logical *and* (and interpreting 1 as true and 0 as false). Finally, to model the projection over the distinguished variables, as the k-tuples for which there exists a consistent extension to an n-tuple, it is enough to assume the additive operation to be the logical *or*. Therefore a CSP is just an SCSP where the c-semiring in the constraint system CS is $S_{CSP} = \langle \{0,1\}, \vee, \wedge, 0, 1 \rangle$. The ordering $\leq_S$ here reduces to $1 \leq_S 0$. As predictable, all the properties related to k-consistency hold. In fact, $\wedge$ is idempotent. Thus the results of Theorems 11 and 13 apply. Also, since the domain of the semiring is finite, the result of Theorem 12 applies as well.

Fuzzy CSPs. Fuzzy CSPs (FCSPs) [6, 25, 26, 27] extend the notion of classical CSPs by allowing non-crisp constraints, that is, constraints which associate a preference level with each tuple of values. Such level is always between 0 and 1, where 1 represents the best value (that is, the tuple is allowed) and 0 the worst one (that is, the tuple is not allowed). The solution of a fuzzy CSP is then defined as the set of tuples of values which have the maximal value. The value associated with n-tuple is obtained by minimizing the values of all its subtuples. Fuzzy CSPs are already a very significant extension of CSPs. In fact, they are able to model partial constraint satisfaction [11], so to get a solution even when the problem is over-constrained, and also prioritized constraints, that is, constraints with different levels of importance [3].

Fuzzy CSPs can be modeled in our framework by choosing the c-semiring $S_{FCSP} = \langle \{x \mid x \in [0,1]\}, max, min, 0, 1 \rangle$. The ordering $\leq_S$ here reduces to the $\geq$ ordering on reals. The multiplicative operation of S_{FCSP} (that is, min) is idempotent. Thus Theorem 11 and 13 can be applied. Moreover, min is AD-closed for any finite subset of [0,1]. Thus, by Theorem 12, any k-consistency algorithm terminates. Thus FCSPs, although providing a significant extension to classical CSPs, can exploit the same kind of local consistency algorithms. An implementation of arc-consistency, suitably adapted to be used over fuzzy CSPs, is given in [27] (although no formal properties of its behavior are proved).

Probabilistic CSPs. Probabilistic CSPs [7] have been introduced to model those situations where each constraint c has a certain independent probability $p(c)$ to be part of the given real problem. This allows one to reason also about problems which are only partially known. The probability of each constraint gives then, to each instantiation of all the variables, a probability that it is a solution of the real problem. This is done by associating with an n-tuple t the probability that all constraints that t violates are in the real problem. This is just the product of all $1 - p(c)$ for all c violated by t. Finally, the aim is to get those instantiations with the maximum probability.

The relationship between Probabilistic CSPs and SCSPs is complicated by the fact that the former contain crisp constraints with probability levels, while the latter contain non-crisp constraints. That is, we associate values with tuples, and not to constraints. However, it is still possible to model Probabilistic CSPs, by using a transformation which is similar to that proposed in [6] to model prioritized constraints via soft constraints in the FCSP framework. More precisely, we assign probabilities to tuples instead of constraints: consider any constraint c with probability $p(c)$, and let t be any tuple of values for the variables involved in c; then we set $p(t) = 1$ if t is allowed by c, otherwise $p(t) = 1 - p(c)$. The reasons for such a choice are as follows: if a tuple is allowed by c and c is in the real problem, then t is allowed in the real problem; this happens with probability $p(c)$; if instead c is not in the real problem, then t is still allowed in the real problem, and this happens with probability $1 - p(c)$. Thus t is allowed in the real problem with probability $p(c) + 1 - p(c) = 1$. Consider instead a tuple t which is not allowed by c. Then it will be allowed in the real problem only if c is not present; this happens with probability $1 - p(c)$.

To give the appropriate value to an n-tuple t, given the values of all the smaller k-tuples, with $k \leq n$ and which are subtuples of t (one for each constraint), we just perform the product of the value of such subtuples. By the way values have been assigned to tuples in constraints, this coincides with the product of all $1 - p(c)$ for all c violated by t. In fact, if a subtuple violates c, then by construction its value is $1 - p(c)$; if instead a subtuple satisfies c, then its value is 1. Since 1 is the unit element of $\times$, we have that $1 \times a = a$ for each a. Thus we get $\Pi(1 - p(c))$ for all c that t violates.

As a result, the c-semiring corresponding to the Probabilistic CSP framework is $S_{prob} = \langle \{x \mid x \in [0,1]\}, max, \times, 0, 1 \rangle$, and the associated ordering $\leq_S$ here reduces to $\geq$ over reals. Note that the fact that P' is α-consistent means that in P there exists an n-tuple which has probability α to be a solution of the real problem.

The multiplicative operation of S_{prob} (that is, $\times$) is not idempotent. Thus neither Theorem 11 nor Theorem 13 can be applied. Also, $\times$ is not closed on any superset of any non-trivial finite subset of [0,1]. Thus Theorem 12 cannot be applied as well. Therefore, k-consistency algorithms do not make much sense in the Probabilistic CSP framework, since none of the usual desired properties hold. However, the fact that we are dealing with a c-semiring implies that, at least, we can apply Theorem 8: if a Probabilistic CSP problem has a tuple with probability α to be a solution of the real problem, then any subproblem has a tuple with probability at least α to be a solution of a subproblem of the real problem. This can be fruitfully used when searching for the best solution in a branch-and-bound search algorithm.

Weighted CSPs. While fuzzy CSPs associate a level of preference with each tuple in each constraint, in weighted CSPs (WCSPs) tuples come with an associated cost. This allows one to model optimization problems where the goal is to minimize the total cost (time, space, number of resources...) of the proposed solution. Therefore, in WCSPs the cost function is defined by summing up the costs of all constraints (intended as the cost of the chosen tuple for each constraint). Thus the goal is to find the n-tuples (where n is the number of all the variables) which minimize the total sum of the costs of their subtuples (one for each constraint).

According to this informal description of WCSPs, the associated c-semiring is $S_{WCSP} = \langle \mathbb{R}^+, \min, +, +\infty, 0 \rangle$, with ordering $\leq_S$ which reduces here to $\leq$ over the reals. This means that a value is preferred to another one if it is smaller.

The multiplicative operation of S_{WCSP} (that is, $+$) is not idempotent. Thus the k-consistency algorithms cannot be used (in general) in the WCSP framework, since none of the usual desired properties hold. However, again, the fact that we are dealing with a c-semiring implies that, at least, we can apply Theorem 8: if a WCSP problem has a best solution with cost α, then the best solution of any subproblem has a cost smaller than α. This can be convenient to know in a branch-and-bound search algorithm. Note that the same properties hold also for the semirings $\langle \mathbb{Q}^+, min, +, +\infty, 0 \rangle$ and $\langle \mathbb{Z}^+, min, +, +\infty, 0 \rangle$, which can be proved to be c-semirings.

Egalitarianism and Utilitarianism: FCSP + WCSP. The FCSP and the WCSP systems can be seen as two different approaches to give a meaning to the notion of optimization. The two models correspond in fact, respectively, to two definitions of social welfare in utility theory [22]: *egalitarianism*, which maximizes the minimal individual utility, and *utilitarianism*, which maximizes the sum of the individual utilities: FCSPs are based on the egalitarian approach, while WCSPs are based on utilitarianism.

In this section we show how our framework allows also for the combination of these two approaches. In fact, we construct an instance of the SCSP framework where the two approaches coexist, and allow us to discriminate among solutions which otherwise would result indistinguishable. More precisely, we first compute the solutions according to egalitarianism (that is, using a $max - min$ computation as in FCSPs), and then discriminate more among them via utilitarianism (that is, using a $max - sum$ computation as in WCSPs). The resulting c-semiring is $S_{ue} = \langle \{\langle l, k \rangle \mid l, k \in [0, 1]\}, \underline{max}, \underline{min}, \langle 0, 0 \rangle, \langle 1, 0 \rangle \rangle$, where $\underline{max}$ and $\underline{min}$ are defined as follows:

$$\langle l_1, k_1 \rangle \underline{max} \langle l_2, k_2 \rangle = \begin{cases} \langle l_1, max(k_1, k_2) \rangle & \text{if } l_1 = l_2 \\ \langle l_1, k_1 \rangle & \text{if } l_1 > l_2 \end{cases}$$

$$\langle l_1, k_1 \rangle \underline{min} \langle l_2, k_2 \rangle = \begin{cases} \langle l_1, k_1 + k_2 \rangle & \text{if } l_1 = l_2 \\ \langle l_2, k_2 \rangle & \text{if } l_1 > l_2 \end{cases}$$

That is, the domain of the semiring contains pairs of values: the first element is used to reason via the *max-min* approach, while the second one is used to further discriminate via the *max-sum* approach. More precisely, given two pairs, if the first elements of the pairs differ, then the $\underline{max} - \underline{min}$ operations behave like a normal $max - min$, otherwise they behave like *max-sum*. This can be interpreted as the fact that, if the first element coincide, it means that the *max-min* criteria cannot discriminate enough, and thus the $max - sum$ criteria is used.

Since $\underline{min}$ is not idempotent, k-consistency algorithms cannot in general be used meaningfully in this instance of the framework.

A kind of constraint solving similar to that considered in this section is the one presented in [8], where Fuzzy CSPs are augmented with a finer way of selecting the preferred solution. More precisely, they employ a lexicographic ordering to improve the discriminating power of FCSPs and avoid the so-called *drowning effect*. We plan to rephrase this approach in our framework (as it is done in the VCSP framework).

N-dimensional SCSPs. Choosing an instance of the SCSP framework means specifying a particular c-semiring. This, as discussed above, induces a partial order which can be interpreted as a (partial) guideline for choosing the "best" among different solutions. In many real-life situations, however, one guideline is not enough, since, for example, it could be necessary to reason with more than one objective in mind, and thus choose solutions which achieve a good compromise w.r.t. all such goals.

Consider for example a network of computers, where one would like to both minimize the total computing time (thus the cost) and also to maximize the work of the least used computers. Then, in the SCSP framework, we would need to consider two c-semirings, one for cost minimization (weighted CSP), and another one for work maximization (fuzzy CSP). Then, one could work first with one of these c-semirings and then with the other one, trying to combine the solutions which are the best for each of them.

However, a much simpler approach consists of combining the two c-semirings and then work with the resulting structure. The nice property is that such a structure is a c-semiring itself, thus all the techniques and properties of the SCSP framework can be used for such a structure as well. More precisely, the way to combine several c-semirings and get another c-semiring just consists of vectorizing the domains and operations of the combined c-semirings.

Definition 14 (composition of c-semirings). Given the n c-semirings $S_i =$

$\langle A_i, +_i, \times_i, 0_i, 1_i \rangle$, for $i = 1, \ldots, n$, we define the structure $Comp(S_1, \ldots, S_n) = \langle\langle A_1, \ldots, A_n \rangle, +, \times, \langle 0_1, \ldots, 0_n \rangle, \langle 1_1 \ldots 1_n \rangle\rangle$. Given $\langle a_1, \ldots, a_n \rangle$ and $\langle b_1, \ldots, b_n \rangle$ such that $a_i, b_i \in A_i$ for $i = 1, \ldots, n$, $\langle a_1, \ldots, a_n \rangle + \langle b_1, \ldots, b_n \rangle = \langle a_1 +_1 b_1, \ldots, a_n +_n b_n \rangle$, and $\langle a_1, \ldots, a_n \rangle \times \langle b_1, \ldots, b_n \rangle = \langle a_1 \times_1 b_1, \ldots, a_n \times_n b_n \rangle$. $\square$

According to the definition of the ordering $\leq_S$ (in Sect. 2.1), such an ordering for $S = Comp(S_1, \ldots, S_n)$ is as follows. Given $\langle a_1, \ldots, a_n \rangle$ and $\langle b_1, \ldots, b_n \rangle$ such that $a_i, b_i \in A_i$ for $i = 1, \ldots, n$, we have $\langle a_1, \ldots, a_n \rangle \leq_S \langle b_1, \ldots, b_n \rangle$ if and only if $\langle a_1 +_1 b_1, \ldots, a_n +_n b_n \rangle = \langle b_1, \ldots, b_n \rangle$. Since the tuple elements are completely independent, $\leq_S$ is in general a partial order, even though each of the $\leq_{S_i}$ is a total order. Thus it is in this instance that the power of a partially ordered domain (as opposed to a totally ordered one, as in the VCSP framework discussed later in the paper) can be exploited.

The presence of a partial order means that the abstract solution of a problem over such a semiring may in general contain an incomparable set of tuples, none of which has $blevel(P)$ as its associated value. In this case, if one wants to reduce the number of "best" tuples (or to get just one), one has to specify some priorities among the orderings of the component c-semirings.

3 Valued Constraint Problems

The framework described in this section is based on an ordered monoid structure. One element of the set of the monoid, a valuation, is associated with each constraint[6] and the monoid operation (denoted $\circledast$) is used to assign a valuation to each assignment, by combining the valuations of all the constraints violated by the assignment. The order on the monoid is assumed to be total and the problem considered is always to minimize the combined valuation of all violated constraints.

We show how this framework can be used to model several existing constraint solving schemes and also to relate all these schemes from an expressiveness and computational complexity point of view. We then try to extend some usual look-ahead backtrack search algorithms to the VCSP framework. In this process, we define an extended version of the arc-consistency property and study the influence of the monoid operation properties on the computational complexity of the problem of checking arc-consistency. The results obtained formally justify the current algorithmic "state of art" for several existing schemes. The content of this section is based on [28].

3.1 Valuation structure

In this section, a classical CSP is defined by a set $V = \{v_1, \ldots, v_n\}$ of variables, each variable v_i having an associated finite domain d_i. A constraint $c = (V_c, R_c)$

---

[6] For the sake of simplicity, the choice was made to associate one valuation with each constraint rather than to use the finer approach where one valuation is associated with each tuple of each constraint. This latter approach is not fundamentally different, as Sect. 4 will show.

is defined by a set of variables $V_c \subseteq V$ and a relation R_c between the variables of V_c *i.e.*, a subset of the Cartesian product $\prod_{v_i \in V_c} d_i$. A CSP is denoted by $\langle V, D, C \rangle$, where D is the set of the domains and C the set of the constraints. A solution of the CSP is an assignment of values to the variables in V such that all the constraints are satisfied: for each constraint $c = (V_c, R_c)$, the tuple of the values taken by the variables of V_c belongs to R_c.

To express the fact that a constraint may eventually be violated, we annotate each constraint with a valuation taken from a set of valuations E equipped with the following structure:

Definition 15 (Valuation structure). A **valuation structure** is defined as a tuple $\langle E, \circledast, \succ \rangle$ such that:

- E is a set, whose elements are called valuations, which is totally ordered by $\succ$, with a maximum element noted $\top$ and a minimum element noted $\bot$;
- $\circledast$ is a commutative, associative closed binary operation on E that satisfies:
  - *Identity*: $\forall a \in E, a \circledast \bot = a$;
  - *Monotonicity*: $\forall a, b, c \in E, (a \succcurlyeq b) \Rightarrow ((a \circledast c) \succcurlyeq (b \circledast c))$;
  - *Absorbing element*[7]: $\forall a \in E, (a \circledast \top) = \top$. $\qquad\qquad\Box$

This structure of a totally ordered commutative monoid with a monotonic operator is also known in uncertain reasoning, E being restricted to $[0, 1]$, as a "triangular co-norm" [5]. In the rest of the paper, we implicitly suppose that the computation of $\succ$ and $\circledast$ are always polynomial in the size of their arguments.

Justification and properties The ordered set E allows different levels of violations to be expressed. Commutativity and associativity guarantee that the valuation of an assignment depends only on the set of the valuations of the violated constraints, and not on the way they are aggregated. The element $\top$ corresponds to unacceptable violation and is used to express *hard* constraints. The element $\bot$ corresponds to complete satisfaction. These maximum and minimum elements can be added to any totally ordered set, and their existence is supposed without any loss of generality. Monotonicity guarantees that the valuation of an assignment that satisfies a set B of constraints will always be as good as the valuation of any assignment which satisfies a subset of B. Two additional properties will be considered later because of their influence on algorithms and computation:

- **Strict monotonicity** $(\forall a, b, c \in E$, if $(a \succ c), (b \neq \top)$ then $(a \circledast b) \succ (c \circledast b))$ guarantees that any modification in a set of valuations that does not contain $\top$ passes on the aggregation, via $\circledast$, of these valuations: the fact that something can be locally improved can not be globally ignored. This type of property is usual in multi-criteria theory, namely in social welfare theory [22].

---

[7] Actually, the "absorbing element" property can be inferred from the other axioms: since $\bot$ is the identity, $(\bot \circledast \top) = \top$; since $\bot$ is minimum, $\forall a \in E, (a \circledast \top) \succcurlyeq (\bot \circledast \top) = \top$; since, $\top$ is maximum, $\forall a \in E, (a \circledast \top) = \top$

- **Idempotency** $(\forall a \in E, (a \circledast a) = a)$ is fundamental in all CSP algorithms that enforce k-consistency since it guarantees that a constraint that is satisfied by all the solutions of a CSP can be added to the CSP without changing its meaning.

Theorem 16. *In a valuation structure* $\langle E, \circledast, \succ \rangle$ *such that* $\circledast$ *is idempotent and* $|E| > 2$, *the operator* $\circledast$ *is not strictly monotonic.* $\qquad\square$

Proof. From identity, it follows that $\forall a \in E, (a \circledast \perp) = a$, then for any $a, \perp \prec a \prec \top$, strict monotonicity implies that $(a \circledast a) \succ a$ and idempotency implies that $(a \circledast a) = a$. $\qquad\square$

Theorem 17. *In a valuation structure* $\langle E, \circledast, \succ \rangle$, *if* $\circledast$ *is idempotent then* $\circledast =$ max.

Proof. This result is well known for t-conorms. From monotonicity and idempotency, we have $\forall b \preccurlyeq a, (a \circledast \perp) = a \preccurlyeq (a \circledast b) \preccurlyeq a = (a \circledast a)$ and therefore $a \circledast b = a$. $\qquad\square$

3.2 Valued CSP

A valued CSP is then simply obtained by annotating each constraint of a classical CSP with a valuation denoting the impact of its violation[8] or, equivalently, of its rejection from the set of constraints.

Definition 18 (Valued CSP). A valued CSP is defined by a classical CSP $\langle V, D, C \rangle$, a valuation structure $S = (E, \circledast, \succ)$, and an application φ from C to E. It is denoted by $\langle V, D, C, S, \varphi \rangle$. $\varphi(c)$ is called the valuation of c. $\qquad\square$

An assignment A of values to some variables $W \subset V$ can now be simply evaluated by combining the valuations of all the violated constraints using $\circledast$:

Definition 19 (Valuation of assignments). In a VCSP $\mathcal{P} = \langle V, D, C, S, \varphi \rangle$ the valuation of an assignment A of the variables of $W \subset V$ is defined by:

$$\mathcal{V}_\mathcal{P}(A) = \underset{\substack{c \in C, V_c \subset W \\ A \text{ violates } c}}{\circledast} [\varphi(c)]$$

$\qquad\square$

---

[8] The finer approach chosen in the SCSP scheme, which associates a valuation with each tuple of a relation, allows the expression of gradual violation of a constraint. However, since $\perp$ is the identity, and since domains are finite, such a gradual relation may be simply expressed as a conjunction of annotated constraints and the restriction to annotated constraints is made without any loss of generality. See sect. 4.

The semantics of a VCSP is a distribution of valuations on the assignments of V (potential solutions). The problem considered is to find an assignment A with a *minimum* valuation. The valuation of such an optimal solution will be called the CSP valuation. It provides a *gradual* notion of inconsistency, from $\bot$, which corresponds to consistency, to $\top$, for complete inconsistency.

Our notion of VCSP is equivalent to [11] view of partial consistency. Indeed, a VCSP defines a relaxation lattice equipped with a distance measure:

Definition 20 (Relaxation of a VCSP). Given a VCSP $\mathcal{P} = \langle V, D, C, S, \varphi \rangle$, a relaxation of $\mathcal{P}$ is a classical CSP $\langle V, D, C' \rangle$, where $C' \subset C$. $\quad\square$

Relaxations are naturally ordered by inclusion of constraint sets. Obviously, the consistent inclusion-maximal relaxations are the classical CSP which can not get closer to the original problem without loosing consistency. It is also possible to order relaxations by extending the valuation distribution to relaxations:

Definition 21 (Valuation of a relaxation). In a VCSP $\mathcal{P} = \langle V, D, C, S, \varphi \rangle$, the valuation of a relaxation $\langle V, D, C' \rangle$ of $\mathcal{P}$ is defined as:

$$\mathcal{V}_{\mathcal{P}}(\langle V, D, C' \rangle) = \underset{c \in C - C'}{\circledast} [\varphi(c)]$$

$\square$

The valuation of the top of the relaxation lattice, the CSP $\langle V, D, C \rangle$, is obviously $\bot$. The valuations of the other relaxations can be understood as a distance to this ideal problem. The best assignments of V are the solutions of the closest consistent problems of the lattice. The monotonicity of $\circledast$ ensures that the order on problems defined by this valuation distribution is consistent with the inclusion order on relaxations.

Theorem 22. *Given a VCSP* $\mathcal{P} = \langle V, D, C, S, \varphi \rangle$, *and* $\langle V, D, C' \rangle$, $\langle V, D, C'' \rangle$, *two relaxations of* $\mathcal{P}$:

$$C' \subsetneq C'' \Rightarrow \mathcal{V}_{\mathcal{P}}(\langle V, D, C' \rangle) \succcurlyeq \mathcal{V}_{\mathcal{P}}(\langle V, D, C'' \rangle)$$

If $\circledast$ *is strictly monotonic, the inequality becomes strict if the valuation of* $\mathcal{P}$ *is not* $\top$. $\quad\square$

This last result shows that strict monotonicity is indeed a highly desirable property since it guarantees that the order induced by the valuation distribution will respect the *strict* inclusion order on relaxations (if the VCSP valuation is not equal to $\top$). In this case, optimal consistent relaxations are always selected among inclusion-maximal consistent relaxations, which seems quite *rational*.

Since idempotency and strict monotonicity are incompatible as soon as E has more than two elements, idempotency can be seen as an undesirable property, at least from the rationality point of view. Using an idempotent operator, it is possible for a consistent non inclusion-maximal relaxation to get an optimal valuation.

There is an immediate relation between optimal assignments and optimal consistent relaxations. Indeed, we can associate the classical *consistent* CSP $[A]_\mathcal{P}$, obtained by excluding the constraints violated by A, with any assignment A of V.

Definition 23. Given a VCSP $\mathcal{P} = \langle V, D, C, S, \varphi \rangle$ and an assignment A of the variables of V, we denote $[A]_\mathcal{P}$ the classical consistent CSP $\langle V, D, C' \rangle$ where $C' = \{c \in C, A \text{ satisfies } c\}$. $[A]_\mathcal{P}$ is called the consistent relaxation of $\mathcal{P}$ associated with A. □

Obviously, $\mathcal{V}_\mathcal{P}(A) = \mathcal{V}_\mathcal{P}([A]_\mathcal{P})$ and $[A]_\mathcal{P}$ is among the optimal problems that A satisfies. It is equivalent to look for an optimal assignment or for an optimal consistent relaxation.

3.3 Instances of the framework

We consider some extensions of the CSP framework as VCSP. Most of these instances have already been described as SCSP in the sect. 2.4 and we just give here the valuation structure needed to cast each instance.

Classical CSP or ∧-VCSP. In a classical CSP, an assignment is considered unacceptable as soon as one constraint is violated. Therefore, classical CSP correspond to the trivial boolean lattice $E = \{t, f\}$, $t = \bot \prec f = \top$, $\circledast = \land$ (or max), all constraints being annotated with $\top$. The operation $\land$ is both idempotent and strictly monotonic (this is the only case where both properties may exist simultaneously in a valuation structure).

Possibilistic, Fuzzy CSP or Max-VCSP. Possibilistic CSPs [27] are closely related to Fuzzy CSPs [6, 25, 26]. Each constraint is annotated with a priority (usually a real number between 0 and 1). The valuation of an assignment is defined as the maximum valuation among violated constraints. The problem defined is therefore a min-max problem [31], dual to the max-min problem of Fuzzy CSP.

Possibilistic CSPs [27] are defined by the operation $\circledast = \max$. Traditionally, $E = [0, 1]$, $0 = \bot$, $1 = \top$. The annotation of a constraint is interpreted as a priority degree. A preferred assignment minimizes the priority of the most important violated constraint. The idempotency of max leads to the so-called "drowning-effect": if a constraint with priority α has to be necessarily violated then any constraint with a priority lower then α is simply ignored by the combination operator $\circledast$ and therefore such a constraint can be rejected from any consistent relaxation without changing its valuation. The notion of lexicographic CSP has been proposed in [8] to overcome this apparent weakness.

Obviously, a classical CSP is simply a specific possibilistic CSP where the valuation $\top$ alone is used to annotate the constraints. Note that finite fuzzy CSP [6] can easily be cast as possibilistic CSP and vice-versa (see Sect. 4).

Weighted CSP or Σ-VCSP. Here, we try to minimize the weighted sum of the valuations associated with violated constraints. Weighted CSP correspond to the operation $\circledast = +$ in $\mathbb{N} \cup \{+\infty\}$, using the usual ordering $<$. First considered in [30], weighted CSP have been considered as *Partial CSP* in [11], all constraint valuations being equal to 1. The operation is strictly monotonic.

Probabilistic CSP or Π-VCSP. Probabilistic CSP have been defined in [7] to enable the user to represent ill-known problems, where the existence of constraints in the real problem is uncertain. Each constraint c is annotated with its probability of existence, all supposed to be independent. The probability that an assignment that violates 2 constraints c_1 and c_2 will not be a solution of the real problem is therefore $1 - (1 - \varphi(c_1))(1 - \varphi(c_2))$. Therefore, probabilistic CSP correspond to the operation $x \circledast y = 1 - (1-x)(1-y)$ in $E = [0, 1]$. The operation is strictly monotonic.

Lexicographic CSP. Lexicographic CSP offer a combination of weighted and possibilistic CSP and suppress the "drowning effect" of the latter [8]. The idea is that the valuation of an assignment will not simply be defined by the maximum valuation among the valuations of violated constraints but will depend on the number of violated constraints at each level of priority, starting from the most prioritary to the least prioritary.

Here, a valuation is either a designated element $\top$ or a multiset (elements may be repeated) of elements of $[0, 1[$ (any other totally ordered set may be used instead of $[0, 1[$).

The operation $\circledast$ is simply multi-set union, extended to treat $\top$ as an absorbing element (the empty multi-set is the identity $\bot$). The order $\succ$ is the lexicographic (or alphabetic) total order induced by the order $>$ on multisets and extended to give $\top$ its role of maximum element: let v and v' be two multisets and α and α' be the largest elements in v and v', $v \succ v'$ iff either $\alpha > \alpha'$ or ($\alpha = \alpha'$ and $v - \{\alpha\} \succ v' - \{\alpha'\}$). The recursion ends on $\varnothing$, the minimum multi-set.

This instance can be compared to the "FCSP+WCSP" instance considered in sect. 2.4: since the number of constraints violated at each level of priority are used to discriminate assignments in the lexicographic CSP approach, it is finer than the "FCSP+WCSP" which simply relies on the number of constraints at one level.

3.4 Relationships between instances

In order to compare the previous VCSP classes, that relies on different valuation structures, we introduce the following notion:

Definition 24 (Refinement of a VCSP). A VCSP $\mathcal{P} = \langle V, D, C, S, \varphi \rangle$ is a *refinement* of the VCSP $\mathcal{P}' = \langle V, D, C', S', \varphi' \rangle$ if for any pair of assignments A, A' of V such that $\mathcal{V}_{\mathcal{P}'}(A) \succ \mathcal{V}_{\mathcal{P}'}(A')$ in S' then $\mathcal{V}_{\mathcal{P}}(A) \succ \mathcal{V}_{\mathcal{P}}(A')$ in S. $\mathcal{P}$ is a *strong refinement* of $\mathcal{P}'$ if the property holds when A, A' are assignments of subsets of V. □

The main point is that if $\mathcal{P}$ is a *refinement* of $\mathcal{P}'$, then the set of optimal assignments of $\mathcal{P}$ is included in the set of optimal assignments of $\mathcal{P}'$; the problem of finding an optimal assignment of $\mathcal{P}'$ can be reduced to the same problem in $\mathcal{P}$.

Definition 25 (Equivalence of two VCSP). Two VCSP $\mathcal{P} = \langle V, D, C, S, \varphi \rangle$ and $\mathcal{P}' = \langle V, D, C', S', \varphi' \rangle$ will be said *equivalent* iff each one is a refinement of the other. They will be said *strongly equivalent* if each one is a strong refinement of the other.

Equivalent VCSP define the same ordering on assignments of V and have the same set of optimal assignments: the problem of finding an optimal assignment is equivalent in both VCSP. Note that this definition of equivalence is weaker than the definition 6 used in SCSP since it does not require that two equivalent VCSP always give the same valuation to the same assignments but only that they order assignments similarly.

Definition 26 (Polynomial time refinement). Given S and S', two valuation structures, a polynomial time refinement from S to S' is a function Φ that:

- transforms any VCSP $\mathcal{P} = \langle V, D, C, S, \varphi \rangle$ in a VCSP $\mathcal{P}' = \langle V, D, C, S', \varphi' \rangle$ such that $\mathcal{P}'$ is a refinement of $\mathcal{P}$;
- is deterministic polynomial time computable.

This notion of polynomial-time refinement is inspired by the notion of polynomial transformation (or many-one reduction) usual in computational complexity [24]. It is finely tuned for optimization problems in valued CSP. This definition is motivated by the fact that the decision problems derived from the optimization problem of finding an optimal assignment are all NP-complete whatever the valuation structure is. As we will see, the notion allows one to bring to light subtle differences between different valuation structures[9].

Considering all previous VCSP classes, presented in the table which follows, we may partition them according to the idempotency of the operator: classical CSP and Possibilistic CSP on one side and Weighted, Probabilistic and Lexicographic CSP on the other. Interestingly, we will now show that this partition is in agreement with polynomial refinement between valuation structures.

---

[9] We have recently discovered that a similar idea has been developped in [16] for optimization problems in general. This work define the class OptP and the notion of polynomial "metric reduction", closely related to the notion of polynomial-time refinement introduced here.

Instance	E	$\circledast$	$\bot$	$\top$	$\succ$	Prop.
$\wedge$-VCSP	$\{t, f\}$	$\wedge = \max$	t	f	$f \succ t$	idemp.
Max-VCSP.	$[0, 1]$	$\max$	0	1	$>$	idemp.
Σ-VCSP	$\overline{\mathbb{N}}$	$+$	0	$+\infty$	$>$	strict. monot.
Π-VCSP	$[0, 1]$	$x + y - xy$	0	1	$>$	strict. monot.
Lex-VCSP	$[0, 1]^* \cup \{\top\}$	$\cup$	$\varnothing$	$\top$	lex.	strict. monot.

We may first consider theVCSP instances with an idempotent $\circledast$ operator: these are Classical and Possibilistic (or Fuzzy) CSP. Note that according to theorem 17, these are the only existing instances.

- a classical CSP is nothing but a specific Possibilistic CSP that uses only the valuation $\top$ and the refinement from classical CSP to Possibilistic CSP is simply the identity.
- the problem of the existence of an assignment of valuation strictly lower than v in a Possibilistic CSP $\langle V, D, C, S, \varphi \rangle$ can easily be reduced to the existence of a solution for the classical CSP $\langle V, D, C' \rangle$ such that $C' = \{c \in C \mid \varphi(c) \geq v\}$: if such a constraint is violated, the assignment valuation is larger than v and conversely. Thus, using binary search, the optimal assignment in a Possibilistic CSP can be found in a logarithmic number of resolution of a classical CSP[10].

We now consider some VCSP instances with a strictly monotonic operator: this are Probabilistic CSP, Weighted CSP and Lexicographic CSP.

- we put Probabilistic CSP aside because the Probabilistic CSP combination operator relies on multiplication of *real* numbers. However, note that if we allow the valuations in Weighted CSP to take values in $\mathbb{R}$ instead of $\mathbb{N}$, then these frameworks are related by a simple isomorphism: a constraint with a probability $\varphi(c)$ of existence can be transformed into a constraint with a weight of $-\log(1 - \varphi(c))$ (and conversely using the transformation $1 - e^{-\varphi(c)}$). The two VCSP obtained in this way are obviously *strongly equivalent*.
- a simple polynomial refinement exists from Weighted CSP to Lexicographic CSP: the valuation $k \in \mathbb{N}$ is transformed in a multiset containing a given element $\alpha \neq 0$ repeated k times (noted $\{(\alpha, k)\}$), where α is a fixed priority and the valuation $+\infty$ is transformed to $\top$. The lexicographic VCSP obtained is in fact *strongly equivalent* to the original weighted VCSP.
- interestingly, a Lexicographic CSP may also be transformed into a *strongly equivalent* Weighted CSP. Let $\alpha_1, \ldots, \alpha_k$ be the elements of $]0, 1[$ that appear in all the Lexicographic CSP annotations, sorted in increasing order. Let n_i be the number of occurrences of α_i in all the annotations of the VCSP. The lowest priority α_1 corresponds to the weight $f(\alpha_1) = 1$, and inductively α_i corresponds to $f(\alpha_i) = f(\alpha_{i-1}) \times (n_{i-1} + 1)$ (this way, the weight $f(\alpha_i)$ corresponding to priority i is strictly larger than the largest possible sum

---

[10] In fact,most of the traditional polynomial classes and problems (k-consistency enforcing...) of classical CSP can be extended to Possibilistic CSP in this way.

of $f(\alpha_j), j < i$. This is immediately satisfied for α_1 and inductively verified for α_i). An initial lexicographic valuation is converted in the sum of the penalties $f(\alpha_i)$ for each α_i in the valuation. The valuation $\top$ is converted to $+\infty$. All the operations involved, sum and multiplication, are polynomial and the sizes of the operands remain polynomial: if k is the number of priorities used in the VCSP and ℓ the maximum number of occurrences of a priority, then the largest weight $f(\alpha_k)$ is in $O(\ell^k)$, with a length in $O(k.\log(\ell))$ while the original annotations used at least space $O(k + \log(\ell))$. Therefore, the refinement is polynomial.

A bridge between idempotent and strictly monotonic VCSP is provided by a polynomial refinement from Possibilistic CSP to Lexicographic CSP: the Lexicographic CSP is simply obtained by annotating each constraint with a multi-set containing one occurrence of the original (possibilistic) annotation if it is not equal to 1, or by $\top$ otherwise. In this case, an optimal assignment of the Lexicographic CSP not only minimizes the priority of the most important constraint violated, but also, the number of constraint violated successively at each level of priority, from the highest first to the lowest. The refinement is obviously polynomial.

These refinements are not only useful for proving that the previous VCSP instances with a non idempotent operator are at least as hard as VCSP instances with an idempotent operator (the translation of the polynomial refinements to "metric reductions" [16] would place this argument in a richer theoretical framework), but are also simple enough to be practically usable: any algorithm dedicated to one of the Probabilistic, Weighted or Lexicographic CSP framework can be used to solve problem expressed in any of these instances (or even in classical and Possibilistic CSP).

Note that the partition between idempotent and strictly monotonic VCSP classes is also made clear at the level of polynomial classes: the existence of an assignment with a valuation lower than v in a (strictly monotonic) binary Weighted CSP *with domains of cardinality two* is obviously NP-hard by restriction to MAX2SAT [12], whereas the same problem is polynomial in all idempotent VCSP classes. One of the few polynomial classes which seems to extend to all classes of VCSP is the class of CSP structured in hyper-tree (see [4, 21, 29]).

3.5 Extending local consistency property

In classical binary CSP (all constraints are supposed to involve two variables only), satisfiability defines an NP-complete problem. k-consistency properties and algorithms [19] offer a range of polynomial time weaker properties: enforcing strong k-consistency in a consistent CSP will never lead to an empty CSP.

¿From the VCSP point of view, strong k-consistency enforcing defines a kind of lower bound of the CSP valuation: if strong k-consistency enforcing yields an empty CSP, then we know that the CSP valuation is greater than $\top$ and therefore equal to $\top$, else it is simply greater than $\bot$, which is always true.

Arc-consistency (strong 2-consistency) is certainly the most prominent level of local consistency and has been extended to Possibilistic/Fuzzy CSP years ago [25]. In Possibilistic CSP, seen as VCSP, arc-consistency can be defined as follows:

Definition 27 (Arc-consistency). A VCSP $\mathcal{P}$ is said to be arc-consistent iff (1) there exists, for each variable, a value that defines an assignment with a valuation strictly lower than $\top$ and (2) any assignment A of one variable can be extended to an assignment A' on two variables with the same valuation ($\mathcal{V}_\mathcal{P}(A) = \mathcal{V}_\mathcal{P}(A')$). $\square$

This is nothing but a specialized form of Definition 7 used for k-consistency in SCSP. Polynomial worst-case time algorithms that enforce this property on Possibilistic CSP are defined in [25, 31, 27]. These algorithms yield an arc-consistent Possibilistic CSP with the same valuation distribution on complete assignments.

Obviously, this definition could also be used in non idempotent VCSP. But it is useless if we can not define the corresponding arc-consistency enforcing algorithms that should compute, in polynomial time, a VCSP $\mathcal{P}'$ which is both arc-consistent and in some sense *"equivalent"* to the original VCSP $\mathcal{P}$. The strongest level of equivalence one could achieve is the equality of the valuations in both VCSP for all complete assignments.

But the generalization of AC enforcing algorithms that consists in using min and ⊛ respectively for projection and combination of constraints are guaranteed to work only on idempotent VCSP instances as it has been shown in the Semiring-CSP framework. Next section considers another possibility for extending local consistency properties to the VCSP framework.

3.6 Extending Look-Ahead Tree Search Algorithms

Following the work in [6, 11, 27, 30], we try to extend some traditional CSP algorithms to the *binary* VCSP framework to solve the problem of finding a provably optimal assignment. The class of algorithms which we are interested in are hybrid algorithms that combine backtrack tree-search with some level of local consistency enforcing at each node. These algorithms have been called look-ahead, prospective or prophylactic algorithms. Some possible instances have been considered in [23]: Backtrack, Forward-Checking , Really Full Look Ahead. As in [23], we consider here that such algorithms are described by the type of local consistency enforcing maintained at each node: check-backward, check forward, arc-consistency or more...

In prospective algorithms, an assignment is extended until either a complete assignment (a solution) is found, or the given local consistency property is not verified on the current assignment: backtrack occurs. The extension of such algorithms to the VCSP framework, where the problem is now an optimization problem, relies on a transformation of the *Backtrack* tree search schema to a *Depth First Branch and Bound* algorithm. DFBB is a simple depth first tree search algorithm, which, like *Backtrack*, extends an assignment until either (1) a complete assignment is reached and a new "better" solution is found or (2) a given lower bound on the valuation of the best assignment that can be found by extending the current assignment exceeds the valuation of the current best solution found and backtrack occurs. The lower bound used defines the algorithm. Our aim is to derive a lower bound from any given local consistency property.

In classical CSP, seen as $\wedge$-VCSP, the actual local consistency property used gives the "lower bound": for example, in *Really Full Look Ahead*, the inexistence of an arc-consistent closure of the CSP guarantees that the valuation of any extension of the current assignment will be greater than $\top$ and therefore equal to $\top$. However, as we pointed out earlier, no arc-consistency enforcing algorithm is available for strictly monotonic VCSP. We will therefore use classical local consistency notions plus the notion of relaxation of a VCSP (which defines classical CSP) to define our class of bounds:

Proposition 28. *Given a classical local consistency property L, a lower bound on the valuation of a given VCSP $\mathcal{P}$ is defined by the valuation α of an optimal relaxation of $\mathcal{P}$ among those that satisfy the local consistency property L used (consistency of the current assignment, absence of domain wipe-out after checkforward or arc-consistency enforcing. . .). In this case, we will say that the VCSP satisfies the property L at the level α.* □

This valuation is a lower bound of the valuation of an optimal assignment since the valuation of an optimal assignment is also the valuation of an optimal *consistent* relaxation and all the relaxations where the "local consistency" property L is not verified are non consistent.

The bounds induced by this definition satisfy two interesting properties:

- they guarantee that the extended algorithm will behave as the original "classical" algorithm when applied to a classical CSP seen as a $\wedge$-VCSP (a classical CSP seen as a $\wedge$-VCSP has only one relaxation with a valuation lower than $\top$: itself);
- a stronger local consistency property will define a better lower bound, leading to a tree search with less nodes but possibly more computation at each node.

This definition also offers an extension of any local consistency property (including k-consistency) to VCSP. If, for example, we consider the arc-consistency property, we get:

Definition 29 (α-arc-consistency). A VCSP will be said α-arc-consistent iff the optimal relaxations among all the relaxations which have a non empty arc-consistent closure have a valuation lower than α. □

This definition offers an extension of local consistency properties which is immediately usable because we do not need anymore "enforcing" algorithms that compute "equivalent" problems (which may not exist) but only algorithms that compute a minimum α and this is obviously always feasible. The main question is the cost of this computation.

Extending Backtrack. *Backtrack* uses the local inconsistency of the current partial assignment as the condition for backtracking. Therefore, the lower bound derived is the valuation of an optimal relaxation in which the current assignment is consistent. This is simply the relaxation which precisely rejects the constraints violated by the current assignment (these constraints have to be rejected or else local inconsistency will occur; rejecting these constraint suffices to restore the consistency of the current assignment in the relaxation). The lower bound is therefore simply defined by:

$$\underset{\substack{c \in C \\ A \text{ violates } c}}{\circledast} \varphi(c)$$

and is obviously computable in polynomial time.

The lower bound can easily be computed incrementally when a new variable x_i is assigned: the lower bound associated with the father of the current node is aggregated with the valuations of all the constraints violated by x_i using $\circledast$.

In the Possibilistic and Weighted instances, this generic VCSP algorithm defined coincides with the "Branch and Bound" algorithms defined for Possibilistic or Weighted CSP in [11, 26, 27]. Note that for Possibilistic CSP, thanks to idempotency, it is useless to test whether constraints whose valuation is lower than the lower bound associated with the father node have to be rejected since their rejection cannot influence the bound.

Extending Forward Checking. *Forward-checking* uses an extremely limited form of arc-consistency: backtracking occurs as soon as all the possible extensions of the current assignment A on any uninstantiated variable are locally inconsistent: the assignment is said non forward-checkable. Therefore, the lower bound used is the minimum valuation among the valuations of all the relaxations that makes the current assignment forward-checkable.

A relaxation in which A is forward-checkable (1) should necessarily reject all the constraints violated by A itself and (2) for each uninstantiated variable v_i it should reject one of the sets $C(v_i, \nu)$ of constraints that are violated if v_i is instantiated with value ν of its domain. Since $\circledast$ is monotonic, the minimum valuation is reached by taking into account, *for each variable*, the valuation of the set $C(v_i, \nu)$ of minimum valuation. The bound is again computable in polynomial time since it is the aggregation of (1) the valuations all the constraints violated by A itself (*i.e.*, the bound used in the extension of the backtrack algorithm, see 3.6) and (2) the valuations of the constraint in all the $C(v_i, \nu)$. This computation needs less than $(e.n.d)$ constraint checks and $\circledast$ operations (e is the number of constraints); all the minimum valuation can be computed with less than $(d.n)$ comparisons and aggregated with less than $n \circledast$ operations. Note that the lower bound derived includes the bound used in the backtrack extension plus an extra component and will always be better than the "*Backtrack*" bound.

The lower bound may be incrementally computed by maintaining during tree search, and for each value ν of every unassigned variable v_i the aggregated valuation $B(\nu, v_i)$ of all the constraints that will be violated if ν is assigned to

v_i given the current assignment. Initially, all $B(\nu, v_i)$ are equal to $\perp$. When the assignment A is extended to $A' = A \cup \{v_j = \mu\}$, the B may be updated as follows:

$$B(\nu, v_i) \leftarrow B(\nu, v_i) \circledast \left(\underset{\substack{c \in C, V_c = \{v_i, v_j\} \\ A' \cup \{v_i = \nu\} \text{ violates } c}}{\circledast} [\varphi(c)] \right)$$

that takes into account all the constraints between v_i and v_j that are necessarily violated if μ is assigned to v_j. Upon backtrack, the B have to be restored to their previous values, as domains in classical *Forward-checking*. Note that the B offer a default value heuristic: choose the value with a minimum B.

The lower bound is simply obtained by aggregating, using $\circledast$, the valuations of all the constraints violated by the assignment and all the minimum $B(\nu, v_i)$ for each unassigned variable. The aggregated valuation $v(A')$, $A' = A \cup \{v_j = \mu\}$), of all the constraints violated by the assignment A' is easily computed by taking the valuation $v(A)$ computed on the father node $\circledast$'ed with $B(\mu, v_j)$.

Additional sophistications include deleting values ν of the domains of non instantiated variables if the aggregated valuation of $v(A')$ and $B(\nu, v_i)$ exceeds the upper bound (see [11]). On the Possibilistic and Weighted VCSP instances, this generic VCSP algorithm coincides roughly with the forward-checking based algorithm for Possibilistic CSP described in [27] or the *Partial Forward-checking* algorithm defined for Weighted CSP in [11]. Note that for Possibilistic CSP, and thanks to idempotency, the updating of B can ignore constraints whose valuation is less than the B updated or than the current lower-bound.

Trying to extend Really Full Look Ahead. *Really Full Look Ahead* maintains arc consistency during tree search and backtracks as soon as the current assignment induces a domain wipe-out: the CSP has no arc-consistent closure. For a VCSP, the bound which can be derived from arc-consistency will be the minimum valuation among the valuations of all the relaxations such that the current assignment does not induces a domain wipe-out.

Let us consider any class $\circledast$-VCSP of the VCSP framework such that $\circledast$ is strictly monotonic and for any $a, b \in E, a, b \prec \top, (a \circledast b) \prec \top$. Let ℓ be any valuation different from $\top$ and $\perp$. The decision problem corresponding to the computation of the lower bound in this class can be formulated as:

*Problem 30 (*MAX-AC-CSP*).* Given such a $\circledast$-VCSP and a valuation α, is there a set $C' \subset C$ such that the relaxation $\langle V, D, C' \rangle$ has a non empty arc-consistent closure and a valuation lower than α?

Note that this problem corresponds also to the verification that the VCSP is at least α-arc-consistent.

Theorem 31. MAX-AC-CSP *is strongly* NP-*complete.* □

The proof is given in appendix A. Therefore, extending *Really Full Look Ahead* seems difficult since computing the lower bound itself is NP-complete. Furthermore, this also proves that the extension of the arc-consistency property to strictly monotonic VCSP such that for any $\forall a, b \in E, a, b \prec \top, (a \circledast b) \prec \top$ looses the quality of being a polynomial time checkable property (if P$\neq$NP).

For idempotent VCSP, this bound may be computed using polynomial time algorithms for enforcing arc-consistency in Fuzzy or Possibilistic CSPs [25, 27, 31] or equivalently the arc-consistency algorithm defined for SCSP, which works fine for idempotent operators: we first apply the extended arc-consistent enforcing algorithm which yields an equivalent problem and then compute the maximum[11] on all variables v_i of the minimum on all values $\nu \in d_i$ of the valuation of the assignment $\{v_i = \nu\}$ in this problem.

3.7 Experiments

The *Forward-Checking* algorithm has been coded and applied to random VCSP generated as follows: a classical random CSP with 16 variables and domains of size 9 is generated as in [13]. A first possibilistic VCSP is obtained by randomly assigning a valuation $\frac{1}{4}, \frac{1}{2}, \frac{3}{4}$ or 1 to each constraint. A lexicographic VCSP is then built simply by using the transformation from possibilistic to lexicographic CSP described in Sect. 3.3. This VCSP is a strong refinement of the original possibilistic CSP.

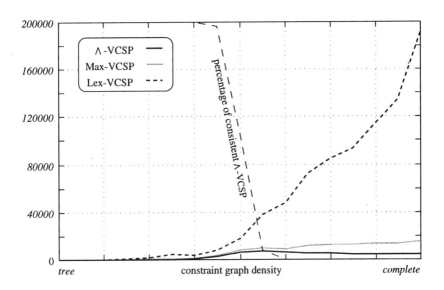

Fig. 1. Number of cc for $\wedge$, max and lex. VCSP

Because of limited space, we only report mean number of constraint checks performed to find an optimal assignment and prove optimality for a slice of

---

[11] Remember that the operator $\circledast$ is necessarily equal to max when it is idempotent.

the random CSP space (see Fig. 1): constraint satisfiability is fixed at 60% and the constraint graph goes from tree structured CSP to a complete graph. At each point 50 classical, possibilistic and corresponding lexicographic CSP are solved with the following heuristics: the variable which minimizes the ratio domain/degree is chosen, the value that minimizes the valuation $B(\nu, x_i)$ (defined in the forward-checking based algorithm) is chosen.

A first conclusion is that solving consistent CSP as VCSP, i.e., uselessly trying to anticipate a possible inconsistency, is relatively inexpensive, even for Lexicographic CSP. On inconsistent CSP, possibilistic CSP are not much harder than classical CSP, but the phase transition is apparently extended to the left. Last, but not least, lexicographic CSP are much more difficult which again shows the computational complexity of strictly monotonic ⊛: rationality seems expensive. Stronger argument could probably be obtained using recent developments in complexity theory, the transformations of Sect. 3.4 defining *metric reductions* between optimization problems [16].

4 Comparison

In this section we will compare the two approaches described above in this paper. In particular, we will show that, if one assumes a total order, there is a way to pass from any SCSP problem to an equivalent VCSP problem, and vice-versa. We also define normal forms both for VSCPs and SCSPs and we discuss their relationship with the transformation functions (from VCSPs to SCSPs and vice-versa).

4.1 From SCSPs to VCSPs

We will consider SCSPs $\langle C, con \rangle$ where *con* involves all variables. Thus we will omit it in the following. A SCSP is thus just a set of constraints C over a constraint system $\langle S, D, V \rangle$, where S is a c-semiring $S = \langle A, +, \times, \mathbf{0}, \mathbf{1} \rangle$, and D is the domain of the variables in V. Moreover, we will assume that the $+$ operation is always *min*, or, in other words, that $\leq_S$ is a total order.

Given a SCSP, we will now show how to obtain a corresponding VCSP, where by correspondence we mean that they associate the same value with each variable assignment (and thus, since they both use the *min* operation, they have the same solution).

Definition 32 (from SCSPs to VCSPs). Given a SCSP P with constraints C over a constraint system $\langle S, D, V \rangle$, where S is a c-semiring $S = \langle A, +, \times, \mathbf{0}, \mathbf{1} \rangle$, we obtain the VCSP $P' = \langle V, D, C', S', \varphi \rangle$, where $S' = \langle E, \circledast, \succ \rangle$, where $E = A$, $\circledast = \times$, and $\prec = <_S$.

For each constraint $c = \langle def, con \rangle \in C$, we obtain a set of constraints $c'_1, \ldots, c'_k$, where k is the cardinality of the range of *def*. That is, $k = |\{a \text{ s.t. } \exists t \text{ with } def(t) = a\}|$. Let us call $a_1, \ldots, a_k$ such values, and let us call T_i the set of all tuples t such that $def(t) = a_i$. All the constraints c_i involve the same variables, which are those involved in c. Then, for each $i = 1, \ldots, k$, we set $\varphi(c'_i) = a_k$, and we define c_i' in such a way that the tuples allowed are those not in T_i. We will write $P' = sv(P)$. Note that, by construction, each tuple t is allowed by all constraints $c_1, \ldots, c_k$ except the constraint c_i such that $\varphi(c_i) = def(t)$. □

Example 1: Consider a SCSP which contains the constraint $c = \langle con, def \rangle$, where $con = \{x, y\}$, and $def(\langle a, a \rangle) = l_1$, $def(\langle a, b \rangle) = l_2$, $def(\langle b, a \rangle) = l_3$, $def(\langle b, b \rangle) = l_1$. Then, the corresponding VCSP will contain the following three constraints, all involving x and y:

- c_1, with $\varphi(c_1) = l_1$ and allowed tuples $\langle a, b \rangle$ and $\langle b, a \rangle$;
- c_2, with $\varphi(c_2) = l_2$ and allowed tuples $\langle a, a \rangle$, $\langle b, a \rangle$ and $\langle b, b \rangle$;
- c_3, with $\varphi(c_3) = l_3$ and allowed tuples $\langle a, a \rangle$, $\langle a, b \rangle$ and $\langle b, b \rangle$.

Figure 2 shows both c and the three constraints c_1, c_2, and c_3. □

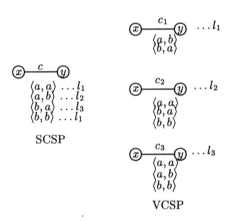

Fig. 2. From SCSP to VCSP.

First we need to make sure that the structure we obtain via the definition above is indeed a valued CSP. Then we will prove that it is equivalent to the given SCSP.

Theorem 33 (from c-semiring to valuation structure). *If we consider a c-semiring $S = \langle A, +, \times, \mathbf{0}, \mathbf{1} \rangle$ and the structure $S' = \langle E, \circledast, \succ \rangle$, where $E = A$, $\circledast = \times$, and $\prec = <_S$ (obtained using the transformation in Definition 32), then S' is a valuation structure.*

Proof. First, $\succ$ is a total order, since we assumed that $\leq_S$ is total and that $\prec = <_S$. Moreover, $\top = \mathbf{0}$ and $\bot = \mathbf{1}$, from the definition of $\leq_S$. Then, since $\circledast$ coincides with $\times$, it is easy to see that it is commutative, associative, monotone, and closed. Moreover, $\mathbf{1}$ (that is, $\bot$) is its unit element and $\mathbf{0}$ (that is, $\top$) is its absorbing element. □

Theorem 34 (equivalence between SCSP and VCSP). *Consider a SCSP problem P and the corresponding VCSP problem P'. Consider also an assignment t to all the variables of P, with associated value $a \in A$. Then $\mathcal{V}_{P'}(t) = a$.*

Proof. Note first that P and P' have the same set of variables. In P, the value of t is obtained by multiplying the values associated with each subtuple of t, one for each constraint of P. Thus, $a = \prod \{def_c(t_c)$, for all constraints $c = \langle def_c, con_c \rangle$ in C and such that t_c is the projection of t over the variables of $c\}$. Now, $\mathcal{V}_{P'}(t) = \prod \{\varphi(c')$ for all $c' \in C'$ such that the projection of t over the variables of c' violates $c'\}$. It is easy to see that the number of values multiplied is this formula coincides with the number of constraints in C, since, as noted above, each tuple violates only one of the constraints in C' which have been generated because of the presence of the constraint $c \in C$. Thus we just have to show that, for each $c \in C$, $def_c(t_c) = \varphi(c_i)$, where c_i is the constraint violated by t_c. But this is easy to show by what we have noted above. In fact, we have defined the translation from SCSP to VCSP in such a way that the only constraint of the VCSP violated by a tuple is exactly the one whose valuation coincides with the value associated with the tuple in the SCSP. □

Note that SCSPs which do not satisfy the restrictions imposed at the beginning of the section, that is, that *con* involves all the variables and that $\leq_S$ is total, do not have a corresponding VCSP.

Corollary 35 (same solution). *Consider a SCSP problem P and the corresponding VCSP problem P'. Then P and P' have the same solution.*

Proof. It follows from Theorem 34, from the fact that both problems use the min operation, and that a solution is just one of the total assignments. □

4.2 From VCSP to SCSP

Here we will define the opposite translation, which allows one to get a SCSP from a given VCSP.

Definition 36 (from VCSP to SCSP). Given the VCSP $P = \langle V, D, C, S, \varphi \rangle$, where $S = \langle E, \circledast, \succ \rangle$, we will obtain the SCSP P' with constraints C' over the constraint system $\langle S', D, V \rangle$, where S' is the c-semiring $\langle E, +, \circledast, \top, \bot \rangle$, and $+$ is such that $a + b = a$ iff $a \preccurlyeq b$. It is easy to see that $\leq_S = \preccurlyeq$. For each constraint $c \in C$ with allowed set of tuples T, we define the corresponding constraint $c' = \langle con', def' \rangle \in C'$ such that con' contains all the variables involved in c and, for each tuple $t \in T$, $def'(t) = \bot$, otherwise $def'(t) = \varphi(c)$. We will write $P' = vs(P)$. □

Example 2: Consider a VCSP which contains a binary constraint c connecting variables x and y, for which it allows the pairs $\langle a, b \rangle$ and $\langle b, a \rangle$, and such that $\varphi(c) = l$. Then, the corresponding SCSP will contain the constraint $c' =$

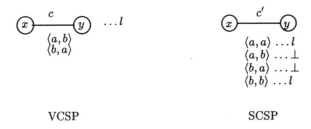

VCSP   SCSP

Fig. 3. From VCSP to SCSP.

$\langle con, def \rangle$, where $con = \{x, y\}$, $def(\langle a, b \rangle) = def(\langle b, a \rangle) = \bot$, and $def(\langle a, a \rangle) = def(\langle b, b \rangle) = l$. Figure 3 shows both c and the corresponding c'. □

Again, we need to make sure that the structure we obtain via the definition above is indeed a semiring-based CSP. Then we will prove that it is equivalent to the given VCSP.

Theorem 37 (from valuation structure to c-semiring). *If we consider a valuation structure $\langle E, \circledast, \succ \rangle$ and the structure $S = \langle E, +, \circledast, \top, \bot \rangle$, where $+$ is such that $a + b = a$ iff $a \preccurlyeq b$ (obtained using the transformation in Definition 36), then S is a c-semiring.*

Proof. Since $\succ$ is total, $+$ is closed. Moreover, $+$ is commutative by definition, and associative because of the transitivity of the total order $\succ$. Furthermore, $\mathbf{0}$ is the unit element of $+$, since it is the top element of $\succ$. Finally, $+$ is idempotent because of the reflexivity of $\succ$, and $\mathbf{1}$ is the absorbing element of $+$ since $\mathbf{1} = \bot$. Operation $\times$ of S coincides with $\circledast$. Thus it is closed, associative, and commutative, since $\circledast$ is Also, $\top$ is its absorbing element and $\bot$ is its identity (from corresponding properties of $\circledast$). The distributivity of $\circledast$ over $+$ can easily be proved. For example, consider $a, b, c \in E$, and assume $b \preccurlyeq c$. Then $a \circledast (b+c) = a \circledast b$ (by definition of $+$) $= (a \circledast b) + (a \circledast c)$ (by the definition of $+$ and the monotonicity of $\circledast$). The same reasoning applies to the case where $c \preccurlyeq b$. □

Theorem 38 (equivalence between VCSP and SCSP). *Consider a VCSP problem P and the corresponding SCSP problem P'. Consider also an assignment t to all the variables of P. The value associated with such an assignment is $A = \mathcal{V}_P(t) = \circledast\{\varphi(c)$ for all $c \in C$ such that the projection of t over the variables of c violates $c\}$. Instead, the value associated with the same assignment in P' is $B = \circledast\{def_{c'}(t_{c'})$, for all constraints $c' = \langle def_{c'}, con_{c'} \rangle$ in C' and such that $t_{c'}$ is the projection of t over the variables of $c'\}$. Then, $A = B$.*

Proof. The values multiplied to produce A are as many as the constraints violated by t; however, the values multiplied to produce B are as many as the constraints in C'. However, by construction, each tuple t_c involving the variables of a constraint $c \in C$ has been associated, in P', with a value which is either $\varphi(c)$ (if t_c violates c), or $\bot$ (if t_c satisfies c). Thus the contribution of t_c to the value of B is important only if t_c violated c in P, because $\bot$ is the unit element for $\circledast$. Thus A and B are obtained by the same number of significant values. Now we have to show that such values are the same. But this is easy, since we have defined the translation in such a way that each tuple for the variables of c is associated with the value $\varphi(c)$ exactly when it violates c. $\quad\square$

4.3 Normal Forms and Equivalences

Note that, while passing from an SCSP to a VCSP the number of constraints in general increases, in the opposite direction the number of constraints remains the same. This can also be seen in Example 1 and 2. This means that, in general, going from a SCSP P to a VCSP P' and then from the VCSP P' to the SCSP P'', we do not get $P = P''$. In fact, for each constraint c in P, P'' will have in general several constraints $c_1, \ldots, c_k$ over the same variables as c. However, it is easy to see that $c_1 \otimes \cdots \otimes c_k = c$, and thus P and P'' associate the same value with each variable assignment.

Example 3: Figure 4 shows how to pass from a SCSP to the corresponding VCSP (this part is the same as in Example 1), and then again to the corresponding SCSP. Note that the starting SCSP and the final one are not the same. In fact, the latter has three constraints between variables x and y, while the former has only one constraint. However, one can see that the combination of the three constraints yields the starting constraint. $\quad\square$

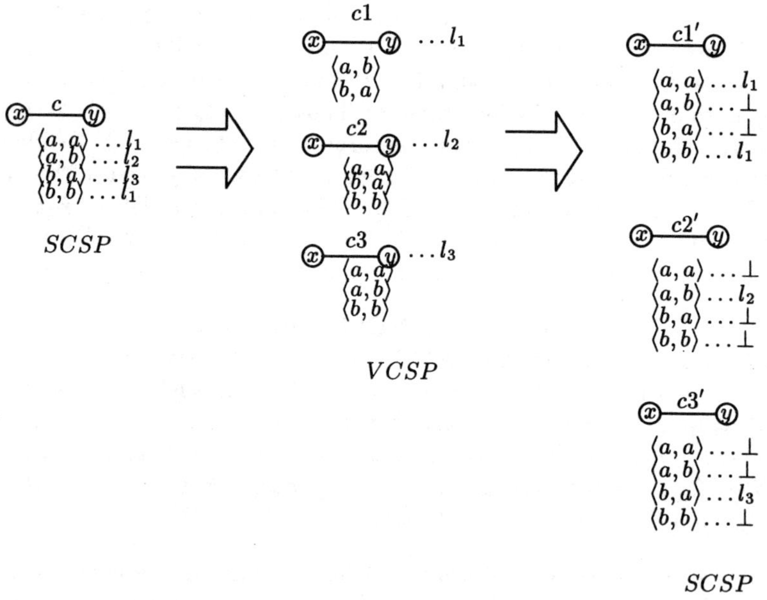

Fig. 4. From SCSP to VCSP and back to SCSP again.

Consider now the opposite cycle, that is, going from a VCSP P to a SCSP

P' and then from P' to a VCSP P''. In this case, for each constraint c in P, P'' has two constraints: one is c itself, and the other one is a constraint with associated value $\bot$. This means that violating such a constraint has cost $\bot$, which, in other words, means that this constraint can be eliminated without changing the behavior of P'' at all.

Example 4: Figure 5 shows how to pass from a VCSP to the corresponding SCSP (this part is the same as in Example 2), and then again to the corresponding VCSP. Note that the starting VCSP and the final one are not the same. In fact, the latter one has two constraints between variables x and y. One is the same as the one in the starting VCSP, while the other one has associated the value $\bot$. This means that violating such constraint yields a cost of value $\bot$. □

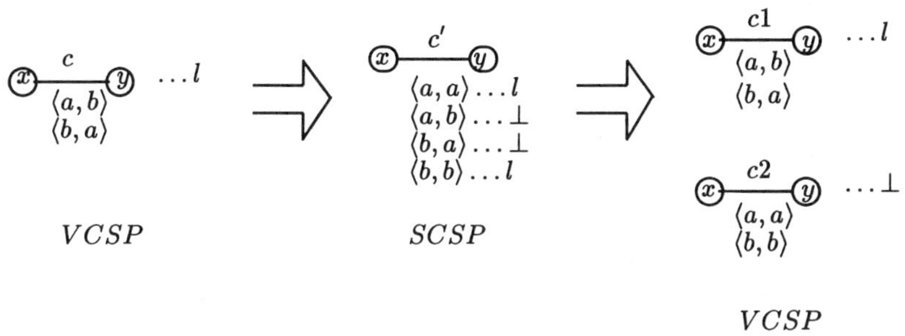

Fig. 5. From VCSP to SCSP, and to VCSP again.

Let us define now normal forms for both SCSPs and VCSPs, as follows. For each VCSP P, its normal form is the VCSP $P' = nfv(P)$ which is obtained by deleting all constraints c such that $\varphi(c) = \bot$. It is easy to see that P and P' are equivalent.

Definition 39 (normal forms for VCSPs). Consider $P = \langle V, D, C, S, \varphi \rangle$, a VCSP where $S = \langle E, \circledast, \succ \rangle$. Then P is said to be in normal form if there is no $c \in C$ such that $\varphi(c) = \bot$. If P in not in normal form, then it is possible to obtain a unique VCSP $P' = \langle V, D, C - \{c \in C \mid \varphi(c) = \bot\}, S, \varphi \rangle$, denoted by $P' = nfv(P)$, which is in normal form. □

Theorem 40 (normal form). *For any VCSP P, P and $nfv(P)$ are equivalent.*

Proof. The theorem follows from the fact that $\forall a, (\bot \circledast a) = a$ and from the definitions of $\mathcal{V}_P(A)$ and $\mathcal{V}_{P'}(A)$, □

Also, for each SCSP P, its normal form is the SCSP $P' = nfs(P)$ which is obtained by combining all constraints involving the same set of variables. Again, this is an equivalent SCSP.

Definition 41 (normal form for SCSPs). Consider any SCSP P with constraints C over a constraint system $\langle S, D, V \rangle$, where S is a c-semiring $S = \langle A, +, \times, \mathbf{0}, \mathbf{1} \rangle$. Then, P is in normal form if, for each subset W of V, there is at most one constraint $c = \langle def, con \rangle \in C$ such that $con = W$. If P is not in normal form, then it is possible to obtain a unique SCSP P', as follows. For each $W \subseteq V$, consider the set $C_W \subseteq C$ which contains all the constraints involving W. Assume $C_W = \{c_1, \ldots, c_n\}$. Then, replace C_W with the single constraint $c = \bigotimes C_W$. P', denoted by $nfs(P)$, is in normal form. $\qquad \square$

Theorem 42 (normal forms). *For any SCSP P, P and $nfs(P)$ are equivalent.*

Proof. It follows from the associative property of $\times$. $\qquad \square$

Even though, as noted above, the transformation from a SCSP P to the corresponding VCSP P' and then again to the corresponding SCSP P'' does not necessarily yield $P = P''$, we will now prove that there is a strong relationship between P and P''. In particular, we will prove that the normal forms of P and P'' coincide. The same holds for the other cycle, where one passes from a VCSP to a SCSP and then to a VCSP again.

Theorem 43 (same normal form 1). *Given any SCSP problem P and the corresponding VCSP $P' = sv(P)$, consider the SCSP P'' corresponding to P', that is, $P'' = vs(P')$. Thus $nfs(P) = nfs(P'')$.*

Proof. We will consider one constraint at a time. Take any constraint c of P. With the first transformation (to the VCSP P'), we get as many constraints as the different values associated with the tuples in c. Each of the constraints, say c_i, is such that $\varphi(c_i)$ is equal to one of such values, say l_i, and allows all tuples which do not have value l_i in c. With the second transformation (to the SCSP P''), for each of the c_i, we get a constraint c_i', where tuples which are allowed by c_i have value $\bot$ while the others have value l_i. Now, if we apply the normal form to P'', we combine all the constraints c_i', getting one constraint which is the same as c, since, given any tuple t, it is easy to see that t is forbidden by exactly one of the c_i. Thus the combination of all c_i' will associate with t a value which is the one associated with the unique c_i which does not allow t. $\qquad \square$

Theorem 44 (same normal form 2). *Given any VCSP problem P and the corresponding SCSP $P' = vs(P)$, consider the VCSP P'' corresponding to P', that is, $P'' = sv(P')$. Then we have that $nfv(P) = nfv(P'')$.*

Proof. We will consider one constraint at a time. Take any constraint c in P, and assume that $\varphi(c) = l$ and that c allows the set of tuples T. With the first transformation (to the SCSP P'), we get a corresponding constraint c' where tuples in T have value $\bot$ and tuples not in T have value l. With the second transformation (to the VCSP P''), we get two constraints: c_1, with $\varphi(c_1) = \bot$, and c_2, with $\varphi(c_2) = l$ and which allows the tuples of c' with value $\bot$. It is easy to see that $c_2 = c$. Now, if we apply the normal form to both P and P'', which implies the deletion of all constraints with value $\perp$, we get exactly the same constraint. This reasoning applies even if the starting constraint has value $\perp$. In fact, in this case the first transformation will give us a constraint where all tuples have value $\perp$, and the second one gives us a constraint with value $\perp$, which will be deleted when obtaining the normal form. $\square$

The statements of the above two theorems can be summarized by the following two diagrams. Note that in such diagrams each arrow represents one of the transformations defined above, and all problems in the same diagram are equivalent (by the theorems proved previously in this section).

$$
\begin{array}{ccc}
VCSP & \xrightarrow{\;vs\;} & SCSP \\
\Big\downarrow{\scriptstyle nfv} & & \Big\downarrow{\scriptstyle sv} \\
VCSP & \xleftarrow[\;nfv\;]{} & VCSP
\end{array}
$$

$$
\begin{array}{ccc}
SCSP & \xrightarrow{\;sv\;} & VCSP \\
\Big\downarrow{\scriptstyle nfs} & & \Big\downarrow{\scriptstyle vs} \\
SCSP & \xleftarrow[\;nfs\;]{} & SCSP
\end{array}
$$

5 Conclusions and Future Work

When we compare the SCSP framework to the VCSP framework, the most striking difference lies in the SCSP framework's ability to represent partial orders whereas the results and algorithms defined in the VCSP framework exploit totally ordered sets of valuations. The ability to represent partial orders seems very interesting for multi-criteria optimization, for example, since the product of two or more c-semirings yields a c-semiring which in general defines a partial order (see last part of Sect. 2.4).

However, it appears that apart from this difference, the assumption of total order gives the two frameworks the same theoretical expressive power. Since SCSPs associate values (that is, costs, or levels of preferences, or else) with tuples, while VCSPs associate these values with constraints, there is a difference in the actual ease of use of each framework. Although it would seem easier in general to use VCSPs, since a problem has less constraints than tuples, in some situations this could lead to a large number of constraints (see the transformation in Sect. 4). But this is more a matter of implementation than a theoretical limitation: it is easy, as Sect. 4 shows, to extend the VCSP framework to the case where valuations are associated with tuples instead of constraints and conversely, it is easy to restrict the SCSP framework to the case where constraints are annotated instead of tuples.

What we get are complementary results. In the SCSP framework, we observe that idempotency of the combination operator guarantees that k-consistency algorithms work (in totally ordered SCSP or equivalently VCSP, the only structure with an idempotent combination operator is the Fuzzy CSP instance). We get the same result in the VCSP framework for arc-consistency, but we also show that strict monotonicity is a dreadful property that turns arc-consistency checking in an NP-complete problem and which also defines harder optimization problems, both from the theoretical and practical point of view.

To solve these difficult problems, we are looking for better lower bounds that could be used in Branch and Bound algorithms to solve {S,V}CSP more efficiently.

Another algorithmic approach of {S,V}CSP which is worth considering, and which is also able to cope with non idempotent combination operators, is the dynamic programming approach, which is especially suited to tree-structured problems.

Finally, we plan to study the possibility of embedding one of the two frameworks (or a merge of them) in the constraint logic programming (CLP) paradigm (see [15]). The presence of such a general framework for constraint solving within CLP would facilitate the use of new or additional constraint systems in the language, and also would allow for a new concept of logic programming, where the relationship between goals is not regulated by the usual *and* and *or* logical connectives, but instead by general $+$ and $\times$ operators. An important issue will be the semantics of the languages defined. This issue has been already addressed for WCSPs [1].

A   Proof of NP-completeness of MAX-AC-CSP

The problem belongs to NP since computing the arc-consistent closure of a CSP can be done in polynomial time and we supposed that $\circledast$ and $\succ$ are polynomial in the size of their arguments.

To prove completeness, we use a polynomial transformation from the NP-complete problem MAX2SAT [12]. An instance of MAX2SAT is defined by a set of n propositional variables $L = \{\ell_1, \ldots, \ell_n\}$, a set Φ of m 2-clauses on L and a positive integer $k \leq m$. The problem is to prove the existence of a truth assignment of L that satisfies at least k clauses of Φ. The problem is known to be strongly NP-complete [12].

Given an instance of MAX2SAT, we built an instance of MAX-AC-CSP defined by a binary CSP (V, D, C). The set V of the CSP variables contains $n+2n.(m+1)$ variables:

1. the first n variables $v_1, \ldots, v_n$ correspond to the n propositional variables of L and have a domain of cardinality two corresponding to the boolean values t and f;

2. the next $2n.(e+1)$ variables will have a domain of cardinality one, containing only the value $\bigstar$. This set of variables is composed of n sets V_i, $1 \leq i \leq n$, of $2e + 2$ variables: $V_i = \{v_{i,1}^t, \ldots v_{i,e+1}^t, v_{i,1}^f, \ldots, v_{i,e+1}^f\}$. Each set V_i is associated with the original variable v_i previously defined.

The constraints that appear in C are composed of three sets:

1. the set C_u is the direct translation of the m 2-clauses of the MAX2SAT instance as CSP constraints. A 2-clause $\phi \in \Phi$ that involves ℓ_i and ℓ_j will be translated to a constraint that involves v_i and v_j and whose relation contains all the truth assignments of $\{\ell_i, \ell_j\}$ that satisfy ϕ;
2. the set C^t contains, for each variable v_i, $1 \leq i \leq n$, and for each variable $v_{i,j}^t$, $1 \leq j \leq m+1$, a constraint involving v_i and $v_{i,j}^t$ authorizing only the pair $(t, \circledast)$;
3. the set C^f contains, for each variable v_i, $1 \leq i \leq n$, and for each variable $v_{i,j}^f$, $1 \leq j \leq m+1$, a constraint involving v_i and $v_{i,j}^f$ authorizing only the pair $(f, \circledast)$;

There is a total of $2n.(m+1) + m$ constraints. All the constraints are annotated with the valuation α, different from $\top$ and $\bot$. For example, Fig. 6 illustrates the micro-structure of the CSP built from the MAX2SAT instance defined by $\Phi = \{v_1 \vee v_2, v_2 \vee v_3, v_1 \vee v_3\}$.

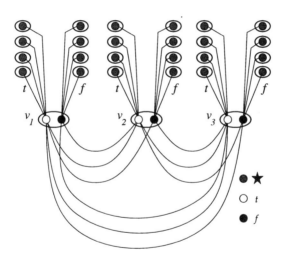

Fig. 6. The micro-structure of an example CSP

The CSP contains $O(m^2)$ constraints, $O(m^2)$ variables and domains of cardinality $O(1)$ and therefore the transformation is clearly polynomial. We shall now prove that the existence of a truth assignment that satisfies at least k clauses of Φ is equivalent to the existence of a relaxation of the VCSP which is non arc-inconsistent and whose valuation is lower than $(\alpha \circledast \cdots \circledast \alpha)$ with $n.(m+1) + (m-k)$ occurrences of α.

We first remark that the CSP (V, D, C) is not arc-consistent: the values t of the variables v_i, $1 \leq i \leq n$, have no support on each of the $m+1$ constraints of C^f that connect v_i to $v_{i,j}^f$, $1 \leq j \leq m+1$. Similarly, the values f of the variables

v_i, $1 \leq i \leq n$, have no support on each of the $m+1$ constraints of C^t that connect v_i to $v_{i,j}^t$, $1 \leq j \leq m+1$. Therefore, a relaxation (V, D, C'), $C' \subset C$, with a non-empty arc-consistent closure will never simultaneously contain constraints from C^t and C^f involving the same variable v_i, $1 \leq i \leq n$. Let us consider a truth assignment ω of the variables of L that satisfies more than k clauses among the m clauses of Φ. Then the following relaxation (V, D, C'), with $|C'| = n.(m+1) + k$ can be defined:

1. we select in C^u all the constraints that correspond to clauses of Φ satisfied by ω;
2. for each propositional variable ℓ_i assigned to t in ω, we also select in C^t the $M + 1$ constraints involving v_i;
3. similarly, for each propositional variable ℓ_i assigned to f in ω, we select in C^f the $M + 1$ constraints involving v_i;

Since $|C'| = n.(m+1) + k$, there are precisely $[2n(m+1) + m] - [n.(m+1) + k] = n(m+1) + (m-k)$ constraints which are rejected, each constraints being annotated with the valuation α. The relaxation has therefore, as we wanted, the valuation $(\alpha \circledast \cdots \circledast \alpha)$ with $n.(m+1) + (m-k)$ occurrences of α. The arc-consistent closure of the CSP defined is non-empty and is obtained by removing of the domain d_i of v_i, $1 \leq i \leq n$, the value t if ℓ_i is assigned to f in ω (because this value is not supported by the constraints in C^f which have been selected) or else the value f (which, in this case, is not supported by the constraint of C^t, which have been selected). The CSP obtained is arc-consistent since all domain have cardinality 1 and the values that appear in the domains satisfy all the constraints selected in C^u, C^t and C^f.

Conversely, if we suppose that there exists a relaxation (V, D, C'), $C' \subset C$, $|C'| \geq n.(m+1) + k$ with a non empty arc-consistent closure. We first show that for all v_i, $1 \leq i \leq n$, at least one constraint among the $2(m+1)$ constraints of C^t and C^f involving v_i belongs to C': let us suppose that for a variable v_j, $1 \leq j \leq n$, no constraint from C^t or C^f involving x_j belongs to C'. Since, as we previously remarked, a maximum of $m + 1$ constraints involving v_i, $1 \leq i \leq n$ can be selected in C^t and C^f without loosing the existence of a non-empty arc-consistent closure, a maximum of $(n-1).(m+1)$ constraints of C' may be selected from C^t and C^f. Since C' can not select more than m constraints in C_u and $(n-1).(M+1) + m < n.(m+1) + k$ since $k > 0$, we get a contradiction. Since no more than $n.(m+1)$ constraints of C^t and C^f appear in C', there is at least k constraints from C^u which appear in C' in order to reach the $n.(m+1) + k$ constraints. Each variable being involved in at least one of the constraints from C^t and C^f, the variables from the arc-consistent closure of the CSP have exactly one value in their domain et these value necessarily satisfy the constraints of C^u since we are considering the arc-consistent closure. Therefore, the truth assignment ω of the variables ℓ_i defined by the assignment of the corresponding variables v_i in this arc-consistent closure satisfy more than k clauses of Φ. This finally shows that MAX2SAT $\propto$ MAX-AC-CSP. $\quad\square$

References

1. S. Bistarelli. Programmazione con vincoli pesati e ottimizzazione (in italian). Dipartimento di Informatica, Università di Pisa, Italy, 1994.
2. S. Bistarelli, U. Montanari, and F. Rossi. Constraint Solving over Semirings. In *Proc. IJCAI95*. Morgan Kaufman, 1995.
3. A. Borning, M. Maher, A. Martindale, and M. Wilson. Constraint hierarchies and logic programming. In Martelli M. Levi G., editor, *Proc. 6th ICLP*. MIT Press, 1989.
4. R. Dechter, A. Dechter, and J. Pearl. Optimization in constraint networks. In R.M Oliver and J.Q. Smith, editors, *Influence Diagrams, Belief Nets and Decision Analysis*, chapter 18, pages 411–425. John Wiley & Sons Ltd., 1990.
5. D. Dubois and H. Prade. A class of fuzzy measures based on triangular norms. a general framework for the combination of uncertain information. *Int. Journal of Intelligent Systems*, 8(1):43–61, 1982.
6. D. Dubois, H. Fargier, and H. Prade. The calculus of fuzzy restrictions as a basis for flexible constraint satisfaction. In *Proc. IEEE Int. Conf. on Fuzzy Systems*. IEEE, 1993.
7. H. Fargier and J. Lang. Uncertainty in constraint satisfaction problems: a probabilistic approach. *Proc. ECSQARU*. Springer-Verlag, LNCS 747, 1993.
8. H. Fargier and J. Lang and T. Schiex. Selecting Preferred Solutions in Fuzzy Constraint Satisfaction Problems. *Proc. 1st European Congress on Fuzzy and Intelligent Technologies* (EUFIT). 1993.
9. E. Freuder. Synthesizing constraint expressions. *CACM*, 21(11), 1978.
10. E. Freuder. Backtrack-free and backtrack-bounded search. In Kanal and Kumar, editors, *Search in Artificial Intelligence*. Springer-Verlag, 1988.
11. E. Freuder and R. Wallace. Partial constraint satisfaction. *AI Journal*, 58, 1992.
12. M. Garey, D. Johnson, and L. Stockmeyer. Some simplified NP-complete graph problems. *Theoretical Computer Science*, 1:237–267, 1976.
13. P. Hubbe and E. Freuder. An efficient cross-product representation of the constraint satisfaction problem search space. In *Proc. of AAAI-92*, pages 421–427, San Jose, CA, 1992.
14. V. Kumar. Algorithms for constraint satisfaction problems: a survey. *AI Magazine*, 13(1), 1992.
15. J. Jaffar and J.L. Lassez. Constraint Logic Programming. Proc. POPL, ACM, 1987.
16. M. Krentel. The complexity of optimization problems. *Journal of Computer and System Sciences*, 36:490–509, 1988.
17. A. Mackworth. Consistency in networks of relations. *AI Journal*, 8(1), 1977.
18. A. Mackworth. *Encyclopedia of AI*, chapter Constraint Satisfaction, pages 205–211. Springer Verlag, 1988.
19. A. Mackworth and E. Freuder. The complexity of some polynomial network consistency algorithms for constraint satisfaction problems. *AI Journal*, 25, 1985.
20. U. Montanari. Networks of constraints: Fundamental properties and application to picture processing. *Information Science*, 7, 1974.

21. U. Montanari and F. Rossi. Constraint relaxation may be perfect. *AI Journal*, 48:143–170, 1991.

22. H. Moulin. *Axioms for Cooperative Decision Making*. Cambridge University Press, 1988.

23. B. A. Nadel. Constraint satisfaction algorithms. *Comput. Intell.*, 5(4):188–224, November 1989.

24. C. M. Papadimitriou. *Computational Complexity*. Addison-Wesley Publishing Company. 1994.

25. A. Rosenfeld, R. Hummel, S. Zucker. Scene Labelling by Relaxation Operations. *IEEE Trans. on Sys., Man, and Cyb.*, vol. 6. n.6, 1976.

26. Z. Ruttkay. Fuzzy constraint satisfaction. *Proc. 3rd Int. Conf. on Fuzzy Systems*, 1994.

27. T. Schiex. Possibilistic constraint satisfaction problems, or "How to handle soft constraints?". *Proc. 8th Conf. of Uncertainty in AI*, 1992.

28. T. Schiex, H. Fargier, and G. Verfaillie. Valued Constraint Satisfaction Problems: Hard and Easy Problems. In *Proc. IJCAI95*. Morgan Kaufmann, 1995.

29. G. Shafer. An axiomatic study of computation in hypertrees. Working paper 232, University of Kansas, School of Business, Lawrence, 1991.

30. L. Shapiro and R. Haralick. Structural descriptions and inexact matching. *IEEE Transactions on Pattern Analysis and Machine Intelligence*, 3:504–519, 1981.

31. P. Snow and E. Freuder. Improved relaxation and search methods for approximate constraint satisfaction with a maximin criterion. In *Proc. of the 8th biennal conf. of the canadian society for comput. studies of intelligence*, pages 227–230, May 1990.

32. L. Zadeh. Calculus of fuzzy restrictions. In K. Tanaka L.A. Zadeh, K.S. Fu and M. Shimura, editors, *Fuzzy sets and their applications to cognitive and decision processes*. Academic Press, 1975.

Defeasible Constraint Solving

Francisco Menezes and Pedro Barahona

Departamento de Informática, Universidade Nova de Lisboa,
2825 Monte da Caparica, PORTUGAL

Abstract. Hierarchical Constraint Solving has been proposed as an adequate scheme to specify over-constrained problems where some of the constraints might remain unsatisfied. However it is often not clear which criteria should be adopted to select the "best" combination of constraints to be relaxed. In previous work we proposed IHCS - an Incremental Hierarchical Constraint Solver - but only with a single criterium. This paper presents IHCS as a general scheme to incrementally handle a hierarchy of constraints for a class of comparators using different criteria. The scheme is further extended with a *satisfaction mode* - given a threshold, this mode enables the search of "satisfactory" solutions to large problems whose optimization is not possible in acceptable time. This scheme can be integrated with different programming environments. In particular, we have integrated it with Prolog to produce an instance of an HCLP language. Because of its portability and incremental nature, IHCS is well suited for reactive systems, allowing the interactive introduction and removal of preferred constraints, illustrated in the examples presented in the paper.

1 Introduction

In many constraint satisfaction applications it is useful to use overconstrained specifications, namely because it enables a more declarative programming style. This requires a defeasible constraint solver which is able to relax some of the constraints in order to solve the problem, the selection of the relaxed constraints being made according to some preference criterium.

This is the case of IHCS, an Incremental Hierarchical Constraint Solver that we have been developing to solve a hierarchy of constraints, i.e., a constraint network where each constraint is assigned a weight or level denoting its relevance to the problem [12]. The higher the weight, the less required is a constraint (weight 0 corresponds to required or mandatory constraints). A comparator defines preferred configurations of the hierarchy, i.e. which constraints are relaxed in the problems solutions, based on the constraint weights. Moreover, it is often not clear which criteria should be adopted to select the "best" combination of constraints to be relaxed, and the choice of a particular criterium is usually application dependent. IHCS has thus been extended in order to use a number of comparators using different criteria, namely *Global-Predicate-Better* and *Global-Weight-Better*.

Since finding optimal solutions to real Constraint Satisfaction Problems is often too difficult due to its intractable nature [6], it is more realistic to search instead non-optimal but simply satisfactory solutions. Therefore, we have also extended IHCS with a *satisfaction mode*, in order to allow users to handle problems that are just too difficult to be optimized in acceptable time. Such solutions may be a base for further refinements leading to better solutions to be found in an incremental and reactive, way.

In addition to the formalization of defeasible constraint solving, we address in this paper possible applications where this scheme can be used. Model-based reasoning and time-tabling are examples of areas where such applications are appropriate. There are many ways of modeling a system, namely a digital circuit [3]. Modeling it as a constraint network, and supplying this model to a constraint solver has been shown to be a declarative and efficient approach, namely to generate test patterns for combinatorial circuits [16]. Within our framework, we show that providing input and output observations made in a faulty circuit, a defeasible constraint solver is able to relax the constraints corresponding to the faulty gates.

In interactive diagnosis one is interested that the constraint solver, is also incremental. Given subsequent observations, discrepancies between observed and predicted behaviors lead to the removal or insertion of further constraints to the initial model of the circuit. It is thus important that the initial diagnostic hypotheses are revised with minimum computing effort, undoing and redoing as little work as possible.

This approach is closely related to the various non monotonic reasoning formalism which have been used in model based diagnosis [13]. Some of these formalism allow the specification of preferences [15, 7], but they are usually non incremental.

An interesting example of interactive behavior is provided by truth maintenance systems (TMS) [5, 4], which associate a theorem prover with a truth maintenance component, where the assumptions or justifications necessary to reach a conclusion are stored. For example in the ATMS [4], contradictions lead to the removal of some assumptions, and a propagation mechanism guarantees that all the dependent conclusions that lack the support of these assumptions are subsequently removed. However, these systems usually maintain all combinations of acceptable assumptions, rather then focussing in preferred combination of assumptions (i.e. preferred diagnoses).

IHCS is also interactive, since it copes with the incremental insertion or removal of constraints in a constraint network, changing pre-existing solutions with minimal re-computations. This is specially useful in time-tabling problems where not only many of the constraints are simple preferences but also their strength may vary during the interactive tuning of the timetable. The incremental nature of IHCS is an important feature that is explored in the integration that we made with an existing Prolog to create a HCLP($\mathcal{FD}, \mathcal{C}$) language [1], where $\mathcal{FD}$ stand for finite domains and $\mathcal{C}$ is a parameter that may be instantiated from a group of built-in comparators. By means of a special mode of the top-level interpreter, partial queries are accepted, i.e., after evaluating some (partial) goal, additional queries may be done using the answer substitutions and constraint stores computed so far. In these additional queries, new constraints may be added or old constraints removed in an incremental way.

This article is organized as follows. Some preliminary definitions about hierarchies are specified in Section 2. The formalization of IHCS by means of a set of transition rules, is described in Section 3. In section 4 we discuss different criteria that can be used to optimize an hierarchy of constraints, identify the kinds of comparators suitable for IHCS, and exemplify these with three different comparators. Section 5 introduces the satisfaction mode. The results obtained with different criteria and solving modes and the exploitation of the the reactive nature are illustrated with some time-table and model-based diagnosis examples in Section 6. Section 7 summarizes the paper and presents some final remarks.

2 Hierarchies and Configurations

A constraint hierarchy $\mathcal{H} = \langle \mathcal{V}, \mathcal{C}, \mathcal{D}om, \text{level}, D \rangle$, is a tuple where $\mathcal{V} = \{v_1, \ldots, v_n\}$ is the set of variables, $\mathcal{C} = \{c_1, \ldots, c_m\}$ the set of constraints, $\mathcal{D}om$ is a finite domain, level : $\mathcal{C} \longrightarrow \mathcal{N}$ is a mapping from constraints to their hierarchical levels, and function $D : \mathcal{V} \longrightarrow \wp(\mathcal{D}om)$ is a mapping from variables to their domains.

In our notation, c or c_i designates any constraint from $\mathcal{C}$. The index indicates the *introduction order* of the constraint in the hierarchy (if $j > i$ then c_j was introduced after c_i)[1]. The index is sometimes omitted in the discussion, whenever the introduction order of c is not relevant.

Definition 1. [Constraint Store] A *constraint store* S is a sequence of constraints from $\mathcal{C}$ ordered by introduction order ($\forall c_i, c_j \in S$, if $i < j$ then c_i precedes c_j in S). Set operations are closed for constraint stores, i.e., introduction order is always preserved.

Given any constraint store S, $S_{[i]}$, $S_{[<i]}$ and $S_{[>i]}$ designate the subsets of S containing all constraints of levels equal, lower and higher then i, respectively. Since $\mathcal{C}$ itself is a constraint store, $\mathcal{C}_{[0]}$ is the set of all required constraints and $\mathcal{C}_{[>0]}$ the set of all non-required constraints.

Furthermore, given any constraint store S, $S_{[IO<i]}$ and $S_{[IO>i]}$ designate the subsets of S containing all constraints whose introduction order (IO) are lower and higher then i, respectively (i.e., all constraints introducced before or after c_i, respectively).

Finally, we will express that a store S is consistent according to some network consistency condition X (e.g. $X = $ AC for Arc-Consistency[9]) by $S \not\vdash_X \bot$. This condition is not required to be a strong consistency condition such as pathconsistency. Although with weak conditions, such as arc-consistency (AC) for finite domains [9], global consistency is not guaranteed, inconsistencies are eventually detected when enough variables become instantiated - it is sufficient to interleave a search method, which assigns each uninstantiated variable with a

---

[1] Do not confuse the introduction order with the hierarchical level of a constraint

value from its domain (a labelling phase). This is a trade off, since AC may be implemented in polynomial time on the number of constraints for the general case [10], and in linear time for a number of important classes of constraints [17], while methods to verify strong K-consistency are exponential for $K > 2$, [8].

Definition 2. [Configuration] A *configuration* Φ of hierarchy $\mathcal{H}$ is a triple of constraint stores $\langle AS \bullet RS \bullet US \rangle$, where AS is the *Active Store*, RS the *Relaxed Store* and US the *Unexplored Store*, with the following properties:

1. $(AS \cup US) \subseteq C$ and $RS \subseteq C_{[>0]}$;
2. $AS \cap RS = \emptyset$ and $AS \cap US = \emptyset$ and $RS \cap US = \emptyset$.

Definition 3. [Total/Partial Configuration] Φ is a *total configuration* if $AS \cup RS \cup US = C$, otherwise it is a *partial configuration*.

A total/partial configuration may be seen as a state/sub-state of the evaluation of a hierarchy where the active store contains all active constraints (those that might have reduced some domains of its variables), the relaxed store is composed by the relaxed constraints (those that cannot be satisfied simultaneously with the active constraints) and the unexplored store is the set of candidates "queuing" for activation. In our notation, any index or prime character used in a configuration symbol, is also used in the correspondent constraint stores. For example, Φ_i or Φ' correspond respectively to the configurations $\langle AS_i \bullet RS_i \bullet US_i \rangle$ and $\langle AS' \bullet RS' \bullet US' \rangle$.

Operations on configurations are defined by the component-wise operation on their constraint stores. For example, $\Phi_i \cup \Phi_j = \langle AS_i \cup AS_j \bullet RS_i \cup RS_j \bullet US_i \cup US_j \rangle$ (the union of two configurations).

Definition 4. [Simplified Configuration] For each configuration Φ, there is a corresponding *simplified configuration* $|\Phi| = \langle NS \bullet RS \rangle$, where $NS = AS \cup US$ is the *Non-relaxed Store*.

Definition 5. [W(S)] Given $S \subseteq C$, let $W(S)$ be the set of all possible simplified configurations $\langle NS \bullet RS \rangle$ of $\mathcal{H}$ such that $NS \cup RS = S$, i.e,

$$\{\langle NS \bullet RS \rangle \mid NS \subseteq S \text{ and } RS \subseteq S_{[>0]} \text{ and} $$
$$NS \cup RS = S \text{ and } NS \cap RS = \emptyset\}$$

Simplified configurations will be used whenever the distinction between active and unexplored constraints is irrelevant, in the sense they are both intended to be active. Operations on simplified configurations are defined in a similar way to those on normal configurations. The word "simplified" will be dropped whenever the context makes it unnecessary. Example 1 shows a set of simplified configurations $W(S)$ which will be used later on.

Example 1. Given $S = \{c_1, c_2, c_3, c_4\}$ such that $S_{[1]} = \{c_1, c_2\}$ and $S_{[2]} = \{c_3, c_4\}$, then $W(S) = \{|\Phi_1|, |\Phi_2|, |\Phi_3|, |\Phi_4|, \cdots, |\Phi_{16}|\}$, where

$$|\Phi_1| = \langle \{c_1, c_2, c_3, c_4\} \bullet \emptyset \rangle \qquad |\Phi_2| = \langle \{c_1, c_2, c_3\} \bullet \{c_4\} \rangle$$
$$|\Phi_3| = \langle \{c_1, c_2, c_4\} \bullet \{c_3\} \rangle \qquad |\Phi_4| = \langle \{c_1, c_2\} \bullet \{c_3, c_4\} \rangle$$
$$|\Phi_5| = \langle \{c_1, c_3, c_4\} \bullet \{c_2\} \rangle \qquad |\Phi_6| = \langle \{c_2, c_3, c_4\} \bullet \{c_1\} \rangle$$
$$|\Phi_7| = \langle \{c_1, c_3\} \bullet \{c_2, c_4\} \rangle \qquad |\Phi_8| = \langle \{c_1, c_4\} \bullet \{c_2, c_3\} \rangle$$
$$|\Phi_9| = \langle \{c_2, c_3\} \bullet \{c_1, c_4\} \rangle \qquad |\Phi_{10}| = \langle \{c_2, c_4\} \bullet \{c_1, c_3\} \rangle$$
$$|\Phi_{11}| = \langle \{c_1\} \bullet \{c_2, c_3, c_4\} \rangle \qquad |\Phi_{12}| = \langle \{c_2\} \bullet \{c_1, c_3, c_4\} \rangle$$
$$|\Phi_{13}| = \langle \{c_3, c_4\} \bullet \{c_1, c_2\} \rangle \qquad |\Phi_{14}| = \langle \{c_3\} \bullet \{c_1, c_2, c_4\} \rangle$$
$$|\Phi_{15}| = \langle \{c_4\} \bullet \{c_1, c_2, c_3\} \rangle \qquad |\Phi_{16}| = \langle \emptyset \bullet \{c_1, c_2, c_3, c_4\} \rangle$$

Definition 6. [Final Configuration] A total configuration Φ is a *final configuration*, denoted FC(Φ), if 1) $AS \not\vdash_X \perp$ and 2) $US = \emptyset$.

Definition 7. [Comparator] A *comparator* $\prec$ is a relation between simplified configurations, such that $\langle W(S), \prec \rangle$ is a complete partial order (cpo) with bottom element $\langle S \bullet \emptyset \rangle$ and top element $\langle S_{[0]} \bullet S_{[>0]} \rangle$.

A comparator provides a criterium to compare alternative configurations of an hierarchy, in order to decide which are the best (according to this criterium). Examples and intended properties of comparators are presented in Section 4.

Definition 8. [Best Configuration] A final configuration Φ is a *best configuration*, denoted BC(Φ), if there is no other final configuration Φ' such that $|\Phi'| \prec |\Phi|$.

Definition 9. [Promising Configuration] A total configuration Φ is a *promising configuration*, denoted PC(Φ), if i) $AS \not\vdash_X \perp$ and ii) there is no final configuration $\Phi' \neq \langle NS \bullet RS \bullet \emptyset \rangle$ such that $|\Phi'| \prec |\Phi|$.

Proposition 10. *If Φ is a promising configuration and $NS \not\vdash_X \perp$, then $\Phi' = \langle NS \bullet RS \bullet \emptyset \rangle$ is a best configuration.*

For convenience, we introduce a total ordering of simplified configurations of an hierarchy (this total order is necessary to assure that the system is sound, complete and only makes finite computations [12]).

Definition 11. [$\prec_t$] Given a comparator $\prec$, let $\prec_t$ be an associated relation such that $\forall C, \langle W(C), \prec_t \rangle$ is a chain (or totally-ordered set) and $\forall \Phi_i, \Phi_j \in W(C), |\Phi_i| \prec |\Phi_j| \Rightarrow |\Phi_i| \prec_t |\Phi_j|$.

$\langle W(C), \prec_t \rangle$ defines the order in which final configurations are searched, enabling a systematic enumeration of configurations even when they are not $\prec$-comparable. Given $\Phi \in W(S)$, succ($\Phi, \prec_t$) denotes the successor of Φ in $\langle W(S), \prec_t \rangle$.

Before presenting the transition rules, the following definition is due.

Definition 12. [next()] Given a configuration Φ, relation next($|\Phi|, \prec_t, |\Phi_{next}|$) holds if and only if there is a successor $|\Phi_{next}|$ of $|\Phi|$ in the $\prec_t$-order.

The relation next($|\Phi|, \prec_t, |\Phi_{\text{next}}|$) is used in several transition rules to denote that it is possible to move from $|\Phi|$ to a successor node $|\Phi_{\text{next}}|$ in the configuration space.

3 Transition Rules of IHCS

IHCS aims at computing best configurations incrementally, according to an arbitrary comparator: given a hierarchy $\mathcal{H}$ with a known best configuration, a new constraint may be inserted, an old constraint may be removed, or an alternative solution may be obtained by performing transitions on the current configuration Φ_0, until a best configuration Φ_n is reached again ($\Phi_0 \longrightarrow \Phi_1 \longrightarrow \cdots \longrightarrow \Phi_n$). This process must be done undoing and redoing as little work as possible.

Conflict Stores

During the search for a best configuration, $\Phi_0 \longrightarrow \Phi_1 \longrightarrow \cdots \longrightarrow \Phi_n$, several conflicts may occur. After a transition $\Phi_{j-1} \longrightarrow \Phi_j$ where a conflict is raised, let $\text{CS}(\Phi_j)$ be a subset of AS_j containing all the constraints pertinent to the conflict raised. $\text{CS}(\Phi_j)$ is a *Conflict Store*, which may be determined by analysing some domain dependent properties of the constraint hierarchy. In this paper we will simple assume that $\text{CS}(\Phi_j)$ exist and can be computed with the properties described above. The actual computation of $\text{CS}(\Phi_j)$ is dealt with in [11, 12], using a Dependency Graph to record dependencies between constraints, adapting techniques previously used for intelligent backtracking [2, 14].

Conflicts may be solved, by the relaxation of some constraints and possible the reactivation of previously relaxed constraints. Conflicts happening in a sequence of transitions may be related among each other. If Φ_j is an inconsistent configuration, $0 \leq j \leq n$, then $\text{CS}(\Phi_j)$ defines its conflict store. Two conflicts are *related* if their conflict stores have common elements. When solving a conflict, any previously related conflict should be considered, since the solution to the current conflict may also be an alternative solution to this related conflict. Constraints relaxed before may thus be reactivated, since alternative constraints involved in the same conflict stores are now being relaxed. This policy will ensure the search for best configurations, where a minimum of constraints are relaxed, according to the criterium defined by the comparator in use. Constraints *pertinent* to a conflict in Φ_j must thus be extended to include not only constraints in $\text{CS}(\Phi_j)$, but also all the constraints in the transitive closure of related conflict sets. The notion of such extended conflict store is captured in Definition 13, which uses the auxiliary function extended:$\mathcal{C} \longrightarrow \mathcal{C}$ defined bellow.

$$\text{extended}(S) \equiv \bigcup_{\forall i} \begin{cases} \text{CS}(\Phi_i) & \text{if } S \cap \text{CS}(\Phi_i) \neq \emptyset \\ \emptyset & \text{otherwise} \end{cases}$$

Given a constraint store S, extended(S) is a constraint store with all constraints belonging to any conflict store that has common elements with S. If

S is a conflict store itself, then extended(S) contains S plus all constraints in related conflict stores. Applying the function again, i.e. extended(S) $\uparrow$ 2, will also add new constraints from conflict stores not related to S, but related to any other conflict store related to S. If this function is applied recursively, a fix point will be reached, extended(S) $\uparrow$ ω, which corresponds to the transitive closure of conflict stores related to S.

Definition 13. [Extended Conflict Store] Given an inconsistent configuration Φ_j, the *extended conflict store* of Φ_j, denoted ECS(Φ_j), is the set of all constraints directly or indirectly pertinent to the conflict in Φ_j (belonging to the transitive closure of conflict stores related to CS(Φ_j)) , i.e.,

$$\text{ECS}(\Phi_j) \equiv \text{extended}(\text{CS}(\Phi_j)) \uparrow \omega$$

Definition 14. [Conflict Configuration] Given an inconsistent configuration Φ_j, the *conflict configuration* of Φ_j, denoted cnfl(Φ_j), is the sub-configuration of $|\Phi_j|$, which contains all the constraints pertinent to the conflict, and rest(Φ_j) denotes the remainder of the configuration, i.e.,

$$\text{cnfl}(\Phi_j) \equiv |\Phi_j| \cap \text{ECS}(\Phi_j)$$
$$\text{rest}(\Phi_j) \equiv |\Phi_j| \setminus \text{cnfl}(\Phi_j)$$

We are now able to introduce the transition rules. Section 3.1 deals with the incremental insertion of constraints (the basic rules). Alternative solutions and the removal of constraints will be dealt in Sections 3.2 and 3.3, respectively.

3.1 The Basic Rules

If a new constraint c is inserted into $\mathcal{H}$ and $\langle AS \bullet RS \bullet \emptyset \rangle$ is a known best configuration, then providing the promising configuration $\Phi_0 = \langle AS \bullet RS \bullet \{c\} \rangle$ to the the following basic rules will lead to a new best configuration, if any, after a finite number n of transitions, $\Phi_0 \longrightarrow \Phi_1 \longrightarrow \cdots \longrightarrow \Phi_n$.

The forward rule (Figure 1) is used to activate a new unexplored constraint if the current configuration is a promising one.

Given an inconsistent configuration Φ_i, the backward rule (c.f. Figure 1) searches for an alternative promising configuration, if there is a successor to the *conflict configuration* cnfl(Φ_i) in the $\prec_t$-order, i.e., relation next(cnfl(Φ_i), $\prec_t$, $|\Phi_{\text{next}}|$) holds. cnfl(Φ_i) represents the sole part of Φ_i worth changing to solve the conflict, by relaxing some of its constraints (the *Relax* set) and re-activating constraints previously relaxed (the *Activate* set). The *Relax* and *Activate* sets are the result of deleting cnfl(Φ_i) from $|\Phi_{\text{next}}|$. This will ensure a complete search on the configuration space, pruning all the intermediate configurations that are the result of relaxing or activating constraints in rest(Φ_i) which are irrelevant to the conflict raised.

The *Reset* set is composed of active constraints that must be temporarily deactivated (moved to the unexplored store), in a backtracking like strategy, where all constraints introduced after any of the constraints being relaxed must

158

Forward rule: $\quad \dfrac{PC(\Phi_i) \qquad\qquad\qquad \exists c \in US_i}{\Phi_i \;\longrightarrow\; \langle AS_i \cup \{c\} \bullet RS_i \bullet US_i \setminus \{c\}\rangle}$
(fwd)

Backward rule: $\quad \dfrac{AS_i \vdash_{\mathcal{X}} \bot \qquad next(\mathrm{cnfl}(\Phi_i), \prec_t, |\Phi_{\mathrm{next}}|)}{\Phi_i \;\longrightarrow\; \Phi_{i+1}}$
(bwd)

$$\text{where} \begin{cases} \langle Activate \bullet Relax\rangle \;\leftarrow\; |\Phi_{\mathrm{next}}| \setminus \mathrm{cnfl}(\Phi_i) \\ Reset \;\leftarrow\; (AS_i \setminus Relax)_{[\mathrm{IO} > \min_{\mathrm{IO}}(Relax)]} \\ AS_{i+1} \;\leftarrow\; AS_i \setminus (Relax \cup Reset) \\ RS_{i+1} \;\leftarrow\; (RS_i \setminus Activate) \cup Relax \\ US_{i+1} \;\leftarrow\; (US_i \setminus Relax) \cup Reset \cup Activate \end{cases}$$

Fig. 1. The basic rules.

be reevaluated[2]. Values returned to current variables domains due to the relaxation of constraints, are checked again for consistency.

Computations made by the basic rules are sound and complete, i.e., starting from a promising configuration Φ_0 there is a finite number of transitions leading either to a best configuration Φ_n such that there is no other best configurations between Φ_0 and Φ_n (in the total order), or to a failure if there is no best configurations after Φ_0 (c.f. [12] for proofs).

3.2 Obtaining Alternative Best Configurations

There may be several best configurations to an hierarchy $\mathcal{H}$, since the $\prec$ ordering is not a total ordering. The system using the rules described so far is not exhaustive in the sense that it does not find all the best configurations, but it rather stops after finding the first of them. This exhaustiveness is achieved by IHCS using a new rule, the *alternative* rule (Figure 2). Given the current best configuration, this rule allows IHCS to find the next promising configuration.

The main difference between the alternative rule and the backward rule, is that the former computes an alternative to a best configuration while the latter computes an alternative to a conflicting configuration. The successor of $|\Phi_i|$ is determined (instead of the successor of some sub-configuration $|\Phi_{\mathrm{cnfl}}|$). The configuration $|\Phi_{i+1}|$ obtained by applying this rule to $|\Phi_i|$ must be no worse then the latter, regarding the $\prec$-order (otherwise Φ_{i+1} would not be a promising configurations since Φ_i is already a final configuration).

The promising configuration Φ_{i+1}, determined by the alternative rule, is subsequently handled by the basic rules in search for the next best configuration. If the next final configuration reached, Φ_j ($j > i + 1$), is not a best configuration, i.e. $|\Phi_{i+1}| \prec |\Phi_j|$, then the search for an alternative best configuration fails. To guarantee the completeness of the system, we have to ensure that all subsequent configurations to Φ_j in the total order are also "worse" then Φ_{i+1}. For

---

[2] The introduction order IO used to compute the *Reset* set is defined in Section 2

Alternative rule: $\dfrac{BC(\Phi_i) \quad next(|\Phi_i|, \prec_t, |\Phi_{next}|) \quad |\Phi_i| \not\prec |\Phi_{i+1}|}{\Phi_i \longrightarrow \Phi_{i+1}}$
(alt)

$$\text{where} \begin{cases} \langle Activate \bullet Relax \rangle \; \leftarrow \; |\Phi_{next}| \setminus |\Phi_i| \\ Reset \; \leftarrow \; (AS_i \setminus Relax)_{[IO > \min_{IO}(Relax)]} \\ AS_{i+1} \; \leftarrow \; AS_i \setminus (Relax \cup Reset) \\ RS_{i+1} \; \leftarrow \; (RS_i \setminus Activate) \cup Relax \\ US_{i+1} \; \leftarrow \; Reset \cup Activate \end{cases}$$

Removal rule: $\dfrac{BC(\Phi_i) \quad |\Phi_{rem}| \subset |\Phi_i|}{\Phi_i \longrightarrow \Phi_{i+1}}$
(rem)

$$\text{where} \begin{cases} Activate \leftarrow (RS_i \setminus RS_{rem}) \cap \text{extended}(AS_{rem}) \uparrow \omega \\ Reset \; \leftarrow \; (AS_i \setminus AS_{rem})_{[IO > \min_{IO}(AS_{rem})]} \\ AS_{i+1} \; \leftarrow \; AS_i \setminus (AS_{rem} \cup Reset) \\ RS_{i+1} \; \leftarrow \; RS_i \setminus (RS_{rem} \cup Activate) \\ US_{i+1} \; \leftarrow \; Reset \cup Activate \end{cases}$$

Fig. 2. Search for alternative best configurations and removal of constraints.

this purpose, an extra requirement is imposed to relation $\prec_t$, and is presented in Definition 11b (that extends Definition 11).

Definition 11b $(\prec_t)$ Given a comparator $\prec$, let $\prec_t$ be an associated relation such that $\forall S \in \mathcal{C}, \langle W(S), \prec_t \rangle$ is a chain and the following conditions hold:

1. $\forall \Phi_i, \Phi_j \in W(S), |\Phi_i| \prec |\Phi_j| \Rightarrow |\Phi_i| \prec_t |\Phi_j|$
2. $\forall \Phi_i, \Phi_j, \Phi_k \in W(S), (|\Phi_i| \prec |\Phi_j| \text{ and } |\Phi_j| \prec_t |\Phi_k|) \Rightarrow |\Phi_i| \prec |\Phi_k|$

3.3 Incremental Removal of a Constraint

Given a best configuration Φ_i of hierarchy $\mathcal{H}$, the removal of some constraints from the hierarchy is handled by the removal rule (Figure 2). Let $|\Phi_{rem}| = \langle AS_{rem} \bullet RS_{rem} \rangle$ be the portion of $|\Phi_i|$ to be removed. The removal of relaxed constraints (RS_{rem}) is straight forward, and it simply amounts to the removal of these constraints from the relax store of Φ_i.

To remove active constraints (AS_{rem}) we have to take similar actions to those taken for the relaxation of constraints in the backward rule. Once more, for backtracking purposes, the $Reset$ set is composed of all remaining active constraints that must be reevaluated, i.e., the ones introduced after any constraint from AS_{rem}. The $Activate$ set contains all remaining relaxed constraints that have previously been in some conflict in the transitive closure of conflicts related to AS_{rem}. These constraints are candidates for activation since the removal of AS_{rem} will hopefully solve some of these previous conflicts. The configuration produced by the removal rule is subsequently handled by the basic rules to search for best configurations to the new hierarchy.

4 Comparators

We have mentioned before that several criteria can be used to define a best configuration. In this section three comparators are presented to illustrate this point. Moreover, we discuss the properties a comparator must have in order to be "suitable" to IHCS.

The comparators presented (globally-predicate- better, locally-predicate-better and globally-weight-better - Definitions 15, 16 and 18 respectively) are adapted from [1]. Figure 3 depicts the cpo's for the first two comparators, using $W(S)$ of example 1.

Definition 15. [Globally-predicate-better]

$$|\Phi| \stackrel{\text{gpb}}{\prec} |\Phi'| \equiv |\Phi| = |\Phi'|$$
$$\text{or}$$
$$\exists k > 0, \forall i < k, \begin{cases} \#RS_{[i]} = \#RS'_{[i]} \\ \#RS_{[k]} < \#RS'_{[k]} \end{cases}$$

Definition 16. [Locally-predicate-better]

$$|\Phi| \stackrel{\text{lpb}}{\prec} |\Phi'| \equiv |\Phi| = |\Phi'|$$
$$\text{or}$$
$$\exists k > 0, \forall i < k, \begin{cases} RS_{[i]} = RS'_{[i]} \\ RS_{[k]} \subset RS'_{[k]} \end{cases}$$

Definition 17. [Global weight] Let n be the maximum hierarchical level of $\mathcal{H}$ and W_i be the weight of level i, such that $\forall i, W_i > 0$ and $\forall j, i < j \rightarrow W_i > W_j$. The global weight of a constraint store S is:

$$\text{weight}(S) = \sum_{i=1}^{n} (W_i \times \#S_{[i]})$$

Definition 18. [Globally-weight-better]

$$|\Phi| \stackrel{\text{gwb}}{\prec} |\Phi'| \equiv \text{weight}(RS) \leq \text{weight}(RS')$$

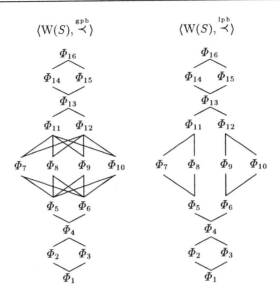

Fig. 3. Global-Predicate-Better versus Local- Predicate-Better.

From $\langle W(S), \overset{gpb}{\prec} \rangle$ of Figure 3, one can easily obtain a chain $\langle W(S), \overset{gpb}{\prec}_t \rangle$ such that $\overset{gpb}{\prec}_t$ meets the requirements of Definition 11b. It is sufficient to adopt the $\overset{gpb}{\prec}$ order and then randomly sort the configurations which are not $\overset{gpb}{\prec}$-comparable. The following chain is an example:

$$\Phi_1, \Phi_3, \Phi_2, \Phi_4, \Phi_5, \Phi_6, \Phi_8, \Phi_7, \Phi_9,$$
$$\Phi_{10}, \Phi_{12}, \Phi_{11}, \Phi_{13}, \Phi_{14}, \Phi_{15}, \Phi_{16}.$$

However, for the case of locally-predicate-better comparisons (c.f. $\langle W(S), \overset{lpb}{\prec} \rangle$ of Figure 3), there is no $\overset{lpb}{\prec}_t$ satisfying requirement 2 of Definition 11b. Take for example Φ_8 and Φ_9 which are not $\overset{lpb}{\prec}$-comparable. Because $\overset{lpb}{\prec}_t$ defines a total order it must be either $\Phi_8 \overset{lpb}{\prec}_t \Phi_9$ or $\Phi_9 \overset{lpb}{\prec}_t \Phi_8$. If $\overset{lpb}{\prec}_t$ is defined such that $\Phi_8 \overset{lpb}{\prec}_t \Phi_9$, then, the above mentioned requirement is violated since ($|\Phi_5| \prec |\Phi_8|$ and $|\Phi_8| \prec_t |\Phi_9|$) $\not\Rightarrow |\Phi_5| \prec |\Phi_9|$. If $\Phi_9 \overset{lpb}{\prec}_t \Phi_8$ instead, the violation remains since ($|\Phi_6| \prec |\Phi_9|$ and $|\Phi_9| \prec_t |\Phi_8|$) $\not\Rightarrow |\Phi_6| \prec |\Phi_8|$.

This example shows that some comparators are not suitable to be used in the IHCS. Based on $\prec$, we define a new relation $\propto$ which is convenient to check whether some $\prec$ is a suitable comparator.

Definition 19. [$\propto$] Given a comparator $\prec$, the associated relation $\propto$ is defined as follow

$$|\Phi| \propto |\Phi'| \equiv |\Phi| = |\Phi'| \text{ or } \begin{cases} |\Phi| \not\prec |\Phi'| \\ |\Phi'| \not\prec |\Phi| \end{cases}$$

As a consequence of its definition, $\propto$ is always reflexive and symmetric. Moreover, $\overset{\text{gpb}}{\propto}$ and $\overset{\text{gwb}}{\propto}$ are transitive and hence they are equivalence relations. In contrast $\overset{\text{lpb}}{\propto}$ is neither transitive nor an equivalence relation. In the example of Figure 3, $\overset{\text{gpb}}{\propto}$ defines the following equivalence classes: $C_1 = \{\Phi_1\}$, $C_2 = \{\Phi_2, \Phi_3\}$, $C_3 = \{\Phi_4\}$, $C_4 = \{\Phi_5, \Phi_6\}$, $C_5 = \{\Phi_7, \Phi_8, \Phi_9, \Phi_{10}\}$, $C_6 = \{\Phi_{11}, \Phi_{12}\}$, $C_7 = \{\Phi_{13}\}$, $C_8 = \{\Phi_{14}, \Phi_{15}\}$ and $C_9 = \{\Phi_{16}\}$. Each configuration of a given class is "as good as" any other configuration of the same class, in the sense that they can not be $\prec$-sorted. A configuration from a class C_i is always better then a configuration from a class C_j if $i < j$. Consequently, a total order relation $\overset{\text{gpb}}{\prec}_t$ must simply keep the $\overset{\text{gpb}}{\prec}$ order between configurations of different classes and establish an arbitrary order between configurations of a same class. For example, we use the introduction order of constraints to sort configurations within each equivalence class (c.f. [12]).

We may now introduce the following necessary and sufficient condition to accept a comparator in IHCS.

Proposition 20. *Given a comparator $\prec$, there exists a $\prec_t$ iff the relation $\propto$ associated to $\prec$ is an equivalence relation.*

In [12] the actual implementation of a comparator $\prec$ (only for the case of Global-Predicate-Better) is addressed. Functions $\text{succ}(\Phi_{\text{cnfl}}, \overset{\text{gpb}}{\prec}_t)$ and $\text{succ}(\Phi_{\text{cnfl}}, \overset{\text{gwb}}{\prec}_t)$ are computable in linear and polynomial time respectively, on the size of the conflict configuration.

5 Satisfaction Mode

The transition rules defined so far aim at optimizing the configuration of an hierarchy according to a given comparator (*optimization mode*).

In this section we present some extensions in order to provide a *satisfaction mode*. Instead of searching a best configuration (where the weight of the relaxed constraints is minimum), a final configuration is sought, whose weight of the relaxed constraints is below a certain *Threshold*. Though this mode is not restricted to be used with a Global-Weight-Better comparator, it is particularly adequate to this type of comparator if the specification of the *Threshold* is made in terms of the sum of constraint weights.

Definition 21. [Extended Configuration] An *extended configuration* $\langle \Phi \bullet DS \rangle$ of hierarchy $\mathcal{H}$ is a 4-tuple of constraint stores $\langle AS \bullet RS \bullet US \bullet DS \rangle$, where $\Phi = \langle AS \bullet RS \bullet US \rangle$ as usual, and DS is the *Discard Store*, with the following properties:

$$
\begin{array}{ll}
\textbf{Discard rule:} & AS_i \vdash_\chi \bot \\
\cdot(\textbf{dsc}) & \dfrac{\exists c \in \mathrm{CS}(\Phi_i)_{[\mathrm{level}>0]}, \ \mathrm{weight}(DS_{i+1}) \leq Threshold}{\langle \Phi_i \bullet DS_i \rangle \longrightarrow \langle \Phi_{i+1} \bullet DS_{i+1} \rangle}
\end{array}
$$

$$
\text{where} \quad
\begin{cases}
Reset \ \leftarrow \ AS_{i[\mathrm{IO}>\mathrm{IO}(c)]} \\
AS_{i+1} \ \leftarrow \ AS_i \setminus (Reset \cup \{c\}) \\
RS_{i+1} \ \leftarrow \ RS_i \\
US_{i+1} \ \leftarrow \ US_i \cup Reset \\
DS_{i+1} \ \leftarrow \ DS_i \cup \{c\}
\end{cases}
$$

$$
\begin{array}{ll}
\textbf{Improve rule:} & \dfrac{AS_i \nvdash_\chi \bot \qquad US_i = \emptyset \qquad \exists c \in DS_i}{\langle \Phi_i \bullet DS_i \rangle \longrightarrow \langle AS_i \bullet RS_i \bullet \{c\} \bullet DS_i \setminus \{c\} \rangle} \\
(\textbf{imp}) &
\end{array}
$$

Fig. 4. Satisfaction mode rules.

1. $(AS \cup US) \subseteq \mathcal{C}$ and $(RS \cup DS) \subseteq \mathcal{C}_{[>0]}$;
2. all the stores are disjoint from each other.

The *Discard Store* is used to accumulate relaxable constraints discarded during the search for solutions, as explained next. The following definition adapts the previously defined rules to the satisfaction mode, i.e., to the use of extended configurations, and two extra transition rules, the Discard and Improve rules, are introduced.

Definition 22. [Extended Rules]
$\forall r \in \{\text{fw,bw,alt,rr,ar,dj,adj}\}$, if $\Phi \xrightarrow{r} \Phi'$ then $\forall DS, \ \langle \Phi \bullet DS \rangle \xrightarrow{r} \langle \Phi' \bullet DS \rangle$.

In satisfaction mode, whenever a conflict is raised, if the threshold is not exceeded (by the total weight of discarded constraints), a relaxable constraint involved in the conflict is simply discarded (moved to the *Discard Store*) by the Discard Rule (c.f. Figure 4). If this is not possible, then the the backward rule is used to find the next promising configuration as before (this rule is extended in order to exclude its applicability when the Threshold is not exceeded).

The *Discard Store* is used to accumulate discarded constraints. Later on, when searching for alternative solutions, the Discard rule is no longer used and discarded constraints are progressively re-introduced by the Improve Rule in order to improve the quality of alternative solutions (c.f. Figure 4). With the re-introduction of discarded constraints, alternative solutions converge towards a best configuration. After all these constraints have been re-introduced, the subsequent search for further alternatives makes use of the alternative rule.

6 Applications

In this section we describe two kinds of applications: i) a simple Time-tabling problem to illustrate the use of different criteria to optimize an hierarchy and the behavior of a satisfaction mode session; ii) a digital circuit diagnostic problem to show the adequacy of a defeasible constraint solver to perform model-based diagnosis interactively.

6.1 Time-tabling

The problem consists of scheduling 6 subjects (A, B, C, D, E and F) in a time-table ranging from monday to friday, between 9am and 13am. Each subject is taught in three blocks per week, except B and F that are only taught twice a week. Each block has one hour duration. The problem amounts to the allocation of a time slot for each block of each subject (the problem variables). The following required constraints must be satisfied in any admissible solution:

1 each block of the same subject must take place at different days;
2 blocks of different subjects must not overlap.

In addition, the following preferred constraints are added to customize the time-table, namely, to leave wednesday free and ensure a uniform distribution of blocks throughout the week:

3 (level 1) blocks should not take place on wednesday;
4 (level 2) two consecutive blocks of a subject should not take place on consecutive days;
5 (level 2) blocks of the same subject should take place at the same hour.

The hierarchy of the different types of preferred constraints, reflects the priority given to each criterion. For example, in case of a conflict it is preferable to place two blocks on consecutive days or different hours rather than placing one of them on wednesday.

Given $W_1 = 3$ and $W_2 = 2$ (the weight of each level), Figure 5 shows two best solutions using different criteria and display the number of constraints relaxed in each level. GPB assumes absolute priority to more "important" constraints (i.e. with lower hierarchical level). In this case, it keeps wednesday free (level 1 constraints) at the cost of 6 relaxed constraints of level 2 (4 of type 4 and 2 of type 5). In contrast, GWB trades off the number and importance of the relaxed constraints. As such it achieves a configuration where the global weight of its relaxed constraints is 10, with a lower number of relaxed constraints (only 4) but leaving wednesday partially occupied.

Figure 6 exemplifies a satisfaction mode session, given a threshold of 16. To improve the first solution found (whose weight of discarded constraints is exactly 16), a type 4 constraint between the first two blocks of subject A discarded in this "satisfactory" solution (thus allowing two blocks of subject A to be scheduled at consecutive days) is re-introduced by the improve rule. Since the new solution produced also satisfies another discarded constraint (type 4 constraint between blocks of subject B), the new global weight for discarded constraints decreases to 12. With two subsequent uses of the improve rule (skipping the third solution),

Global-Predicate-Better

	9-10	10-11	11-12	12-13
Mon	A	C	D	E
Tue	A	B	D	F
Wed				
Thu	B	C	F	E
Fri	A	C	D	E

Relaxed Constraints

level 1	level 2
0	6

Global-Weight-Better

	9-10	10-11	11-12	12-13
Mon	A	C	D	E
Tue	B	F	D	
Wed	A	C		
Thu	B	F		E
Fri	A	C	D	E

Relaxed Constraints

level 1	level 2
2	2

Fig. 5. Example of Global-Predicate-Better versus Global-Weight-Better.

two previously discarded constraints (type 3 constraints on blocks of subjects D and E placed on Wednesday) are re-introduced, producing a fourth solution. The re-introduction of this constraints leads to the relaxation of two weaker type 4 constraints. This solution happens to be a best configuration (already obtained in Figure 5), with a global weight of 10, and further use of the improve rule will relax the remaining discarded constraints. The successive configurations are thus obtained incrementally, allowing a user to stop as soon as he/she is satisfied with the (possibly non-optimal) solution obtained.

6.2 Model Based Diagnosis

A digital circuit may be modeled by means of a constraint network, each gate corresponding to a boolean constraint over its input and output values. Figure 7 shows a simple example of a circuit and its model in the HCLP language developed as the front-end of IHCS. Each gate is represented by one boolean constraint, either and/3 or or/3, both accepting negated arguments (in IHCS, boolean constraints are implemented over finite domains).

IHCS is suitable to interactive model based diagnosis since it is able to suggest preferred diagnoses,and revise them interactively, following additional observations made by the user. From the model specification and the observations made, IHCS suggests preferred diagnoses, where faulty gates correspond to the constraints relaxed by the constraint solver. In the model, all constraints are assigned the same weight (@1), since it is assumed that the gates are all equally prone to faults. Therefore, the preferred diagnoses that IHCS finds with both the Global-Predicate-Better and the Global-Weight-Better comparators, are those with least number of faulty gates.

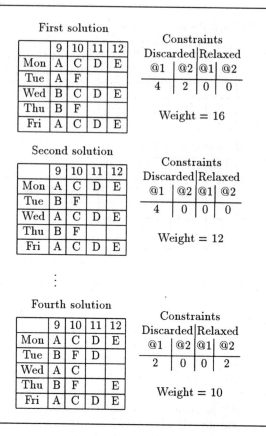

First solution

	9	10	11	12
Mon	A	C	D	E
Tue	A	F		
Wed	B	C	D	E
Thu	B	F		
Fri	A	C	D	E

Constraints

Discarded		Relaxed	
@1	@2	@1	@2
4	2	0	0

Weight = 16

Second solution

	9	10	11	12
Mon	A	C	D	E
Tue	B	F		
Wed	A	C	D	E
Thu	B	F		
Fri	A	C	D	E

Constraints

Discarded		Relaxed	
@1	@2	@1	@2
4	0	0	0

Weight = 12

⋮

Fourth solution

	9	10	11	12
Mon	A	C	D	E
Tue	B	F	D	
Wed	A	C		
Thu	B	F		E
Fri	A	C	D	E

Constraints

Discarded		Relaxed	
@1	@2	@1	@2
2	0	0	2

Weight = 10

Fig. 6. Example of Satisfaction Mode with Global-Weight-Better.

To perform diagnosis of a circuit, it is sufficient to supply a) the model of the circuit, b) the input values, and c) the observed output values. If the system can solve the constraint network relaxing no constraints, then the circuit is assumed to have no faulty gates. Otherwise, relaxed constraints represent a preferred diagnosis. All the other equally preferred diagnoses may be found with the alternative rule.

Example 2. **Preferred Configurations:** Given the values $A = 0, B = 1, C = 1$ and $D = 1$ input to the circuit of figure 7, value 1 is observed at output I (instead of the expected value zero). In this case, the query "?- circuit(0,1,1,1,_,_,_,_,1)." produces the only two diagnoses with a single faulty gate (either $g5$ or $g3$).

In general, diagnosis is an interactive task. Once a certain fault is suspected, it can be tested, in order to be validated (eg. the diagnosis of a faulty gate may be validated by comparing the expected input and output values with those observed). If this validation succeeds, then the diagnosis is assumed correct and

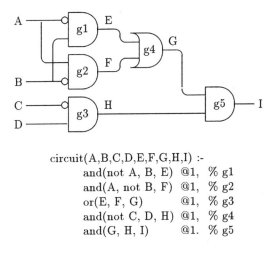

```
circuit(A,B,C,D,E,F,G,H,I) :-
    and(not A, B, E)  @1,  % g1
    and(A, not B, F)  @1,  % g2
    or(E, F, G)       @1,  % g3
    and(not C, D, H)  @1,  % g4
    and(G, H, I)      @1.  % g5
```

Fig. 7. Digital circuit and its specification.

complete, in the sense that the faults completely justify the observations. Otherwise, diagnosis shall proceed taking into account the new information.

Example 3. **Interactive Addition of Constraints:** Given the input values $A = 1$, $B = 1$, $C = 1$ and $D = 1$ to the circuit of figure 7, value 1 is observed at output I (instead of the expected value zero). The following table traces the steps required to identify the faulty gates (a faulty gate corresponds to a relaxed constraint), and the expected values are those assigned by the constraint solver and that justifie the faults. Given that in step 1 the observed values G and H are different from those expected, this new information is entered as two additional constraints. The constraint solver then incrementally solves the extended constraint network. The resulting preferred diagnosis, suggests faults in both gates $g3$ and $g4$. This procedure is repeated given the inconsistency of the observed and expected values of E, and the correct diagnosis of faulty gates $g1$ and $g3$ is eventually computed in the next step.

Query	Faulty gates	Expected values	Observed values
1. ?- circuit(1,1,1,1, E,F,G,H,1).	$g5$	G=0 H=0	G=1 H=1
2. ?- add G=1, H=1.	$g3, g4$	E=0 F=0	E=1 F=0
3. ?- add E=1.	$g1, g3$		

Example 4. **Alternative Configurations:** After step 1 of the previous example, more than one diagnostic solution exists with two faulty gates (but none with a single fault). These alternatives can all be found by the constraint solver, as shown in the following trace.

Query/Action	Faulty Gates
2. ?- add G=1, H=1.	$\{g3, g4\}$
3. retry ? yes	$\{g2, g3\}$
4. retry ? yes	$\{g1, g3\}$
5. retry ? yes	$\langle FAIL \rangle$

Example 5. **Interactive Removal of Constraints:** The previous examples illustrate the incremental behavior of the constraint solver upon the inclusion of additional constraints. One can also take advantage of the incremental nature of the constraint solver upon removal of constraints. For example, $g5$ is suggested as faulty in step 1 of example 2. Nevertheless, if another gate is directly observed as faulty (eg. because it is burned) then the constraint corresponding to that gate may be simply removed as exemplified below (via predicate remConstraint/1). Upon removal of $g3$, the constraint solver finds a diagnostic solution without any further faulty gates, thus reactivating $g5$.

Query	Faulty gates	Observed
1. ?- circuit(0,1,1,1, E,F,G,H,1).	$g5$	g3 is burned
2. ?- add remConstraint(g3).	none	

7 Conclusion

In previous work [11, 12] we presented the formalization and an implementation of IHCS, restricted to a single comparator (global-predicate-better), and proved some of its properties, namely a) soundness (only best solutions are obtained), b) completeness (all such solutions are computed), and c) non redundancy (no repeated solutions).

This paper generalizes IHCS by a) allowing a diversity of comparators to be used and b) introducing the possibility of searching satisfactory solutions to avoid the intractability of the optimization problems. The usefullness of these generalizations are illustrated by means of time-tabling and model-based diagnostic problems, showing the versatility and adequacy of IHCS to handle such problems interactively.

Moreover, the paper describes the properties that a comparator $\prec$ must have, so that the soundness and completeness properties of IHCS proven for globally-predicate-better comparisons, still hold for $\prec$.

Experimental time results are not presented here, since only toy-problems are described (because they are easy to explain), and their solutions are obtained in insignificant time. However, previous experiences [11, 12] have shown us that IHCS is quite efficient in its constraint handling. It uses a very efficient Arc Consistency algorithm which presents time complexity $O(ed)$ to handle linear equations and disequations and $O(e)$ to handle boolean constraints such as the ones used in the examples (where e is the number of constraints and d the medium size of the domains of variables).

The belief revision process necessary to find an optimal configuration is by nature an hard problem, given the exponential size of the configuration space. Nevertheless, a large number of practical problems are efficiently tackled, due to the use of dependencies among constraints to focus the search for optimality on pertinent sub-configurations, and the use of simple redundancy tests to avoid a large number of useless configurations [12]. The scope of IHCS is additionally extended with the introduction of the satisfaction mode. In the future, we intend to research heuristic methods in order to further improve the search for optimal configurations.

Acknowledgments

This work was supported by JNICT, the Portuguese Board for Science and Technology, under grant PBIC/C/TIT/1242/92.

References

1. A. Borning, M. Maher, A. Martingale, and M. Wilson. Constraints hierarchies and logic programming. *Logic Programming: Proceedings of the 6th International Conference*, pages 149–164, Lisbon, Portugal, June 1989. The MIT Press.
2. C. Codognet, P. Codognet, and G. Filé. Yet Another Intelligent Backtracking Method. In *Proceedings of 5th ICLP/SLP*, Seattle, 1988.
3. R. Davis and W. Hamscher. Model-based Reasoning: Troubleshooting. *Exploring Artificial Intelligence*, chapter 8, pages 297–379. Morgan Kaufmann Publisher, INC., 1988.
4. J. de Kleer. Problem Solving with the ATMS. *Artificial Intelligence*, 28:197–224, 1986.
5. J. Doyle. A Truth Maintenance System. *Artificial Intelligence*, 12:231–272, 1979.
6. Michael R. Garey and David S. Johnson. *Computers and Intractability: A Guide to the Theory of NP-Completness*. W.H.Freeman and Company, New York, 1979.
7. U. Junker. Generating Diagnoses by Prioritized Defaults. In *Working Notes of 2nd International Workshop on Principles of Diagnosis*, Milano, 1991.
8. Vipin Kumar. Algorithms for Constraint-Satisfaction-Problems: A Survey. *AI Magazine*, pages 32–44, Spring 1992.
9. Alan K. Mackworth. Consistency in Networks of Relations. *Artificial Intelligence*, 8:99–118, 1977.

10. Alan K. Mackworth and Eugene C. Freuder. The Complexity of Some Polynomial Network Consistency Algorithms for Constraint Satisfaction Problems. *Artificial Intelligence*, 25:65–74, 1985.
11. F. Menezes and P. Barahona. Preliminary Formalization of an Incremental Hierarchical Constraint Solver. In *Proceedings of EPIA '93*, Porto, October 1993.
12. F. Menezes and P. Barahona. An Incremental Hierarchical Constraint Solver. *Principles and Practice of Constraint Programming*, MIT Press, 1994.
13. L. M. Pereira, C. Damásio, and J. J. Alferes. Diagnosis and Debugging as Contradiction Removal. *2nd Int. Ws. on Logic Programming and NonMonotonic Reasoning*, pages 316–330. MIT Press, 1993.
14. Luis Moniz Pereira and M. Bruynooghe. Deduction Revision by Intelligent Backtracking. *Implementations of Prolog*, pages 194–215. J.A. Campbell, 1984.
15. D. Poole. A Logical Framework for Default Reasoning. *Artificial Intelligence*, 36:27–47, 1988.
16. H. Simonis. Test Generation using the Constraint Logic Programming Language CHIP. *Logic Programming: Proceedings of the 6th International Conference*, pages 101–112, Lisbon, Portugal, June 1989. The MIT Press.
17. P. Van Hentenryck, Y. Deville, and C.-M. Teng. A Generic Arc Consistency Algorithm and its Specializations. Technical Report RR 91-22, K.U. Leuven, F.S.A., December 1991.

Transforming Ordered Constraint Hierarchies into Ordinary Constraint Systems*

Armin Wolf

GMD – German National Research Center for Information Technology
GMD FIRST, Rudower Chaussee 5, D-12489 Berlin, Germany
e-mail: `armin@first.gmd.de`

Abstract. In this paper, we propose ordered constraint hierarchies and a non-trivial error function to model and solve over-constrained real-world problems adequately. We substantiate our proposition by an example. Then we present a new method to transform constraint hierarchies into equivalent ordinary constraint systems. For practical applications, we present a modified algorithm based on an incomplete finite domain constraint solver. We conclude with a prototype implementation of the method and some aspects about our future work.

1 Introduction

In real-world applications, constraints represent conditions which have to be fulfilled. With an according constraint solver, we get a useful problem solving system, which has a high degree of declarativity. It distinguishes strongly between the two parts of problem solving algorithms: *logic* and *control* (Kolwalski). The logic is given by the constraints, and the constraint solver is the controller of this kind of logic.

Two wide areas where constraint systems are successfully applied, are computer graphics, especially automatic layout generation [1, 9] and planning, especially scheduling [6]. In practical applications we often are confronted with situations where the problems are over-specified, and a correct solution is not possible, but it is unsatisfying that a system reacts with the output: `no solution` like normal constraint logic programming (CLP) systems do. Well-known over-constrained problems are initial specifications of production planning problems. Other more or less known over-constrained problems arise in graphic systems during interaction [8] and during derangements in production processes, e.g. a break-down of a machine or a delay of delivery terms. In both cases not only the current layout/plan but also the new requirements have to be considered in the new layout/plan. This means that it is necessary to satisfy the constraints of the new problem specification, and it is desirable to preserve the previous layout/plan as well as possible.

---

* The research in this paper is a result of the project WISPRO funded by the *German Federal Minister for Education, Science, Research, and Technology* (BMBF) under grant no. 01 IW 206.

We want to show the arising difficulties by an example. Let us consider two tasks in a task schedule of a production plan. Their starting times are $S \leftarrow 100$ and $T \leftarrow 100$ (w.r.t. a given reference time). Suppose we have to satisfy the additional constraints $S < T$ (task S starts before task T) and $S \neq 100$ (task S must not start at time 100) because there is a derangement, and we also want to consider the previous plan as well as possible. This means that we have to take the constraints $S = 100$ and $T = 100$ into account, too, resulting in an over-constrained problem. Its intuitively "best" solution (in the integer domain) is a schedule with $S \leftarrow 99$ and $T \leftarrow 100$. To get this solution, we use an appropriate "constraint-hierarchy" where we reduce the strength of the constraints created from the old plan. Details will be shown later in this paper.

In the next section we present the state-of-the-art. After this we present a brief overview of the main aspects of constraint hierarchies. After the definitions we comment on some unsatisfying properties of the definition. We then present ordered constraint hierarchies combined with a non-trivial error function. We continue with proposing a method to transform such hierarchies into ordinary constraint systems, which may be solved with well-known, available constraint solvers. The presented method has a flaw: One prerequisite for its application is the existence of a complete constraint solver for linear constraints like the Simplex algorithm integrated in $\text{CLP}(\mathcal{R})$ [7]. In general, complete solvers are exponential in time, thus, we prefer incomplete solvers, like finite domain solver based on local propagation techniques. For incomplete solvers we propose a modified transformation method. We conclude with implementation details, some results, and our future research activities.

2 State-of-the-Art

The formal foundations of *constraint hierarchies* are presented in [2]. Further developments are made by Molly Ann Wilson in her PhD thesis [14], especially the extension of constraint logic programming (s.a. [15]). For constraint hierarchies based on the trivial error function, a lot of algorithms to solve such hierarchies are known [10, 11, 13]. The first solution uses method graphs, the second is a model theoretic approach of the problem, and the third is based on deductive database techniques. The only algorithm known to the author for constraint hierarchies with a non-trivial error function[2] is based on the constraint solver DeltaStar [5] and the Simplex algorithm for linear constraints over the reals. The basics of the algorithm are sketched in Wilson's PhD thesis.

3 Constraint Hierarchies

In this section we give a brief, thus incomplete overview on constraint hierarchies. For more details see [14] or [15]. We will then critically discuss some aspects of the current theory.

---

[2] In this paper the same error function is considered.

3.1 Definitions

A *constraint hierarchy* H consists of a finite sequence of *constraint levels* $H_0, \ldots, H_n$. Each constraint level is a finite vector of constraints $(c_1, \ldots, c_k)$ – in the meantime in some arbitrary order. The level H_0 is the vector of all required constraints. H_1 is the vector of the strongest non-required constraints. They have to be satisfied as well as possible with highest priority. H_2 is the vector of all constraints with the second-highest priority, and so forth through the weakest level H_n with the lowest priority.

A *valuation* of a constraint is a substitution of all free variables, which maps the free variables to values in their domains. The symbols $\sigma, \theta, \ldots$ will be used for substitutions. The application of a valuation σ to a constraint c is written as $c\sigma$. The application to a constraint vector $C = (c_1, \ldots, c_k)$ is defined by $C\sigma := (c_1\sigma, \ldots, c_k\sigma)$.

An *error function* e maps valuated constraints to non-negative real numbers. It holds that $e(c\sigma) = 0$ if and only if c is satisfied by σ. The generalization of an error function e is a function E which maps constraint vectors to real-valued vectors. For a constraint vector $C = (c_1, \ldots, c_k)$ it is defined by

$$E(C\sigma) := [e(c_1\sigma), \ldots, e(c_k\sigma)] \ .$$

An *error sequence* of a hierarchy $H = (H_1, \ldots, H_n)$ w.r.t. an extended error function E and a valuation σ is the sequence $[E(H_1\sigma), \ldots, E(H_n\sigma)]$.

A *combining function* g combines the errors in an error sequence. The result is a kind of "total error" in a not further specified set. However, there must be two binary relations associated with the function g in this set: $<>_g$ and $<_g$. It is required that $<_g$ is irreflexive, antisymmetric and transitive (an irreflexive partial ordering) and that $<>_g$ is reflexive and symmetric. The generalization of a combining function g for error sequences is a function G defined by

$$G\left([E(H_1\sigma), \ldots, E(H_n\sigma)]\right) := [g(E(H_1\sigma)), \ldots, g(E(H_n\sigma))] \ .$$

The result $[g(E(H_1\sigma)), \ldots, g(E(H_n\sigma))]$ is a *combined error sequence*.

A lexicographic ordering $<_G$ on combined error sequences is defined in the obvious manner. Its definition is based on the ordering $<_g$ and the relation $<>_g$:

$$(x_1, \ldots, x_n) <_G (y_1, \ldots, y_n) \quad \text{if and only if}$$
$$\text{there exists } k \leq n, \text{ such that}$$
$$x_i <>_g y_i \text{ for } i = 1, \ldots, k-1 \text{ and } x_k <_g y_k.$$

A *comparator* $better_{e,g}$ based on an error function e and a combining function g with its associated relations $<>_g$ and $<_g$ is defined by

$$better_{e,g}(\sigma, \theta, H) \equiv G([E(H_1\sigma), \ldots, E(H_n\sigma)]) <_G G([E(H_1\theta), \ldots, E(H_n\theta)]) \ ,$$

where σ and θ are valuations and H is a constraint hierarchy. If the lexicographic ordering is well-defined, this means that it is really an irreflexive ordering, and if we have a fixed hierarchy H, then the comparator will define an irreflexive ordering on the valuations of the hierarchy. It is used to compare the solutions of the required constraints in H_0. A solution is one of the "best" solutions, if there is no better solution. The "best" solutions are the solutions of the whole hierarchy. More formally, the solution set S is defined by:

$$S_0 := \{\theta \mid \forall c \in H_0 : e(c\theta) = 0\}$$
$$S := \{\theta \mid \theta \in S_0 \wedge \forall \sigma \in S_0 : \neg better_{e,g}(\sigma, \theta, H)\} \ .$$

Here S_0 is the solution set of the required constraints.

3.2 Critical Remarks

In the following we give some critical remarks about the current constraint hierarchy theory.

Finding Solutions. Theoretically, the solutions of a constraint hierarchy are well-defined, but in general the given definition does not yield a practical algorithm to compute these solutions. Following the definition, we have to compute all solutions of the required constraints and compare them. If we do not know all solutions, we will be unable to decide whether a solution in S_0 is a solution of the whole hierarchy. We do not know whether there is a better solution or not. We are only able to filter solutions which are no "candidates" for the solution set S. We will have this problem if the solution set S_0 is very large or infinitely. – This problem will arise, if we consider the previously introduced example: $H_0 = \{S < T, S \neq 100\}, H_1 = \{S = 100, T = 100\}$.

Trivial and Non-Trivial Error Functions. An *error function* defines a "measure" of the degree how good or bad a valuation σ satisfies a constraint c. Satisfaction will be perfect if $c\sigma$ holds and the error value is zero. The greater the value the worse the satisfaction. However, if the trivial error function e_t is given with either $e_t(c\sigma) = 0$ ($c\sigma$ holds) or $e_t(c\sigma) = 1$ ($c\sigma$ is unsatisfied), there will be no discrimination between unsatisfying valuations. They have the same error value.

In practical applications, this coarse grained discrimination is not acceptable. For example, suppose the constraint $S = 100$ and the valuations $S \leftarrow 99$ and $S \leftarrow 220$. Using the trivial error function, both valuations yield the same error value, and both valuations are solutions of a hierarchy with $S \neq 100$ required and $S = 100$ non-required. If S specifies the starting time of a task and the values are interpreted as minutes, it will be of minor importance in practice if the task starts one minute earlier. However, if the tasks starts two hours later it may cause great delays in delivery times etc.

An non-trivial error function which better reflects the degree of the "real" error in this case is based on the absolute function. We define it for equations and inequations over the integers, especially linear equations and inequations:

$$e_{|.|}(a = b) := |b - a|$$

$$e_{|.|}(a \leq b) := \begin{cases} 0 & \text{if } a \leq b \\ a - b & \text{otherwise} \end{cases}$$

$$e_{|.|}(a < b) := e_{|.|}(a \leq b - 1)$$

Using this function, the error of the valuated constraint $(S = 100)\{S \leftarrow 99\}$ is 1 and the error of $(S = 100)\{S \leftarrow 220\}$ is 120. This error is identical with the "distance" between the ideal and the selected value, which matches more the intended meaning of the notion "error".

Comparators Respecting the Hierarchies. The property that a comparator *respects* a hierarchy is of practical interest:

Definition 1 [15]. A comparator $better_{e,g}$ *respects* a constraint hierarchy $H = (H_1, \ldots, H_n)$ if there is a valuation θ in S_0 which satisfies all constraints up to level k $(k \leq n)$ then all solutions of the hierarchy will satisfy all constraints up to level k, too. More formally:

$$\exists \theta \in S_0 \forall p \in H_1 \cup \ldots \cup H_k : p\theta \Rightarrow \forall \sigma \in S \, \forall p \in H_1 \cup \ldots \cup H_k : p\sigma$$

It is obvious that the solutions resulting from a comparator "without respect" may be surprising. Thus we claim for practical applications that a comparator has to respect *every* hierarchy.

A sufficient condition for comparators respecting the hierarchies is given by the following lemma.

Lemma 2. *Let $better_{e,g}$ be a comparator for constraint hierarchies. The comparator respects every hierarchy if the two conditions*

1. *if* $g(x_1, \ldots, x_l) <>_g g(0, \ldots, 0)$ *then* $(x_1, \ldots, x_l) = (0, \ldots, 0)$,
2. *if* $(x_1, \ldots, x_l) \neq (0, \ldots, 0)$ *then* $g(x_1, \ldots, x_l) >_g g(0, \ldots, 0)$

hold.

Proof. Assume there be a comparator $better_{e,g}$ and g fulfills both conditions but the comparator does not respect any hierarchy H. This means that there is a valuation $\theta \in S_0$, which satisfies all constraints up to level k, and there is another valuation $\sigma \in S$, which satisfies all constraints at the levels $1, \ldots, l - 1$ but there is at least one unsatisfied constraint at level l $(1 \leq l \leq k)$. As a consequence $E(H_l\theta) = (0, \ldots, 0)$ and $E(H_l\sigma) \neq (0, \ldots, 0)$ holds.

Further $[g(E(H_1\theta)) <>_g g(E(H_1\sigma))], \ldots, [g(E(H_{l-1}\theta)) <>_g g(E(H_{l-1}\sigma))]$ holds, because both valuations satisfy all constraints up to level $l - 1$. At level l it holds that $E(H_l\theta)) = (0, \ldots, 0)$ and $E(H_l\sigma) \neq (0, \ldots, 0)$. Using the first condition the error combinations $g(E(H_l\theta))$ and $g(E(H_l\sigma))$ are not comparable w.r.t. $<>_g$ (otherwise $E(H_l\sigma) = (0, \ldots, 0)$ using the first condition) but w.r.t. $<_g$:

Using the second condition it holds that $g(E(H_l\theta)) <_g g(E(H_l\sigma))$. Obviously $better_g(\theta, \sigma, H)$ holds. It follows directly that σ is no solution of the hierarchy H. This contradicts the assumption, so the comparator $better_{e,g}$ respects every hierarchy, especially H. □

Note 3. The second condition will be sufficient if the lexicographical ordering $<_G$ is well-defined or – more general – if either $x <_g y$, $y <_g x$, or $x <>_g y$ holds for any two error combinations x and y.[3]

The Influence of the Constraint Ordering. In the definition of constraint hierarchies the remark is missing that the ordering of the constraints in the hierarchy levels may strongly influence the solution set of a hierarchy which we proof by an example. We choose the hierarchy consisting of the levels $H_0 =$ (**true**) and $H_1 = (X = 0, X = 1)$. We choose the comparator $g(v) = v_1$, a special instance of *weighted-sum-better* (cf. [14]): $g(v) = \sum_{i=1}^{|v|} w_i v_i$ where $<_g \equiv <$ and $<>_g \equiv =$ on the reals, and we choose the trivial error function.

If we prefer the ordering $X = 0$ before $X = 1$, the solution set will consist of the valuation $X \leftarrow 0$. Preferring the ordering $X = 1$ before $X = 0$ the solution of the hierarchy is $X \leftarrow 1$. In this case both solution sets are disjoint.

It depends on the selected comparator whether the ordering of the constraints influences the solution set. In the case of influence, we have to specify an ordering strategy. Possible strategies may be *first-in-first-considered* and *last-in-first-considered*.

4 Ordered Constraint Hierarchies

In this section we present a special type of constraint hierarchies. In the following section we introduce a method to transform these hierarchies into ordinary constraint systems which can be solved with well-known methods.

In the given theory "constraint hierarchy" is a generic concept. Types of constraint hierarchies are further determined by the constraints from which they are build up, and the used comparators. A comparator is uniquely determined by the used error function and error combining function with its associated relations.

4.1 The Constraints

We want to consider one special type of constraint hierarchies. Its constraints are linear equations and inequations over the integers. Without loss of generality, we assume that the equations and inequations are in normal form (e.g. preprocessing with an appropriate reduction system). These normal forms are

$$\sum_{i=1}^{n} a_i X_i = b, \quad \sum_{i=1}^{n} a_i X_i \leq b, \quad \text{and} \quad \sum_{i=1}^{n} a_i X_i < b$$

---

[3] In Wilson's PhD thesis this is intended, because there is the hint ... *The symbol $<>_g$ indicates that two valuations cannot be ordered using $<_g$.* ...

where the a_i's and b's are integer values and the X_i's are integer variables.

4.2 The Comparator

For specified constraint hierarchies we use a comparator which is a modification of *locally-better* (cf. [14]). The comparator $better_{e_{|.|},l}$ is based on the previously defined non-trivial error function $e_{|.|}$ and the simple error combining function l with

$$l(x) = x \qquad \text{where } x \text{ is an error sequence.}$$

The associated relation $<_l$ is the usual lexicographic (irreflexive) ordering on real-valued vectors, based on the irreflexive ordering $<$ and the equality $=$ over the reals:

$$(x_1, \ldots, x_n) <_l (y_1, \ldots, y_n) \quad \text{if and only if}$$
$$\text{there is a } k \ (1 \le k \le n) \text{ with}$$
$$x_1 = y_1, \ldots, x_{k-1} = y_{k-1} \text{ and } x_k < y_k.$$

The relation $<>_l$ is the usual equality over real-valued vectors:

$$(x_1, \ldots, x_n) <>_l (y_1, \ldots, y_n)$$
$$\text{if and only if } x_1 = y_1, \ldots, x_n = y_n.$$

Obviously, the comparator $better_{e_{|.|},l}$ respects every hierarchy. The order of the constraints in the hierarchy levels influences the solution sets because the defined ordering $<_l$ is lexicographic. In the following, we speak about *ordered constraint hierarchies* to notify the importance of the constraint ordering.

The irreflexive ordering $<_l$ is a total ordering on error sequences. As a consequence two valuations σ and θ either result in the same error sequence/combination or either $better_{e_{|.|},l}(\sigma, \theta, H)$ or $better_{e_{|.|},l}(\theta, \sigma, H)$ holds w.r.t. an ordered hierarchy H.

4.3 Transformation into Ordinary Constraint Systems

The transformation of an ordered constraint hierarchy is an iterative process. The transformation method works on "partial" hierarchies.

Definition 4. Let $H = H_0, H_1 \ldots, H_n$ be a constraint hierarchy. A *partial hierarchy* of H is a constraint hierarchy $H_{i,j} := H_0, H_1, \ldots, H_{i-1}, H_i'$ $(0 \le i \le n)$ with $H_i' := (c_1^i, \ldots, c_j^i)$ if $H_i = (c_1^i, \ldots, c_{m_i}^i)$ $(1 \le j \le m_i)$.

The *extension* of a partial hierarchy $H_{i,j}$ is either the hierarchy $H_{i,j+1}$ (if $j < m_i$) or the hierarchy $H_{i+1,1}$ (if $j = m_i$) or undefined (if $H_{i,j} = H$). For $H_{i,j}$ $(\ne H)$ we define the *extending constraint* $c_{i,j}$, which is either c_{j+1}^i or c_1^{i+1} respectively.

It is remarkable that the system H_0 of the required constraints and the whole hierarchy H are partial hierarchies of H.

The comparator $better_{e_{|.|},l}$ is "order preserving". This means the solution set of the extension of a partial hierarchy is a subset of the solution set of the partial hierarchy.

Lemma 5. *Let J be the extension of a partial hierarchy $H_{i,j}$ of some hierarchy H. It holds that*

1. $better_{e_{|.|},l}(\sigma, \theta, H_{i,j})$ *implies* $better_{e_{|.|},l}(\sigma, \theta, J)$
   for any two valuations σ and θ.
2. *The solution set S_J of J is a subset of the solution set $S_{H_{i,j}}$ of $H_{i,j}$:*

$$S_J \subseteq S_{H_{i,j}}.$$

Proof. It will be shown that the first proposition holds. The second proposition is a direct consequence of the first one.

$better_{e_{|.|},l}(\sigma, \theta, H_{i,j})$ implies that there is a hierarchy level $k(k \leq i)$ so that $[E(H_1\sigma) <>_l E(H_1\theta)], \ldots, [E(H_{k-1}\sigma) <>_l E(H_{k-1}\theta)]$, and $[E(H_k^\star\sigma) <_l E(H_k^\star\theta)]$ hold. $H_k^\star$ is either $H_k(k < i)$ or $H_k'(k = i)$. Further it follows that there is an index $m \leq j$ so that $[e(c_1^k\sigma) = e(c_1^k\theta)], \ldots, [e(c_{m-1}^k\sigma) = e(c_{m-1}^k\theta)]$, and $[e(c_m^k\sigma) < e(c_m^k\theta)]$ hold. Independent from the values $e(c_{i,j}\sigma)$ and $e(c_{i,j}\theta)$ it holds that $better_{e_{|.|},l}(\sigma, \theta, J)$. □

The comparator $better_{e_{|.|},l}$ is "satisfaction invariant". The following lemma guarantees that all partial hierarchies of a constraint hierarchy are satisfiable if (and trivially only if) its required constraints are satisfiable.

Lemma 6. *Let H be one of the determined constraint hierarchies with the comparator $better_{e_{|.|},l}$.*

If the constraint system H_0 is satisfiable, every partial hierarchy of H – especially H itself – is satisfiable, too. In short terms:

$$S_0 \neq \emptyset \Rightarrow S \neq \emptyset.$$

Proof. Induction base: The partial hierarchies $H_{0,1}, \ldots, H_{0,m_0}$ of H are satisfiable because every subsystem of a satisfiable constraint system is satisfiable, too.

Induction step: The partial hierarchy $H_{i,j} \neq H$ is satisfiable by induction hypothesis. The solution set of this hierarchy, called $S_{i,j}$, is non-empty. Now let us consider the extension of the partial hierarchy $H_{i,j}$, called J, and the added extending constraint, called c, for simplicity.

The minimum of the set $\{e_{|.|}(c\sigma) \mid \sigma \in S_{i,j}\}$ is well-defined because the set is closed and has the trivial lower bound 0. With other words, there is (at least one) valuation $\theta \in S_{i,j}$ with $e_{|.|}(\theta) = \min\{e_{|.|}(c\sigma) \mid \sigma \in S_{i,j}\}$. Per definition of the comparator it holds that $\neg better(\theta, \sigma, J)$ for every $\sigma \in S_{i,j}$. The valuation is one solution of the partial hierarchy J. □

Transformation with Complete Constraint Solvers. Now we show in detail how to transform a constraint hierarchy into an equivalent ordinary constraint system. This means that the solution sets of the "flattening" system and the hierarchy are identical.

Initially, we consider the partial hierarchy H_0. It is an ordinary constraint system which is equivalent to itself. The whole hierarchy is satisfiable if the system H_0 is satisfiable (see Lemma 6). In the other case the hierarchy is equivalent to every inconsistent system, especially to false.

Suppose the system H_0 be satisfiable. Inductively we assume that we have already transformed the partial hierarchy $H_{i,j}$ into an equivalent constraint system $C_{i,j}$. If $H_{i,j} = H$ the system we are looking for is $C_{i,j}$. In all other cases we consider the extending constraint $c_{i,j}$.

For the sake of simplicity, we assume that $c_{i,j} \equiv X = b$, where X is a variable and b is an integer value (other cases are considered later). This constraint has to be fulfilled as well as possible. With other words, we are looking for valuations σ satisfying the system $C_{i,j}$ and minimizing the error $e((X = b)\sigma)$. If the system $C_{i,j}$ extended by $X = b$ is satisfiable, all solutions of the extended system will minimize the error. In this case the error value is zero. In general, especially if we do not know whether the extended system is satisfiable, the following considerations will be helpful: The smallest error is defined by

$$e_{\min}(X = b) := \min\{e(X = b) \mid \{X \leftarrow u\} = \sigma\mid_X \text{ and } C_{i,j}\sigma\} \, \P.$$

The solutions of the extended partial hierarchy satisfy the constraint system $C_{i,j}$ and in addition the "constraint"

$$e(X = b) = e_{\min}(X = b)$$

For representing this "constraint" by constraints in the normal sense, we have to evaluate the two values

$$u_0 := \min\{u \mid \{X \leftarrow u\} = \sigma\mid_X, C_{i,j}\sigma \text{ and } u \geq b\}$$
$$u_1 := \max\{u \mid \{X \leftarrow u\} = \sigma\mid_X, C_{i,j}\sigma \text{ and } u \leq b\}.$$

In systems like $\text{CLP}(\mathcal{R})$ [7] which include complete constraint solvers these evaluations are simple. They are supported by the projection operation dump/3. If one of the two considered sets is empty, the corresponding value will be undefined. At most one set is empty and at most one value is undefined: Assume both sets be empty, the system $C_{i,j}$ and the system H_0 are not satisfiable. However, this contradicts the hypothesis that H_0 is satisfiable.

Knowing the values of u_0 and u_1 we distinguish three different cases:

1. u_0 and u_1 are defined, it holds that $e_{\min}(X = b) = \min\{u_0 - b, b - u_1\}$.
2. u_0 is defined but u_1 is undefined, it holds that $e_{\min}(X = b) = u_0 - b$.
3. u_0 is undefined but u_1 is defined, it holds that $e_{\min}(X = b) = b - u_1$.

So we have three different representations of $e(X = b) = e_{\min}(X = b)$:

$\P$ $\sigma\mid_X$ is the valuation σ restricted to the variable X.

1. u_0 and u_1 are defined and $u_0 - b = b - u_1$ holds: The disjunction $X = u_0 \vee X = u_1$ represents the constraint.
2. either both u_0 and u_1 are defined and $u_0 - b < b - u_1$ holds or u_1 is undefined: The equation $X = u_0$ represents the constraint.
3. either both u_0 and u_1 are defined and $u_0 - b > b - u_1$ holds or u_0 is undefined: The equation $X = u_1$ represents the constraint.

In the following, we call these representations *representatives*. Obviously the extension of the partial hierarchy $H_{i,j}$ is equivalent to the system which we will get if we extend the system $C_{i,j}$ by the appropriate representative.

It is remarkable that this general transformation includes the special case, where the extension of $C_{i,j}$ (extended by $X = b$) is satisfiable. In this case u_0 and u_1 are both defined and equal. Their common value is b. Consequently the constraint $e(X = b) = e_{\min}(X = b)$ is correctly represented by $X = b$.

In general we have the extending constraint $a_1 X_1 + \ldots + a_n X_n = b$ which has to be fulfilled as well as possible. This means that we are looking for an appropriate representative of the "constraint"

$$e(\sum a_i X_i = b) = e_{\min}(\sum a_i X_i = b)$$

To solve this problem we reduce the general case to the special case mentioned above: We extend the system $C_{i,j}$ by the constraint $\sum a_i X_i = Y$ resulting in the system $C'_{i,j}$. The variable Y is a "new" internal variable, which is not used in any constraint in the whole hierarchy. Trivially each solution of the extended system is a solution of the original system, and every solution of the extended system is obviously extensible to a solution of the extended system. Thus, we are looking for solutions of the constraint system

$$C'_{i,j} \wedge e(Y = b) = e_{\min}(Y = b) \text{ with } C'_{i,j} := C_{i,j} \wedge \sum a_i X_i = Y \ ,$$

where the representation of $e(Y = b) = e_{\min}(Y = b)$ – now considering $C'_{i,j}$ instead of $C_{i,j}$ – is clear.

In the case where the extending constraint has the form $X \le b$ we evaluate

$$w := \min\{w \mid \{X \leftarrow w\} = \sigma \mid_X, C_{i,j}\sigma\}$$

and distinguish between two different cases:

1. $w \le b$ holds: $(w \le) X \le b$ represents $e(X \le b) = e_{\min}(X \le b)$.
2. $w > b$ holds: $X = w$ represents $e(X \le b) = e_{\min}(X \le b)$.

The value of w is well-defined because the system $C_{i,j}$ is satisfiable by hypothesis.

It is remarkable that in the first case the system $C_{i,j} \wedge X \le b$ is satisfiable. The part in parentheses is redundant but it makes the implicitly given restriction $w \le X$ explicit.

In general, when we have a constraint of the form $\sum a_i X_i \le b$, we extend the system $C_{i,j}$ by a constraint $\sum a_i X_i = Y$ in the introduced manner. The problem reduces itself to the problem where we have to consider the constraint $Y \le b$.

Constraints of the form $\sum a_i X_i < b$ are equivalent to constraints of the form $\sum a_i X_i \le b - 1$ because the variables have integer domains.

We conclude: The introduced method describes an algorithm to transform partial constraint hierarchies step by step into ordinary constraint systems.

Lemma 7. *Let $C_{i,j}$ be the constraint system which is equivalent to the partial hierarchy $H_{i,j}$ ($\ne H$). Let $c_{i,j}$ be the extending constraint of $H_{i,j}$ and let $r_{i,j}$ be its appropriate representative. Then the constraint system $C_{i,j} \wedge r_{i,j}$ is equivalent to the extension of $H_{i,j}$.*

Proof. The proof follows directly from the previous considerations. □

Finally, the method yields a transformation algorithm for ordered constraint hierarchies.

Theorem 8. *Let H be one of the determined constraint hierarchies with the comparator $better_{e_{|.|},l}$. The result of the successive application of the introduced transformation method results in finite steps in a constraint system C, which is equivalent to H.*

Proof. The number of steps we need to generate the system C is at most the number of constraints in the hierarchy H. Initially we start with the constraint system H_0. Assume this system consists of n_0 constraints. The proof is given by induction over the number of constraints in the hierarchy. The case $n = n_0$: If we have no other hierarchy levels than H_0 then the hierarchy is identical to H_0. The case $n \to n + 1$: By induction hypothesis we have a partial hierarchy consisting of n constraints and its equivalent constraint system. We are able to transform the extension of this partial hierarchy – which consists of $n + 1$ constraints – into an equivalent constraint system (see Lemma 7). If the hierarchy consists of $n_0 + k$ constraints, we will get a system C after k steps, which is equivalent to the hierarchy H. □

4.4 Transformation with Incomplete Solvers

There are only a few constraint (logic) programming systems which integrate complete constraint solvers, for example the CLP($\mathcal{R}$) system [7] and the ECL^{i}PS^e system [3]. They use a combination of the Gaussian elimination method and the Simplex algorithm. The Simplex algorithm is exponential in time (worst case), needs a lot of space for "slack" variables (for details see [12]) and is restricted to linear programming problems.

When we have discrete problem domains and finite value domains like in task scheduling (discrete time axis) and layout generation (pixel matrix or grid), it is more convenient to use finite domain solvers (FD solvers). They are able to deal with non-linear constraints. These are delayed until they become linear. So it is possible to consider polynomial equations and inequations instead of linear ones and the following solution strategy is also applicable to polynomial constraints.

FD solvers are correct but incomplete and more efficient than complete solvers. *Correct* means that the solution set of a constraint system is a subset of the solution set of the FD solver. *Incomplete* means that the solution set of a constraint system is in general a proper subset of the evaluated solution set. Normally we have to search for solutions of the constraint systems in the solution sets of the solver. The search space is given by the Cartesian product of the value domains of the variables. They are properly restricted by the solver using (local) propagation techniques which are polynomial in time.

Let $D_C(X)$ be the value domain of the variable X w.r.t. the constraint system C. This domain is a superset of all admissible values of the variable X w.r.t. C:

$$\{u \mid \{X \leftarrow u\} = \sigma \mid_X, C\sigma\} \subseteq D_C(X)$$

if we use an efficient but incomplete constraint solver.

Based on set inclusion and the properties of the minimum and maximum functions it holds that:

$$u_0 := \min\{u \mid \{X \leftarrow u\} = \sigma \mid_X, C_{i,j}\sigma \text{ and } u \geq b\}$$
$$\geq \min\{u \mid u \in D_{C_{i,j}}(X) \text{ and } u \geq b\} =: v_0$$
$$u_1 := \max\{u \mid \{X \leftarrow u\} = \sigma \mid_X, C_{i,j}\sigma \text{ and } u \leq b\}$$
$$\leq \max\{u \mid u \in D_{C_{i,j}}(X) \text{ and } u \leq b\} =: v_1$$
$$w := \min\{w \mid \{X \leftarrow w\} = \sigma \mid_X, C_{i,j}\sigma\}$$
$$\geq \min\{w \mid w \in D_{Ci,j}(X)\} =: m$$

Without knowledge about the values of u_0, u_1, and w, we replace in the representatives of the equations $X = u_0$ by $X \geq v_0$, $X = u_1$ by $X \leq v_1$ (in the disjunctions, too). In the representatives of the inequations we replace $w \leq X \leq b$ or $X = w$ respectively by $m \leq X$. These replacements weaken the generated representatives, because $u_0 \geq v_0$, $u_1 \leq v_1$, and $w \geq m$.

In all cases where v_0 and v_1 are defined, it is impossible to distinguish between the three cases $u_0 - b = b - u_1$, $u_0 - b < b - u_1$, and $u_0 - b > b - u_1$. The representative of an equation have to be replaced by a disjunction $X \geq v_0 \vee X \leq v_1$. This disjunction is a weakening of the representatives in all cases.

The weakenings in the constraint systems of all partial hierarchies of the partial hierarchy $H_{i,j}$ result in a constraint system $\tilde{C}_{i,j}$. It is obvious that the solution set of the weakened system $\tilde{C}_{i,j}$ is a superset of the solution set of the system $C_{i,j}$.

For the special partial hierarchy H the replacement results in a constraint system $\tilde{C}$ which is not equivalent to the hierarchy H, but a weakening of it. However, the solution set of the system $\tilde{C}$ is a subset of S_0, the solution set of the required constraints in H_0. So the solution set of $\tilde{C}$ is a better starting point for a naive solving of the hierarchy than the solution set S_0.

In the following, we propose a solution strategy different from a naive "generate-and-compare" strategy. By integration of found solutions into additional constraints, we follow the goal to strengthen the constraint system $\tilde{C}$ iteratively, until we finally get the system C.

Improving the Representatives of the Equations. Let $X \geq v_0$ be a replacement in a representative of an equation (eventually in a disjunction). If we have found a solution of the system $\tilde{C}$ with $X \leftarrow e_0$ ($e_0 \geq v_0$), we replace the weaker version $X \geq v_0$ by $e_0 \geq X \geq v_0$ in $\tilde{C}$ without loosing any solution of the hierarchy. The reason for this is given by the following considerations: Per definition it holds that $e_0 \geq u_0 \geq v_0$. Both inequations reduce themselves to one equation, namely $X = v_0$ if $e_0 = v_0$. This is completely correct, because in this special case $v_0 = u_0$ holds.

Let the replacement $X \leq v_1$ and a solution with $X \leftarrow e_1$ ($e_1 \leq v_1$) be given. Then, we replace $X \leq v_1$ by $e_1 \leq X \leq v_1$. Analogously, we do not loose any solution of the hierarchy and in the special case, where $e_1 = v_1$ holds, it holds that $v_1 = u_1$. The weakened representative is replaced by $X = u_1$ correctly.

If we have replacements of the form $e_0 \geq X \geq v_0$ (or $e_1 \leq X \leq v_1$) in a strengthened version of $\tilde{C}$ and some additional solutions $X \leftarrow f_0$ with $f_0 < e_0$ ($X \leftarrow f_1$ with $e_1 < f_1$), we will replace the weaker versions by $f_0 \geq X \geq v_0$ (or $e_1 \leq X \leq v_1$ respectively).

Reducing Disjunctions. Under certain circumstances it is possible to reduce weakened versions of disjunctions of the form $(e_0 \geq)X \geq v_0 \vee (e_1 \leq)X \leq v_1$. These circumstances are:

- $e_0 = b$ or $e_1 = b$ holds. It follows (by definition) that $u_0 = b = u_1$ and the disjunction reduces itself to $X = b$. This means that the original equation is satisfiable w.r.t. the corresponding partial hierarchy.
- $e_0 - b < b - v_1$ holds and implies $u_0 - b < b - u_1$. The second disjunctor has no consequences and will be removed (e.g. by assertion of the additional constraint $X > b$).
- $b - e_1 < v_0 - b$ holds and implies $u_0 - b > b - u_1$. The first disjunctor has no consequences and will be removed (e.g. by assertion of the additional constraint $X \leq b$).

Now we consider the cases where the values of u_0 or u_1 are known. Under certain circumstances reductions of disjunctions are possible, too:

- The disjunction has the form $X = u_0 \vee (e_1 \leq) X \leq v_1$ and additionally $b - v_1 > u_0 - b$ holds. The disjunction reduces itself to $X = u_0$, because $b - u_1 \geq b - v_1 > u_0 - b$ holds.
- The disjunction has the form $(e_0 \geq) X \geq v_0 \vee X = u_1$ and additionally $b - u_1 > v_0 - b$ holds. The disjunction reduces itself to $X = u_1$, because $b - u_1 < v_0 - b \leq u_0 - b$ holds.

Improving the Representatives of the Inequations. Let $m \leq X$ ($m \leq b$) be a replacement of a representative of an inequation $X \leq b$. If we have found a solution of the system $\tilde{C}$ with $X \leftarrow e_0$ ($m \leq e_0 \leq b$), the inequation is satisfiable, and we replace the weaker constraint $m \leq X$ by the stronger version ($m \leq)X \leq b$ – for example by adding the constraint $X \leq b$.

Let $m \leq X$ ($b \leq m$) be a replacement of a representative of an inequation $X \leq b$. If we have found a solution of the system $\tilde{C}$ with $X \leftarrow e_1$ ($m \leq e_1$), we replace the weaker constraint $m \leq X$ by the stronger version $m \leq X \leq e_1$ (e.g. by adding the constraint $X \leq e_1$). If $m = e_1$ holds, $m \leq X \leq e_1$ will be replaced by $X = m$ and it will hold that $m = w$.

General Improvement Strategy. The introduced improvements have to be carried out w.r.t. the constraint order starting with the constraint c_1^1 of highest priority. It is necessary to repeat the improvements of the considered constraint until the correct representative is found, before we continue with the improvements of the representatives of the remaining constraints. Only this general strategy guarantees that the resulting constraint system is equivalent to the constraint hierarchy.

5 Implementation

A prototype of the introduced transformation method is implemented in $\mathrm{ECL^iPS^e}$ [4] – the ECRC common logic programming system – using the incomplete constraint solver in its finite domain library [3]. We use directed search to increase the improvements mentioned above. The search is activated by the assertion of the "next" extending constraint and stops after the values of of u_0, u_1, or w, respectively, are found. This strategy guarantees that the systems $\tilde{C}_{i,j}$ converges to the desired systems $C_{i,j}$ in finite replacement steps.

Starting with H_0 we consider the constraint system which is equivalent to the partial hierarchy and the weakened system of the extension of this partial hierarchy. Then we strengthen the weakened system until we get a system which is equivalent to the extension of the partial hierarchy and so forth until we get the system which is equivalent to the whole hierarchy H.

It is also possible to delay the search and the improvements until all constraints in the hierarchy are known. However in the generation phase of the constraint hierarchy only the constraints at level 0 are propagated, and possible restrictions yielding additional information (e.g. control information) do not take place.

5.1 Directed Search for Increasing Improvements

The main goal of our directed search is the replacement of the weakened representatives by the correct representatives. To reach this goal the search concentrates its activities on the "estimations" v_0, v_1, and m for the values of u_0, u_1, and w.

If the valuation $X \leftarrow v_0$ has been extended to a solution of the weakened system, then $u_0 = v_0$ holds. Otherwise we replace v_0 by $v_0 + 1$, because for all solutions it holds that $X > v_0$. Analogously, $u_1 = v_1$ $(w = m)$ holds if $X \leftarrow v_1$ $(X^{\cdot} \leftarrow m)$ has been extended to a solution and v_1 (m) is replaced by $v_1 - 1$ $(m + 1)$ if this was not the case.

In all cases we also use the previously introduced improvement methods, especially the reduction of disjunctions, and repeat the search with its improvements until the correct representatives are found. The search terminates because we consider discrete value domains and non-negative integer values a, b, and c with $u_0 = v_0 + a$, $u_1 = v_1 - b$, and $w = m + c$ are existing.

5.2 Examples

We choose two small examples to show the activation and results of our prototype implementation. In the first example we consider the constraint hierarchy H consisting of $H_0 = \{S < T, S \neq 100\}$ and $H_1 = \{S = 100\}$:

```
[eclipse 2]: S #< T, S ## 100,
             adding([S, T], iseq(S, 100), _).
```

The evaluated results are correct (the upper bound of T is a default value):

```
S = S{[99, 101]}
T = T{[100..10000000]}
```

```
Delayed goals:
    T{[100..10000000]} - S{[99, 101]}#>=1
```

```
yes.
```

In the second example, we consider the constraint hierarchy H consisting of $H_0 = \{S < T, S \neq 100\}$, $H_1 = \{S = 100\}$, and $H_2 = \{T = 100\}$:

```
[eclipse 3]: S #< T, S ## 100,
             adding([S, T], iseq(S, 100), _),
             adding([S, T], iseq(T, 100), _).
```

The prototype evaluates the correct results:

```
S  = 99
T  = 100
```

```
yes.
```

6 Results and Future Work

The proposed method is useful to handle "static" over-constrained systems. It may be the base for maintaining dynamic systems. So there are two main topics in our future research activities:

- incremental solving of dynamically changing hierarchies,
- more sophisticated search strategies.

In interactive environments, especially in planning, scheduling, and graphical applications, the problem arises how to manage dynamically changing constraint hierarchies. The application of the proposed method requires that the constraints are given in the order of the hierarchy. Our future effort will be investigated in an incremental transformation method being able to manage arbitrary insertions and removals of constraints in a hierarchy.

It is not clear whether the proposed easy-but-rigid search strategy is more efficient than search strategies using all formerly generated weaker representatives. An advantage of such strategies may be the possibility to consider the variables in the representatives in a flexible order, and the use of heuristics. How to organize the search and how to guarantee correctness and termination of such strategies are open problems.

The proposed constraint hierarchies are "unfair": The first constraint of a hierarchy level is satisfied as well as possible without consideration of the other constraints at the same level. Currently, it is impossible to consider more than one constraint with the same priority during the minimization of the error. In many practical applications this is not acceptable. In the future we will minimize the errors at the hierarchy levels more globally. We have the idea that it is possible to transform any constraint hierarchy into an equivalent hierarchy having at most one constraint at each level, which is trivially ordered.

References

1. Alan Borning and Robert Duisberg. Constraint-based tools for building user interfaces. *ACM Transactions on Graphics*, 5(4):345–374, October 1986.
2. Alan Borning, Robert Duisberg, Bjorn Freeman-Benson, Axel Kramer, and Michael Woolf. Constraint hierarchies. In *Proceedings of the OOPSLA*, 1987.
3. ECRC, München. *ECL^{i}PS^e 3.5, Extensions User Manual*, 1995.
4. ECRC, München. *ECL^{i}PS^e 3.5, User Manual*, 1995.
5. Bjorn Freeman-Benson, Molly Wilson, and Alan Borning. DeltaStar: A general algorithm for incremental satisfaction of constraint hierarchies. In *Proceedings of the 11th Annual IEEE Phoenix Conference on Computers and Communication*, pages 561–568. IEEE, March 1992.
6. H.-J. Goltz. Reducing domains for search in CLP(FD) and its application to job-shop scheduling. In Ugo Montanari and Francesca Rossi, editors, *Proceedings of the First International Conference on Principles and Practice of Constraint Programming 1995*, number 976 in Lecture Notes in Computer Science, pages 549–562. Springer Verlag, 1995.
7. Nevin C. Heintze, Spiro Michaylov, Peter J. Stuckey, and Roland H.C. Yap. *The CLP(R) Programmer's Manual, Version 1.2*. IBM Thomas J. Watson Research Center, PO Box 704, Yorktown Heights, NY 10598, U.S.A., September 1992.
8. G. Nelson. Juno, a constraint-based graphics system. *Computer Graphics, ACM SIGGRAPH*, pages 235–243, 1985.
9. Francis Newberry Paulisch and Walter F. Tichy. Edge: An extendible graph editor. *Software — Practice and Experience*, 20(S1):63–88, June 1990.

10. Michael Sannella. The SkyBlue constraint solver and its applications. In Vijay A. Saraswat and Pascal van Hentenryck, editors, *Proceedings of the 1993 Workshop on Principles and Practice of Constraint Programming*. The MIT Press, 1994.

11. Ken Satoh and Akira Aiba. Computing soft constraints by hierarchical constraint logic programming. Technical Report 610, Institute for New Generation Computer Technology (ICOT), 1991.

12. A. Schrijver. *Theory of Linear and Integer Programming*. John Wiley & Sons, 1986.

13. Fujio Tsusumi. An efficient algorithm of logic programming with constraint hierarchy. In Jean-Pierre Jouannaud, editor, *Proceedings of the First International Conference on Constraints in Computational Logics*, number 845 in Lecture Notes in Computer Science, pages 170–182. Springer Verlag, 1994.

14. Molly Ann Wilson. *Hierarchical Constraint Logic Programming*. PhD thesis, Dept. of Computer Science and Engineering, University of Washington, 1993.

15. Molly Ann Wilson and Alan Borning. Hierachical constraint logic programming. *The Journal of Logic Programming*, 16(3 & 4):277–318, July and August 1993.

A Compositional Theory of Constraint Hierarchies (Operational Semantics)

Michael Jampel

Department of Computer Science, City University, Northampton Square, London
EC1V 0HB, U.K. Email `jampel@cs.city.ac.uk`.

Abstract. We propose a variant of the Hierarchical Constraint Logic Programming (HCLP) scheme of Borning, Wilson, and others. We consider compositionality and incrementality in Constraint Logic Programming, introduce HCLP, and present Wilson's proof that it is non-compositional. We define a scheme which uses bags (multisets) called BCH (Bags for the Composition of Hierarchies) for composing together solutions to individual hierarchies; it calculates a superset of the solutions expected from HCLP. We prove that BCH is compositional. We then define FGH (Filters, Guards, Hierarchies), a non-compositional scheme which removes precisely those BCH solutions which are unacceptable to HCLP. Thus we separate HCLP into two parts, one compositional and one non-compositional.

1 Background

1.1 Introduction

The Hierarchical Constraint Logic Programming (HCLP) scheme of Borning, Wilson, and others [2, 11, 13] greatly extends the expressibility of the general CLP scheme [5]. There is also related work by Satoh [10]. A semantics has been defined for HCLP [11, 12] and some instances of it have been implemented [9, 11]. However, the semantics is not as natural as one might hope, and the implementations are inherently less efficient than those of CLP. We believe that these two issues may be related, and suggest that splitting HCLP into two parts, one of which has a clean compositional semantics, and the second which introduces the non-monotonicity necessary to mimic HCLP, may overcome both these issues. Our earlier work in this area [7] did not include the idea of splitting HCLP into two schemes, which we now feel was a limiting factor.

We propose a decomposition of the semantics of HCLP into two parts, which we call BCH (Bags for the Composition of Hierarchies) and FGH (Filters, Guards, Hierarchies). BCH is compositional and so is amenable to efficient implementation. It composes individual hierarchies together, giving a superset (in fact, super-bag) of the solutions that would be obtained in HCLP from combining the programs and starting from scratch. The reason we need FGH is to remove those extra solutions calculated by BCH, so that we end up with exactly the same answers as would be obtained from HCLP. The main mathematical tool we use is the theory of bags; we augment it with a new relational operator and a new class of functions.

Motivation: it is good for constraint languages to be compositional. HCLP is non-compositional, so we have defined a compositional variant which we call BCH. Unfortunately BCH calculates too many solutions, and so we describe a scheme called FGH which removes precisely those solutions which we do not want.

Structure: We begin by explaining the terms 'incremental' and 'compositional' which are important for motivating our work (Sect.1.2). In Sect.1.3 we introduce HCLP and present a proof of its non-compositional nature. We discuss bags in Sect.1.4. In Sect.2 we define BCH, including a definition of guards; we discuss BCH's relationship to HCLP and its complexity and incrementality. Section 3 contains FGH, the second part of our scheme, and includes the description of a particular filter function and a discussion of its complexity. Section 4 begins with an example which includes a comparison with HCLP, and then proves the equivalence of FGH and HCLP. Finally we draw conclusions and mention future directions.

1.2 Incrementality and Compositionality

Standard HCLP is beautifully expressive, but the efficiency of its implementations may be poor, and its semantics lacks certain desirable properties. Our proposal in this paper involves splitting HCLP into two parts to gain efficiency and more tractable semantics. In constraint logic programming, efficiency is discussed with reference to 'incrementality', whereas in discussing semantics one characteristic that we look for is compositionality; these are the concerns of the rest of this section.

There is no precise algebraic characterisation of compositionality: we say that a function or operator is compositional if it preserves certain structure that is relevant in a given situation. For example, set union is commutative and associative, but does not preserve number of occurrences. So union is compositional when we consider, say, a collection of raindrops merging together (as one drop plus another drop is still just one drop), but not when we consider the mass of water involved. In the context of incremental implementations of CLP systems, we can say that a theory is compositional if the solution to the combination of two problems is the same as the combination of the solutions to the problems separately. More formally, if we have some kind of solution function or proof system ρ, if we can combine *problems* using $\cup_\rho$, and if we can combine *solutions* using $\circ_\rho$, then

$$\rho(A \cup_\rho B) = \rho(A) \circ_\rho \rho(B)$$

For our purposes, proving the compositionality of the ρ system will entail proving the associativity and commutativity of $\circ_\rho$, at least.

In a CLP context, compositionality is a desirable property for a system to have because it shows that the semantics of the system can be modelled by a mathematical system with formal properties, and because it suggests that implementations will be efficient. This is because compositionality implies *decompositionality*, i.e. we can solve a complex problem by splitting it into simpler parts, solving them, and then composing the results into a complete solution.

Whereas compositionality is a property of formal systems, incrementality is a (desirable) property of Constraint Logic Programming *implementations*. There is no precise definition, but what it means is that the work required to add an extra constraint to the solution of a large set of constraints and check its satisfiability is proportional to the complexity of the addition, and not related to the size of the initial set. If a system is not incremental, then adding one more constraint to the solution of, say, 20 constraints, involves as much work as solving the system of 21 constraints from scratch.

(In fact, even in an 'incremental' system the amount of work required to deal with an additional constraint will probably depend on more than just the constraint itself: the number of variables in the original set may be relevant, as well as other factors.)

We wish to suggest that compositionality is weaker than incrementality: a theory which has compositional semantics may have a non-incremental implementation, but a truly incremental implementation of a non-compositional formalism is difficult to define. What is more important than the distinction between incrementality and compositionality is the distinction between having both these properties and having neither, which is related to the distinction between 'sufficiently efficient' and 'unusably inefficient'. Both logic programming and constraint logic programming are in principle sufficiently efficient, and they have compositional theories (see next section), and so compositionality is assumed to hold in general, almost without being mentioned. Therefore, the focus in previous CLP work has tended to be on incrementality alone, rather than on its relationship with compositionality.

Query Composition in Logic Programming and CLP. Consider the two programs in Fig.1. They may be logic programs or constraint logic programs. The subscripts 1,2,3 are not part of the predicate names; they indicate that the query ?- p(X) will have up to three solutions {P_1, P_2, P_3}, one from each of the three clauses of p. The composed query ?- p(X) & q(X) will have up to six solutions, arising from ⊛, the composition, in some sense, of the elements of the cartesian product of the two sets of solutions. We could treat each of the {P_i ⊛ Q_j} combinations independently, and make a distinction between this multiplicity and multiple solutions all arising from one branch[1], but in this paper we will just consider the collection of solutions as a whole, ignoring how they arose.

In standard logic programming, the composition of the solutions to two or more queries is the most general unifier (m.g.u.) of those solutions. Calculating the m.g.u. of two solutions takes time related to the size of the solutions, and not related to the size of the original programs. Similarly, in constraint logic programming, solving the conjunction of two output sets of constraints will depend

---

[1] For example, '$X > 4$' over the finite domain $0 \ldots 10$ will give rise to 6 solutions.

```
p₁(X) :- ...                                q₁(X) :- ...

p₂(X) :- ...                                q₂(X) :- ...

p₃(X) :- ...

?- p(X).                                    ?- q(X).
{P₁, P₂, P₃}                                {Q₁, Q₂}

          ?- p(X) & q(X).
          {Pᵢ ⊛ Qⱼ | i = 1,2,3, j = 1,2}
```

Fig. 1. Composing queries in logic programming

on the size of the output, not on the size or complexity of the input constraints. (These two operations, of finding the m.g.u. and conjoining the constraints, are intensional, i.e. they deal with the abstract representation of the solutions. If we consider extensional solutions (models), then in both cases composition is just intersection.)

Hierarchical CLP (HCLP) is not compositional, and so incremental implementations have to make assumptions which may then need to be retracted [9]. But before we can demonstrate this non-compositionality, it is necessary to provide an overview of HCLP.

1.3 Hierarchical Constraint Logic Programming

A good introduction to HCLP can be found in Molly Wilson's PhD thesis [11, chapter 4] or in the early reference [2]; here is a brief overview. Just as Logic Programming can be extended to CLP, so CLP can be extended to a Hierarchical CLP scheme including both 'hard' and 'soft' constraints. The HCLP scheme is parameterised not only by the constraint domain $\mathcal{D}$ but also by the 'comparator' $\mathcal{C}$, which is used to compare and select from the different ways of satisfying the soft constraints.

An HCLP rule has the form

$$p(\mathbf{t}) :- q_1(\mathbf{t}), \ldots, q_m(\mathbf{t}), l_1 c_1(\mathbf{t}), \ldots, l_n c_n(\mathbf{t}).$$

where $\mathbf{t}$ is a list of terms, $p, q_1, \ldots, q_m$ are atoms and $l_1 c_1, \ldots, l_n c_n$ are labelled constraints. A program is a bag (multiset) of rules, and a query is a bag of atoms. The strengths of the different constraints are indicated by a non-negative integer label. Constraints labelled with a zero are *required* (hard), while constraints labelled j for some $j > 0$ are optional (soft), and are preferred over those labelled k, where $k > j$. (A program can include a list of symbolic names, such as *required, strongly-preferred*, etc., for the strength labels, which will be mapped to the natural numbers by the interpreter. If the strength label on a constraint is omitted, it is assumed to be *required*.)

Goals are executed as in CLP, except that initially non-required constraints are accumulated but otherwise ignored[2]. If there is more than one solution to a goal, the accumulated hierarchy of optional constraints is solved, thus refining the valuations. The method used to solve the non-required constraints will vary from domain to domain, and for different comparators.

The constraint store σ (a set) is partitioned into the set of required constraints S_0 and the set of optional ones S_i. (Solutions in standard HCLP are generally described using sets, not bags.) The solution set for the whole hierarchy is a subset of the solution set of S_0, such that no other solution could be 'better', i.e. for all levels up to k, S_k is completely satisfied, and for level S_{k+1} this solution is better than all others, in terms of some comparator. Backtracking and incomparable hierarchies give rise to multiple possible solution sets, each a subset of the solution to S_0.

Certain comparators can be used with any domain. For example, a 'predicate' comparator prefers one solution to another if it satisfies more constraints at some level (and an equal number of constraints at all previous levels). However if the domain has a metric, for example the real numbers, it is possible to ask how far from the preferred answer a solution is, in which case one might prefer fewer constraints to be exactly satisfied if the distance of the answer from a given point can be minimised[3]. Weights can be used within a particular level of the hierarchy in order to influence the solution, but a heavily weighted constraint in a given level is completely dominated by the lightest constraint in any more important level. Wilson calls this property 'respecting the hierarchy' [11].

In this paper, we shall only consider the *unsatisfied-count-better* (UCB) comparator, which is quite simple to understand and which can be defined over any domain. Basically, one valuation is better than another if it leaves fewer constraints unsatisfied (or equivalently, if it satisfies more constraints).

Definition: a solution θ is *unsatisfied-count-better* than a solution σ if it satisfies as many constraints as σ does in levels $1 \ldots k - 1$, and at level k it satisfies strictly more constraints than σ. See [11].

Quite a lot of the other work on HCLP considers a different comparator, *locally-predicate-better*, which is slightly less discriminating than *UCB* but is easier to implement in certain situations. Locally-predicate-better is concerned not just with the *numbers* of constraints satisfied by a particular valuation, but by the particular constraints themselves. For the sake of comparison, its definition is given here:

---

[2]This is an implementation detail but, while it is not essential, it helps us to understand the differing roles of required and optional constraints. Menezes et al. use an alternative 'optimistic' strategy [9].

[3]For a complete understanding of comparators it is necessary to consider 'error sequences' [11, 13], but they are quite complicated and are only used briefly in this paper, so we omit them here for reasons of space.

Definition: a solution θ is *locally-predicate-better* than a solution σ if it satisfies every constraint that σ does in levels $1 \ldots k-1$, and at level k it satisfies a strict superset of the constraints satisfied by σ. (If θ and σ satisfy different constraints then they are incomparable and both will appear in the solution set.) See [11, 13].

The Disorderly Property of HCLP. Wilson discusses a very simple but powerful example in her PhD thesis [11], which shows the non-monotonic, hence non-compositional, nature of any variant of HCLP which respects the hierarchy.

Wilson defines the 'orderly' property as follows: let P and Q be constraint hierarchies, and let $S_{\{P\}}(\mathcal{C})$ be the set of solutions to the hierarchy P when comparator $\mathcal{C}$ is used. Then $\mathcal{C}$ is *orderly* if $S_{\{P \cup Q\}}(\mathcal{C}) \subseteq S_{\{P\}}(\mathcal{C})$. A comparator which is not orderly is *disorderly*.

Proposition 1. *Any comparator which respects hierarchies over a non-trivial domain $\mathcal{D}$ is disorderly.*

Proof. (Wilson [11, Sect.2.5.3]): Let $P = \{\texttt{weak X = a}\}$ and $Q = \{\texttt{strong X = b}\}$ for two distinct elements of $\mathcal{D}$, a and b. (The existence of distinct elements is what makes $\mathcal{D}$ non-trivial.) $S_{\{P\}}(\mathcal{C})$ will contain the valuation which maps X to a, and if $\mathcal{C}$ respects the hierarchy then $S_{\{P \cup Q\}}(\mathcal{C})$ will contain the valuation which maps X to b. So $S_{\{P \cup Q\}}(\mathcal{C}) \not\subseteq S_{\{P\}}(\mathcal{C})$, so $\mathcal{C}$ is disorderly. □

Non-compositionality: We can extend this proof as follows[4]. Consider $R = \{\texttt{required X = a}\}$. Then $S_{\{R\}}(\mathcal{C}) = S_{\{P\}}(\mathcal{C})$ but $S_{\{R \cup Q\}}(\mathcal{C}) \neq S_{\{P \cup Q\}}(\mathcal{C})$, which is a proof of the absence of monotonicity, an important compositionality property.

1.4 Bags

Bags, sometimes called multisets, are like sets except that duplicate elements are allowed i.e. $\langle a, a \rangle \neq \langle a \rangle$ (we use $\langle, \rangle$ to denote bags). In this section, we define various properties of bags, treating as basic the notion of number of occurrences of an element in a bag (see Gries & Schneider [4]).

By definition, an element which occurs a times in a bag A, and b times in B, occurs $a + b$ times in the additive-union $A \uplus B$, $\max(a,b)$ times in $A \cup B$, which is the equivalent of set-theoretic union, and $\min(a,b)$ times in the intersection $A \cap B$. We will not use $A \cup B$ in this paper, and so we feel free to refer to $A \uplus B$ as union, rather than the more clumsy 'addition' or 'additive-union'.

Let $e \# B$ denote the natural number n of occurrences of the element e in the bag B.[5]

---

[4]Adapted from a suggestion by Michael Maher [Private Communication].

[5]Schemes based on bags seem more appropriate for finite domains. But they can also be considered for domains where an extensional view would be impractical, such as the reals, if we allow the 'extension' of a constraint to be written as the union of disjoint

Example. $a\# \{a, a\} = 2.$ $b\# \{a, a\} = 0.$

A shorthand is to label elements with a superscript indicating the number of occurrences.

Example. $\{a, a\} = \{a^2\}.$ Of course $a\#\{a^k\} \equiv k.$

Note that bag intersection is a 'lifted' version of set intersection. In other words, if A and B are both bags with no duplicate elements ('set-like' bags), then so is $A \cap B$; we can define a function **set** which makes bags set-like as follows:

Definition.

$$e\#\mathbf{set}(A) = \begin{cases} 1, \text{ if } e\#A \geq 1 \\ 0, \text{ otherwise} \end{cases}$$

The bag equivalent of the subset operation may be called sub-bag. We will use the symbol $\subseteq$, and define it in terms of $\#$ as follows:

Definition. $A \subseteq B \Leftrightarrow \forall e \cdot e\#A \leq e\#B$

2 BCH — Bags for the Composition of Hierarchies

2.1 Guards

Later in the paper we need an alternative to bag intersection, with certain other properties, and so we now define a new infix binary operator $/\!\!/$ ('guarded by'). If A and B are both bags, then $A /\!\!/ B$ contains only those elements of A which are also in B, with the same multiplicity as in A. (For set-like bags, $/\!\!/$ is the same as intersection.)

Definition.

$$e\#(A /\!\!/ B) = \begin{cases} e\#A, \text{ if } e\#B > 0 \\ 0, \quad \text{ if } e\#B = 0 \end{cases}$$

Example.

$$\{a, a, c\} /\!\!/ \{a, b\} = \{a, a\}$$
$$\{a, a, c\} /\!\!/ \{b\} \quad = \{\}$$

We claim that BCH, to be defined below using guards, has a strong mathematical basis, including the fact that it is compositional. In order to defend this view, we need to prove that $/\!\!/$ has various properties (such as associativity, distributivity through $\uplus$, etc.). This is done in the longer version of this paper [6].

---

ranges. For example, the 'extensional solution' of the constraints 'weak $X > 5$, weak $X < 10$' is '$(X \leq 5)$, $(5 < X < 10)$, $(5 < X < 10)$, $(10 \geq X)$' or, equivalently, '$(X \leq 5)$, $(5 < X < 10)^2$, $(10 \geq X)$'. Then $\#$ denotes the number of occurrences of a *range*, rather than a single element.

2.2 BCH Solutions to Compositions of Hierarchies

We now define an associative, commutative composition operator $\circ$ in terms of its effect on solutions to individual hierarchies $p_1, p_2, \ldots, p_k$, which we assume have each been calculated separately by some method to be discussed below. This operator defines what we mean when we refer to BCH or $\mathrm{BCH}_\circ$:

Definition (BCH Compose).

$$S_0(\circ\, p_i) =_{\text{def}} \bigcap_i S_0(p_i)$$

$$S_n(\circ\, p_i) =_{\text{def}} \left(\biguplus_i S_n(p_i) \right) /\!/ S_0(\circ\, p_i) \qquad\qquad (n > 0)$$

Note that all these levels are guarded by the solution to level 0, i.e. the required constraints. Solutions to p_i's optional constraints which contradict p_j's required constraints are removed by the guard, thus any valuation found lower down the hierarchy will be acceptable to all the required constraints. But certain valuations are present which are not as preferred as others — these rules have no analogue for comparators in HCLP (but see below, Sect.3). Consequently, as each remaining element of the S_i's satisfies the required constraints we can say that the rules 'respect the requireds'. But as elements may be present in S_2 which are not in S_1, say, these rules do *not* 'respect the hierarchy'.

It is clear that $\circ$ is commutative, due to the commutativity of intersection and union of bags. To show that $\circ$ preserves all that we require under composition, it is necessary to prove that it is associative.

Proposition 2. *Compose Associativity* $(p \circ q) \circ r = p \circ (q \circ r)$

Proof. **Case S_0:** obvious, by the associativity of $\cap$.
Case S_n $(n > 0)$: In the following, let P be an abbreviation for $S_0(p)$, let p mean $S_n(p)$ except inside expressions such as $S_n(p \circ q)$ on the first and last lines, and similarly for Q, R, q, and r.

$$e\#(S_n((p \circ q) \circ r))$$

$$
\begin{aligned}
&= e\#(((p \uplus q)/\!/(P \cap Q)) \uplus r)/\!/(P \cap Q \cap R) && \{\text{by defn. of } \circ\} \\[4pt]
&= \begin{cases} 0, & \text{if } e\#P = 0 \text{ or } e\#Q = 0 \text{ or } e\#R = 0 \\ e\#((p \uplus q)/\!/(P \cap Q)) + e\#r, & \text{otherwise} \end{cases} && \{\text{by defn. of } /\!/\} \\[4pt]
&= \begin{cases} 0, & \ldots \\ e\#p + e\#q + e\#r \end{cases} && \{\text{as } e\#P \neq 0,\ e\#Q \neq 0\} \\[4pt]
&= \begin{cases} 0, & \ldots \\ e\#p + e\#((q \uplus r)/\!/(Q \cap R)) \end{cases} && \{\text{as } e\#Q \neq 0,\ e\#R \neq 0\} \\[4pt]
&= e\#(((q \uplus r)/\!/(Q \cap R)) \uplus p)/\!/(P \cap Q \cap R) && \{\text{by defn. of } /\!/\} \\[4pt]
&= e\#(S_n(p \circ (q \circ r))) && \{\text{by defn. of } \circ\}
\end{aligned}
$$

(where the third and fourth explanatory comments, beginning 'as $\ldots$', refer to the non-zero case). $\qquad\qquad\qquad\qquad\qquad\qquad\qquad\qquad\qquad\qquad\qquad\qquad\square$

Empty Solutions for a Level. In HCLP, the key difference between required and non-required constraints is that the former can cause failure to occur. In other words, the required constraints may have an empty solution set, but no weaker constraints can cause a failure if stronger constraints have been satisfied[6]. In our composition rules, in $A /\!/ B$, A represents the solutions for a weaker level than B, and yet the definition of $/\!/$ allows the possibility of $A /\!/ B$ being empty even if B is not (in the case that $A \cap B = \emptyset$). Thus our rules appear to allow failure to arise from optional constraints. In fact, as we define the solution S to be the tuple $\langle S_0, S_1, \ldots, S_n \rangle$ and not just its final element S_n, it is not a problem if one of the elements of S is empty: the solution that is offered to the user is no longer the final element of the tuple, but all the non-empty elements. Thus this aspect of HCLP is not present in our logic, but is left until the interpretation[7].

2.3 BCH Solutions to Individual Hierarchies

The previous section was concerned with composing solutions to hierarchies; but where do these solutions come from in the first place? One possibility is to assume that HCLP has been used initially, with BCH being invoked to compose the previously calculated answers in order to avoid starting HCLP from scratch on the composed program texts. All the theory in this paper is well-defined for this situation (treating the various solution sets $S_1, \ldots, S_n$ as set-like bags), however in the rest of this section we discuss the second possibility, namely that BCH is used for everything. When we use BCH to solve an individual hierarchy we label the rules $\text{BCH}_{\text{indiv}}$; when used to compose solutions the terms BCH or $\text{BCH}_{\circ}$ are used.

Note that the answers for a single hierarchy defined by $\text{BCH}_{\text{indiv}}$ will not be the same as those calculated by HCLP (the latter will be a subset of the former, as discussed below). The answer to a composition of solutions of course depends on the individual solutions; as these will differ depending on whether $\text{BCH}_{\text{indiv}}$ or HCLP was used to find them, the answers to the composition will also differ. But given the solutions to the sub-problems, then the solution to their composition calculated by BCH is completely determined.

We now define what we mean by $\text{BCH}_{\text{indiv}}$: it is the method of calculating the solution to an individual hierarchy, defined as follows. The solution to a hierarchy P is defined to be a tuple $\langle S_0(P), S_1(P), \ldots, S_n(P) \rangle$ where each S_i is the solution to the constraints at level i, guarded by S_0. To solve the required constraints

---

[6]In fact we need a slightly stronger condition in HCLP; if S_0 is non-empty *and finite* then we can be certain that S_n will not be empty. The extra condition is necessary to deal with certain pathological cases, such as the hierarchy 'required $X > 0$, strong $X = 0$' with a metric comparator. See [1, pp. 233–234] for more details.

[7]An alternative is to define a binary infix 'choice' operator $\diamond$ such that $A \diamond B = A$ if $A \cap B \neq \emptyset$, and $A \diamond B = B$ otherwise. Then every occurrence of $A /\!/ B$ in our rules would be replaced by $(A /\!/ B) \diamond B$. Thus if the optional constraints contradict the required ones, the solution is just the solution to the required constraints. This shows that we *can* model this aspect of HCLP inside BCH, but at the cost of complicating the rules.

and find S_0, we invoke the appropriate CLP solver. To 'solve' the optional constraints, take the (bag- additive-) union of the extensions of the individual constraints. Then guard each of the S_i, $i > 0$, with S_0, thus removing solutions which violate the required constraints.

This definition of the solution to an individual hierarchy may seem extravagant, but it becomes less so in the context of filter functions such as $\mathbf{f}_{max}$ (Sect.3.2).

Relationship between HCLP and BCH$_{indiv}$. In the above discussion each of the non-required levels contains the union of the extensions of the constraints at that level. HCLP would also produce at most the union of the extensions (if all the constraints were consistent with all stronger levels, but also mutually contradictory). At the other extreme, if all the constraints were mutually consistent, then the HCLP solution would be the *intersection* of the extensions. If we removed the labels from all the constraints at a given level and solved them using CLP, the result would be the (possibly empty) intersection of the extensions. Therefore the solutions for each level satisfy the following relation:

$$CLP \subseteq HCLP \subseteq B\dot{C}H_{indiv}$$

No Constraints at a Level. If there are no constraints at the *required* level, the solution set is the entire domain (the cartesian product of the domains of all the variables) which we denote by saying $S_0 = \mathcal{U}$. It can be seen that guarding any bag with this universal set will not have any effect: $\forall A \cdot A /\!/ \mathcal{U} = A$. If there are no constraints at one of the optional levels, the BCH solution for that level will be $\mathcal{U} /\!/ S_0 = S_0$. Therefore, we can see that the rules we have defined here will work for hierarchies with arbitrarily many or few constraints at any given level.

The Orderly Property of BCH. The particular example used in the proof of the disorderliness of HCLP in Sect.1.3 does not apply to BCH. This is encouraging, as is the proof of the associativity of $\circ$, but is not a proof of orderliness; we have yet to develop a formal proof.

Equivalence of BCH$_{indiv}$ and BCH$_\circ$.

Proposition 3. $BCH_{indiv}(A) \circ BCH_{indiv}(B) = BCH_{indiv}(A \cup B)$
As long as all levels contain constraints, and where $\cup$ represents program text union.

Proof. The definition of BCH$_{indiv}$ requires the extensions of each constraint at a level to be combined using union ($\uplus$). The definition of $\circ$ also combines equivalent levels from two hierarchies using union. The distributivity of $\uplus$ through $/\!/$ guarantees that the order in which the various combinations are done is unimportant. Hence the proposition is true. Note that the left-hand-side of the proposition defines the BCH_o solution (i.e. using $\circ$ to compose two previously calculated hierarchies) and so we can see that the difference between BCH_{indiv} and BCH_o is that BCH_{indiv} defines the meaning of $\langle \_ \rangle$, whereas BCH_o composes $\langle \_ \rangle$'s, however they are defined. $\square$

Relationship between HCLP and BCH_o. By a similar argument to that on the relationship between HCLP and BCH_{indiv} (see Sect.2.3 above), and considering Proposition 3, we can see that $HCLP \subseteq BCH_o$.

Complexity of BCH. (The following is a summary; for detailed calculations, see the longer technical report version of this paper [6].) Using a naïve representation, the complexity of composing k hierarchies each with n solutions to their required constraints (*not* necessarily the number of solutions from n constraints) and l levels of optional constraints is $O(lkn^2)$ in the worst case. Using a more sensible representation, and noting that k and l are likely to be fixed at small values, the total complexity of BCH is $O(n\log n)$.

This result is expressed in terms of basic comparison operations. Comparison with the complexity of HCLP is difficult, partly because no analysis of the latter exists in the literature; but note that whatever results might be obtained are likely to be in terms of constraint checks, an inherently more difficult operation than simple element comparison.

Incrementality of BCH. More interesting than standard complexity results is the question of incrementality. If we have already composed k hierarchies using BCH, what is the extra work required to compose one additional hierarchy?

Let us assume that the previously composed system has a total of n elements in S_0, and hence at most n distinct elements in each of the optional levels. Let us assume that the new hierarchy has at most m elements per level. Let l be the maximum of the number of optional levels in the two systems. Calculating the new S_0 requires the intersection of two sets, of size m and n. This will take $O(mn)$ at worst, and the resulting set will have size at most $\max(m,n)$. For each optional level, finding the union of two bags takes $O(m + n)$, and then guarding with S_0 takes at most $O(\max(m, n).(m + n))$. If we make the assumption that m is as large as n, this expression becomes $O(n^2)$.

Therefore when we include the calculations on l optional levels the total complexity is $O(mn + l.\max(m, n).(m+n)) = O(l.\max(m, n).(m+n)) = O(ln^2)$ (setting $m = n$). This could be improved by guarding each level of the previously composed system with the required level of the new hierarchy *before* taking the union of the optional levels, and by other methods. But even so, it is clear that the work required will be approximately quadratic rather than linear in m. Therefore, perhaps, we should not claim that BCH is incremental. But note that k, the number of systems which were originally composed together, before the addition of the new hierarchy, is not mentioned in these results. In other words, the work done by BCH may depend on the number of *solutions* in the previously composed system, but not on the number of original hierarchies. We are able to avoid starting from scratch, re-calculating the composition of k systems, which is not the case in HCLP. So we have achieved the goal set in Sect.1.2 of this paper.

3   FGH — Filters, Guards and Hierarchies

3.1   Filter Functions

The second half of this paper describes FGH, which takes the results of BCH and uses them to calculate solutions equivalent to those of HCLP. Unlike BCH, FGH respects the hierarchy, and so is disorderly and non-monotonic, hence non-compositional.

FGH uses an extra mathematical construct, a class of filters, functions from bags to bags, which remove some of the elements from the input bag. It is necessary to introduce them here because we will use them in the next section to define rules for removing less preferred valuations from BCH solutions. Particular examples are discussed later, but in general filter functions will be denoted by $\mathbf{f}$, and $\mathbf{F}$ will be used to denote the raised version (i.e. $\mathbf{f}$ applied to each member of a tuple of bags). Note that the guard operator $/\!\!/$ is concerned with the relationship between *different strength levels* in a hierarchy, whereas filter functions select solutions *within* a given level. (Similarly, in HCLP the comparators compare solutions within a given level.) More details can be found in the next section, where we examine one particular filter at length.

We now define an operation $\mathbf{F}$ which removes elements from bags in a tuple, forming new bags. Let P be a program with a BCH solution $\langle S_0(p), S_1(p), \ldots \rangle$. Note that $S_1(p) \subseteq S_0(p)$, $S_2(p) \subseteq S_0(p)$, etc., but it is not necessarily the case that $S_2(p) \subseteq S_1(p)$, etc. As we have already used $p \circ q$ to refer to a BCH operation we will describe the FGH extension as $\mathbf{F}(p \circ q)$, where $\mathbf{F}(p)$ is defined as follows:

Definition (F).

$$\begin{aligned}
\mathbf{F}\langle S_0(p), S_1(p), \ldots, S_n(p) \rangle \\
= \langle \mathbf{F}(S_0(p)), \mathbf{F}(S_1(p)), \ldots, \mathbf{F}(S_n(p)) \rangle \\
\mathbf{F}(S_0(p)) = S_0(p) \\
\mathbf{F}(S_n(p)) = \mathbf{f}\left(S_n(p) /\!\!/ \mathbf{F}(S_{n-1}(p))\right) \qquad (n > 0)
\end{aligned}$$

where $\mathbf{f}$ is a filter function, applied to the result of guarding a level with the previous level (itself filtered). Note that we overload the symbol $\mathbf{F}$, allowing it to operate both on tuples and on members of tuples. In fact, instead of $\mathbf{F}(S_n(p)) = \mathbf{f}\left(S_n(p) /\!\!/ \mathbf{F}(S_{n-1}(p))\right)$, we should really write $\mathbf{F}(S_n(p), S_{n-1}(p)) = \mathbf{f}\left(S_n(p) /\!\!/ \mathbf{F}(S_{n-1}(p))\right)$, as it is a function of two inputs, not one.

The reason for filtering each level is to remove solutions which are less preferred than others. The reason for guarding S_n with level S_{n-1} is to remove solutions which are acceptable to level n, but less preferred by the stronger constraints of level $n-1$. Of course, we also need to remove solutions which are less preferred by levels $n-2$, $n-3$, etc., but this is dealt with by the fact that S_{n-1} has itself been guarded by S_{n-2}, and so it is not necessary to guard S_n by S_{n-2} explicitly.

Note that in the definition of **F** we have not included an application of **set**. We consider it to be the very last action that should be performed on the bags of solutions, perhaps just treated as an aspect of pretty-printing.

3.2 The Filter $\mathbf{f}_{max}$

The rules defined in the previous section are parameterised by filter functions. We now define one particular filter function $\mathbf{f}_{max}$, which is the most interesting when compared to HCLP's 'unsatisfied-count-better' comparator [11]. $\mathbf{f}_{max}$ removes those elements of a bag which do not occur a maximal number of times. In other words, if some elements occur once in a given bag, and some elements occur twice, $\mathbf{f}_{max}$ defines the bag containing only those elements occurring twice.

Definition.

$$e \; \# \; \mathbf{f}_{max}(A) = \begin{cases} e\#A, & \text{if } e\#A = \max\{i\#A \cdot i \in A\} \\ 0, & \text{otherwise} \end{cases}$$

Example.

$$\mathbf{f}_{max}\{a, a, b\} = \{a, a\}$$
$$\mathbf{f}_{max}\{a, b\} = \{a, b\}$$

$\mathbf{F}_{max}$ is defined as the raised version of $\mathbf{f}_{max}$ in the obvious way.

FGH Example: **F** was defined in terms of a single hierarchy p, but p may arise from the BCH composition of two or more hierarchies. For this example, let us assume that we have two hierarchies q and r with the following solution sets: $S_0(q) = \{a, b, c\}$ and $S_0(r) = \{b, c, d\}$. Let $S_1(q) = \{b, c\}$ and $S_1(r) = \{b, d\}$. Then $S_0(q \circ r) = \{b, c\}$ and $S_1(q \circ r) = \{b, b, c, d\}$.

$$\mathbf{F}_{max}(S_1(q \circ r)) = \mathbf{f}_{max}(\{b, b, c, d\} /\!/ \{b, c\})$$
$$= \mathbf{f}_{max}(\{b, b, c\})$$
$$= \{b, b\}$$

So the BCH solution is $\langle \{b, c\}, \{b, b, c, d\} \rangle$ and the FGH($\mathbf{F}_{max}$) solution is $\langle \{b, c\},$ $\{b, b\} \rangle$; therefore the preferred solution to p is the value b as it is the only element of the final non-empty bag in the solution tuple, where the final bag has been affected by all the levels of optional constraints. Thus BCH allows us to explore the *space* of solutions, while FGH produces its best element.

For simplicity, the example above concerned possible values for a single variable (i.e. $X = a$ or $X = b$, for some variable X). But it is clear that everything can be defined in terms of possible values for a vector or sequence $\mathbf{X} = [X_1, X_2, \ldots, X_n]$. In this case, # counts the number of occurrences of a particular vector and not simply a particular value for a single variable. For example, the bag $\{\langle a, b \rangle, \langle a, c \rangle\}$ contains two distinct 'elements', corresponding to $[X_1 = a, X_2 = b]$ and $[X_1 = a, X_2 = c]$.

Non-incrementality of FGH. We can give a simple example why $\mathrm{FGH}(\mathbf{F}_{\max})$ is not compositional as follows: consider two bags $\lbrace a, b\rbrace$ and $\lbrace b, c\rbrace$. Composing them with BCH gives $\lbrace a, b, b, c\rbrace$, and then filtering with $\mathbf{f}_{\max}$ will result in $\lbrace b, b\rbrace$. If the third bag contains $\lbrace a, c\rbrace$, then composing it with $\lbrace b, b\rbrace$ followed by filtering will give the incorrect answer $\lbrace b, b\rbrace$. However, if we had composed it at the BCH stage (getting $\lbrace a, a, b, b, c, c\rbrace$), filtering would have resulted in a different answer. So FGH is not associative, hence not compositional.

Therefore if we wish to add a constraint to the results of filtering a tuple of bags, we cannot do it directly. We must compose the BCH solution tuple for the new constraint (which will be very simple to calculate) with the BCH solution of the original hierarchy, and then re-filter. Thus FGH is not really incremental in its own right. Of course, BCH *is* incremental, and it is relatively cheap to re-filter using FGH, and so we still have some gain compared to HCLP, which must restart right from the very beginning.

The Disorderly Property of FGH. Let us consider again the example which was used in the proof of the disorderly nature of HCLP (Sect.1.3). The two hierarchies were $P = \lbrace\text{weak X = a}\rbrace$ and $Q = \lbrace\text{strong X = b}\rbrace$ and the solutions calculated by BCH were $\mathrm{BCH}(P) = \langle \mathcal{U}_X, \mathcal{U}_X, \lbrace a\rbrace\rangle$, $\mathrm{BCH}(P \circ Q) = \langle \mathcal{U}_X, \mathcal{U}_X \uplus \lbrace b\rbrace, \mathcal{U}_X \uplus \lbrace a\rbrace\rangle$ and $\mathrm{BCH}(P \cup Q) = \langle \mathcal{U}_X, \lbrace b\rbrace, \lbrace a\rbrace\rangle$.

The FGH solutions are: $\mathbf{F}_{\max}(P) = \langle \mathcal{U}_X, \mathcal{U}_X, \lbrace a\rbrace\rangle$, $\mathbf{F}_{\max}(P \circ Q) = \langle \mathcal{U}_X, \lbrace b, b\rbrace, \lbrace b\rbrace\rangle$ and $\mathbf{F}_{\max}(P \cup Q) = \langle \mathcal{U}_X, \lbrace b\rbrace, \lbrace\rbrace\rangle$. So for this example $\mathbf{F}_{\max}(P \cup Q) \subseteq \mathbf{F}_{\max}(P) \not\subseteq \mathbf{F}_{\max}(P \circ Q)$, which does not yet demonstrate the disorderliness of FGH.

Now consider a different hierarchy $P' = \lbrace\text{weak X = a}, \text{strong X = c}\rbrace$. Then $\mathrm{BCH}(P') = \langle \mathcal{U}_X, \lbrace c\rbrace, \lbrace a\rbrace\rangle$ and $\mathbf{F}_{\max}(P') = \langle \mathcal{U}_X, \lbrace c\rbrace, \lbrace\rbrace\rangle$. Also $\mathrm{BCH}(P' \circ Q) = \langle \mathcal{U}_X, \lbrace b, c\rbrace, \mathcal{U}_X \uplus \lbrace a\rbrace\rangle$ and $\mathrm{BCH}(P' \cup Q) = \langle \mathcal{U}_X, \lbrace b, c\rbrace, \lbrace a\rbrace\rangle$. Therefore $\mathbf{F}_{\max}(P' \circ Q) = \langle \mathcal{U}_X, \lbrace b, c\rbrace, \lbrace b, c\rbrace\rangle$ and $\mathbf{F}_{\max}(P' \cup Q) = \langle \mathcal{U}_X, \lbrace b, c\rbrace, \lbrace\rbrace\rangle$. As $\lbrace b, c\rbrace \not\subseteq \lbrace c\rbrace$, $\mathbf{F}_{\max}(P' \cup Q) \not\subseteq \mathbf{F}_{\max}(P')$, thus demonstrating the disorderliness of FGH.

Complexity of FGH. It is clear that $\mathbf{F}$ requires l filtering steps for a hierarchy with l levels, using any given $\mathbf{f}$.

If $\mathbf{f}_{\max}$ is used, for each level it is necessary to sort by number of occurrences (greatest first). If there are up to k occurrences of each of n distinct elements in a level, sorting will take $O(kn \log kn)$ simple comparison operations (or $O(n \log n)$ if a sensible representation is used). Keeping a note of the maximum number of occurrences of an element found so far, it is easy to select the first part of the output bag, i.e. those elements occurring precisely k times. This process must be repeated for each level, and so the complexity of FGH using $\mathbf{f}_{\max}$ is $O(lkn \log kn)$, where in practice k and l are likely to be fixed at quite small values, leading again to a complexity of $O(n \log n)$.

(All the above is in addition to the complexity of the BCH step, which was found to be $O(\log k . n \log n + lkn)$, or $O(n \log n)$ under some sensible assumptions. See Sect.2.3.)

3.3 Other Filter Functions

Although the previous section has concentrated on $\mathbf{f}_{max}$, other filter functions are possible. It is interesting to define filter functions which mimic the various HCLP comparators; whereas $\mathbf{f}_{max}$ is derived from *unsatisfied-count-better* one can imagine other filters based on metric comparators such as *least-squares-better (lsb)*. These might not simply filter existing elements from bags, but could actually create new ones. For example, if $A = \{2, 5, 5\}$, $\mathbf{f}_{max}(A) = \{5, 5\}$, but $\mathbf{f}_{lsb}(A) = \{3\}$ (or perhaps $\{3, 3, 3\}$).

4 Discussion

4.1 Example — Comparison with HCLP

Figure 2 contains an example of the use of both BCH and FGH, and a comparison with HCLP. Part of the example (P) is taken from Wilson's thesis [11, Sect.2.2]. Note that BCH does not discriminate enough, by itself, to capture the behaviour of standard HCLP. This is not surprising, and motivates the use of filter functions. Also note that the combination of FGH and BCH does indeed give precisely the same answers as HCLP.

4.2 Relationship between HCLP and FGH

HCLP is parameterised by a comparator $\mathcal{C}$, and FGH is parameterised by a particular filter function. Therefore it is not possible to state a relationship as general as, say, "The HCLP solution set for a hierarchy is identical to its FGH solution"[8].

Notwithstanding the above, we *can* claim the following, where *UCB* stands for the *unsatisfied-count-better* comparator:

Proposition 4. $\text{FGH}(\mathbf{F}_{max})(H) = \text{HCLP}(UCB)(H)$

Proof. Reminder: a solution θ is *unsatisfied-count-better* than a solution σ if it satisfies as many constraints as σ does in levels $1 \ldots k - 1$, and at level k it satisfies strictly more constraints than σ. See [11].

As discussed in Sect.2.3, we assume that the the members of the BCH solution tuple will contain only those elements which are acceptable to the required constraints, and will contain at least all the elements which are preferred by one or more optional constraints.

$\mathbf{f}_{max}$ selects all elements occurring at least n times, for some n, and filters out all elements occurring less then n times. An element appears n times at a given optional level because it is preferred by n constraints at that level. An element only appears less than n times if it is preferred by less than n constraints. Therefore it would not have been selected by *UCB*. □

---

[8]More precisely, we cannot state "The HCLP solution set for a hierarchy H is identical to the result of applying **set** to the final non-empty bag of the FGH tuple of solutions for H".

	P	Q	R
required	$X > 0$	$0 < X < 20$	$0 < X < 23$
strong	$X < 10$	$X > 5$	$X > 15$
weak	$X = 4$	X div 7^a	X div 8
Sols.^{b}			
S_0	$X > 0$	$0 < X < 20$	$0 < X < 23$
S_1	$0 < X < 10$	$5 < X < 20$	$15 < X < 23$
S_2	$X = 4$	$X = 7, X = 14$	$X = 16$

c	HCLP^d	BCH	FGH^e
S_0	$0 < X < 20$	$0 < X < 20$	$0 < X < 20$
S_1	$5 < X < 10$ $15 < X < 20$	$(0 < X \leq 5)^1$ $(5 < X < 10)^2$ $(10 \leq X \leq 15)^1$ $(15 < X < 20)^2$	$(5 < X < 10)^2$ $(15 < X < 20)^2$
S_2	$X = 7$, $X = 8$, $X = 16$	$X = 4$, $X = 7$, $X = 8$, $X = 14$, $X = 16$	$X = 7$, $X = 8$, $X = 16$

^{a}i.e. X is divisible by 7, shorthand for the domain constraint $X::[0,7,14,\ldots]$
^{b}HCLP solutions for each hierarchy individually
^{c}Solutions for the combined hierarchies
^{d}Calculated taking into account error sequences [11]
^{e}Using $\mathbf{f}_{\max}$

Fig. 2. Compositions using HCLP, BCH, and FGH

Let us define a CLP(all) solution to be one which satisfies *all* constraints. Now any BCH solution satisfies at least one of the optional constraints. An HCLP solution satisfies as many as possible of the optional constraints, as does an FGH solution. CLP(required) refers to the solutions found by ignoring the optional constraints, hence it satisfies all the required constraints but none of the optional ones. Given these definitions, and combining the result of Proposition 4 with that of Sect.2.3, we can say:

$$\text{CLP(all)} \subseteq \mathbf{set}(\text{FGH}) = \text{HCLP}(UCB)$$
$$\subseteq \mathbf{set}(\text{BCH}) \subseteq \text{CLP(required)}$$

5 Conclusions

We have developed a compositional variant of HCLP based on bags, which stores the intermediate solutions to a hierarchy in a tuple $\langle S_0, S_1, \ldots, S_n \rangle$. We are able to avoid invoking the constraint solver to recalculate solutions from scratch, due to the simplicity and elegant mathematical properties of our scheme. This scheme allows the exploration of the solution space, and can be implemented in an incremental manner.

We have defined a new binary infix relation over bags, called 'guard'. We have also defined a class of filter functions over bags, and placed them in a non-compositional framework which respects the theory of HCLP. We have examined one filter function in detail, and shown that it can be used to calculate the same solutions to a hierarchy as would be obtained by HCLP using the unsatisfied-count-better comparator. Thus we have separated HCLP into its compositional and non-compositional parts.

Most of our presentation has been couched in terms of finite domains of integers, but it is clear that our work can be extended to any of the usual constraint domains, such as reals.

In general, constraint satisfaction is of exponential complexity, compared to which guarding and filtering are cheap. In addition to being efficient, these operations are simple to understand, and calculate the answers we would obtain from HCLP while avoiding its computational expense and complex semantics.

6 Further Work

In the future we wish to extend this scheme to take account of error sequences, and hence bring different (metric) comparators within the scope of FGH. We would like to investigate constraint deletion and dynamic constraint satisfaction, which is difficult in most formalisms but should be made easier by the modular nature of our tuple representation of solutions.

We believe that our bag-based approach to HCLP will facilitate the integration of hierarchical techniques with Freuder's Partial Constraint Satisfaction method, as we are able to move the strength label from a constraint to each member of its extension without the loss of preference information that would occur in a set-based scheme, and hence work at this lower tuple level as required by all CSP techniques. (We have already made progress towards such an integration using a slightly different approach. See [8].)

We are also investigating inconsistencies arising from the composition of CLP programs i.e. programs without hierarchical strength labels. We wish to explore the parallels between composition of these unlabelled systems and composition of the labelled systems discussed in this paper.

More generally, we would like to investigate the links between HCLP and default logic, and other paradigms which allow preferences to be overridden. We believe that fuzzy logic may also provide interesting comparisons with HCLP.

Acknowledgements. Thanks to Sebastian Hunt, Alan Borning, Michael Maher, Rob Scott and David Gilbert for helpful discussions on CLP, HCLP, and previous versions of this paper.

References

1. Alan Borning, Bjorn Freeman-Benson, and Molly Wilson. Constraint Hierarchies. *Lisp and Symbolic Computation*, 5:223–270, 1992.
2. Alan Borning, Michael Maher, Amy Martindale, and Molly Wilson. Constraint Hierarchies and Logic Programming. In *ICLP'89*, Lisbon, 1989.
3. Michel Gangnet and Burton Rosenberg. Constraint Programming and Graph Algorithms. In *2nd International Symposium on Artificial Intelligence and Mathematics*, January 1992.
4. David Gries and Fred Schneider. *A Logical Approach to Discrete Math*. Springer-Verlag, 1994.
5. Joxan Jaffar and Jean-Louis Lassez. Constraint Logic Programming. In *POPL'87*, Munich, 1987.
6. Michael Jampel. A Compositional Theory of Constraint Hierarchies. Technical Report TCU/CS/1995/5, Department of Computer Science, City University, London, March 1995.
7. Michael Jampel and Sebastian Hunt. Composition in Hierarchical CLP. In *IJCAI'95*, Montreal, August 1995.
8. Michael Jampel. *Over-Constrained Systems*. PhD thesis, Department of Computer Science, City University, 1996. (Draft, January 1996. Final version to be submitted in June 1996.).
9. Francisco Menezes, Pedro Barahona, and Philippe Codognet. An Incremental Hierarchical Constraint Solver. In *PPCP'93*, 1993.
10. Ken Satoh. Formalizing Soft Constraints by Interpretation Ordering. In *ECAI'90*, 1990.
11. Molly Wilson. Hierarchical Constraint Logic Programming. Technical Report 93-05-01, University of Washington, Seattle, May 1993. (PhD Dissertation).
12. Molly Wilson and Alan Borning. Extending Hierarchical Constraint Logic Programming: Nonmonotonicity and Inter-Hierarchy Comparison. In *NACLP'89*, 1989.
13. Molly Wilson and Alan Borning. Hierarchical Constraint Logic Programming. *Journal of Logic Programming*, 16(3):277–318, July 1993.

Heuristic Methods for Over-Constrained Constraint Satisfaction Problems*

Richard J. Wallace and Eugene C. Freuder

University of New Hampshire, Durham, NH 03824 USA

Abstract. Heuristic repair methods have successfully solved constraint satisfaction problems (CSPs) and satisfiability problems (SAT) that are too large to be solved by complete algorithms. In this paper we develop methods for testing the efficiency and quality of solution returned by these methods when applied to overconstrained CSPs and SAT. The key strategy is to test heuristic methods on problems of moderate size with known optimal distances (number of constraint violations), as determined with complete algorithms. This allows us to determine whether heuristic methods find optimal distances and allows us to carry out more incisive analyses of efficiency when different strategies are incorporated into these methods and parameter values are varied. The present work tested the min-conflicts algorithm with CSPs, either alone or in combination with walk, reset or tabu strategies. SAT was tested with GSAT and walk-SAT. The best results for min-conflicts were found with the walk strategy, when the probability of random assignment was set at 0.10 or 0.15. Both GSAT and walk-SAT readily found optimal solutions for 3-SAT, the latter being somewhat faster overall.

1 Introduction

Constraint satisfaction problems (CSPs) involve finding an assignment of values to variables that satisfies a set of constraints between these variables. In many important applications the problems may be overconstrained, so that no complete solution is possible. In these cases, partial solutions may still be useful if a sufficient number of the most important constraints are satisfied. An example of such partial constraint satisfaction is the maximal constraint satisfaction problem (MAX-CSP), in which the goal is to find assignments of values to variables that satisfy the maximum number of constraints.

An important type of CSP is the satisfiability problem (SAT), in which the goal is to find a truth assignment that satisfies a CNF formula in propositional logic. A generalization of satisfiability that corresponds to MAX-CSP is the maximum satisfiability problem (MAX-SAT) in which the object is to find an truth assignment that satisfies the maximum number of clauses in the formula.

In recent years methods have been developed for solving constraint satisfaction problems which are based on local improvements of an initial assignment

---

* This material is based on work supported by the National Science Foundation under Grant Nos. IRI-9207633 and IRI-9504316.

that violates an unspecified number of constraints, rather than on incremental extensions of a fully consistent partial solution. In some cases these local or heuristic repair methods have solved problems that are too large to be solved by complete methods that involve backtrack search [3] [5].

In view of these results, it is worthwhile to apply these heuristic techniques to overconstrained problems, in particular to MAX-CSPs and MAX-SAT problems. Since these are optimization problems, the basic question is whether heuristic methods will return optimal (here, maximal) or near-optimal solutions to these problems after a reasonable amount of time.

To answer this question, one must be able to assess the quality of solutions found by these methods. For problems with complete solutions such evaluation is easy, since an optimal solution is one that is complete, i.e., it must satisfy all the constraints in the problem. Quality can then be judged in terms of the difference between the solution found and a complete solution. In this case, the number of violated constraints or the sum of the weights of these constraints is a straightforward measure of quality.

In contrast, with overconstrained problems one cannot determine optimality *a priori*. However, rigorous assessment is possible if complete methods can be used that return guaranteed optimal solutions. Solutions found by heuristic repair methods can then be compared with those found by complete methods, and differences in quality can be assessed in the same way as for problems with complete solutions.

As indicated by the remarks above, the size of problems for which this is possible will be limited by restrictions on the capacity of complete methods to solve large problems. However, complete methods are now available for solving some classes of MAX-CSPs with up to 60 variables (and easier problems in some classes with up to about 100 variables) and moderately large MAX-SAT. This allows assessment of solutions to problems that are large enough to be interesting in some applications, and may also allow some assessment of trends with increasing problem size (as well as changes in other parameters such as density and constraint tightness).

The complete methods used in the present work are branch and bound algorithms developed to solve MAX-CSPs [1] and MAX-SAT problems [6]. These were used to evaluate several important heuristic methods. For MAX-CSPs, heuristic methods are based on the min-conflicts procedure [3] and include versions that incorporate tabu search procedures [2]. For MAX-SAT, heuristic methods are variants of GSAT [5].

The next section reviews basic concepts and describes the algorithms. Section 3 describes our experimental methodology. Section 4 gives results for MAX-CSPs and MAX-SAT problems. Section 5 gives some conclusions.

2 Algorithms

A constraint satisfaction problem (CSP) involves assigning values to *variables* that satisfy a set of *constraints* among subsets of these variables. The set of values that can be assigned to one variable is called the *domain* of that variable. In the present work all constraints are binary, i.e., they are based on the Cartesian product of the domains of two variables. A binary CSP is associated with a constraint graph, where nodes represent variables and arcs represent constraints. If two values assigned to variables that share a constraint are not among the acceptable value-pairs of that constraint, this is an *inconsistency* or constraint violation.

Heuristic repair procedures for CSPs begin with a complete assignment and try to improve it by choosing alternative assignments that reduce the number of constraint violations. In the min-conflicts procedure, the first assignment is made by choosing values that minimize the number of constraint violations with values already chosen [3]. Then random methods are used to choose a variable whose assignment conflicts with another assignment and, from the domain of that variable, a value that has the minimal number of conflicts. GSAT, a repair procedure for SAT problems, begins with a random truth assignment and then alters ("flips") the assignment of a variable that leads to the greatest net increase in satisfiable clauses. After a certain number of flips, a new random assignment is chosen (called a new "try") [5]. A variation of GSAT, here called walk-SAT, chooses a variable at random with a probability, p, (the "walk probability") and flips its value, while the normal greedy procedure is followed with probability $1 - p$. Two features of this procedure were also tested with min-conflicts: retries after a certain number of new assignments and the random walk strategy, in which, after choosing a variable with a conflicting value as before, a new value is chosen at random for assignment, with probability, p, while the usual min-conflicts procedure is followed with probability, $1 - p$. In initial testing, values for retry and walk were derived from published sources: here, the number of assignments before a retry was five times the number of variables, while the probability of a random choice in the walk procedure was always 0.35. Since these values had originally been used for SAT problems, there was no reason to think that they were optimal for CSPs. Hence, later testing included systematic variation of the values for assignments before retry and for the walk probability. A further version of min-conflicts incorporated some of the techniques for tabu search: (i) a list of candidate variable/value pairs (values chosen randomly from the domains of conflicting variables), (ii) a tabu list, which was a simple queue of domain values that had been switched with other values.

In the present work branch and bound versions of CSP and SAT algorithms were used to determine the optimal number of constraint violations in the problems. These algorithms are described in [1] [6].

3 Experimental Methods

Random constraint satisfaction problems were generated using a "probability of inclusion" (PI) model of generation (cf. [1]). The number of variables was fixed, as well as the maximum domain size. Each possible domain element, constraint and constraint value pair was then chosen with a specified probability. PI problems had either 30 or 100 variables; domains were either variable, with an expected value of 5 or were fixed at 5 or 10, for different problem sets. All problems with less than 100 variables could be run to completion with the branch and bound methods used, so that quality of solution could be evaluated by comparison with a known optimal distance as well as with solutions obtained by other methods.

3-SAT problems were produced with a program obtained from Bell Laboratories. This generates random SAT problems with a specified number of variables and fixed-length clauses. Problems had 50 variables and either 225 or 300 clauses. In the first case the problems are near the transition region described above; about 25 percent of the sample problems had solutions. None of the 300-clause problems had solutions. Problems with these features can be solved to completion with branch and bound methods, so that the quality of solution returned by the heuristic methods could be evaluated in relation to the known optimum, as well as in relation to the complete method.

For all sets of problems, the sample size was 25. Anytime curves reported here are based on one run with each problem. (In some cases ten successive runs were made through the entire sample to determine the stability of the mean for number of inconsistencies after k sec. The standard deviations of the mean was about 5% of the mean, for each point tested.) Means reported in Section 4.2 are based on ten runs per problem.

4   Results for MAX-CSPs

4.1   Anytime Curves

Figure 1 shows anytime curves with some typical characteristics for two versions of min-conflicts and for a branch and bound algorithm based on forward checking with dynamic search rearrangement. PI problems were used in this experiment, with 30 variables, expected density = 0.33, average domain size of 8, and expected tightness = 0.30. The mean optimal distance was 0.92. (Eight problems had complete solutions.)

Using the min-conflicts preprocessing procedure, both heuristic methods begin with a much better solution than the branch and bound algorithm, in this case one based on forward checking. (It is not clear from the figure, but the mean for the initial distance is 14.00 with the heuristic methods, while the initial mean for forward checking is 48.4; these are obtained in about 0.01 and 0.05 seconds, respectively.) The simple min-conflicts procedure then quickly improves on the initial solution, so that at very short durations (here about 10 seconds or less) it is superior to the walk version as well as the branch and bound algorithm, which is obviously a poor third. However, simple min-conflicts quickly becomes 'stuck'; in most cases it reaches a local minimum in about five seconds and remains at this level for the duration of the experiment. (In fact, it does not seem to ever get 'unstuck' with these problems, as indicated by experiments with much greater cutoff times than the one used here [100 seconds].) The walk version continues to improve, and eventually surpasses the basic version of min-conflicts. Branch and bound improves more slowly, and is still lagging at the time of cutoff. (For these problems, this algorithm required up to about 6000 seconds to find an optimal solution in the hardest case.) As shown below, the value for the walk probability used here is not in fact the best. Nonetheless, the same pattern of results is usually found: simple min-conflicts finds better solutions at very short intervals but gets stuck before finding an optimal solution, while mincon-walk catches up and surpasses the basic version in a fairly short period of time. The variants of min-conflicts based on the retry strategy and tabu search also surpass the basic version. (For these problems the results for tabu search [using a candidate set of 5 and a tabu list of length 7] were very similar to those for mincon-walk with probability 0.35, while the retry [resetting after each 150 tries] was appreciably better.)

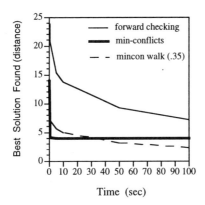

Fig. 1. Averaged anytime curves for branch and bound and heuristic repair algorithms. 30-variable problems, exp. density $= 0.33$, exp. tightness $= 0.30$, mean optimal distance $= 0.92$.

To summarize, each of the strategies for breaking out of local minima, random selection, resetting and using a tabu strategy, is successful to some extent with these problems, so that eventually the means for these techniques fall below the mean for the basic min-conflicts. The question of whether these methods can actually find an optimal solution in a reasonable amount of time is answered in the following subsections.

4.2 Parametric Analysis of Mincon-Walk

Further analysis with problems like those used in the last section has indicated that the walk strategy gives the best results of the variants tested to date–under the best parameter settings. For this reason, the walk strategy is emphasized in the present section.

The data in this section are based on PI problems with 30 variables. In the first series of problem sets, the domain size was fixed at 5. An inclusion probability for constraint pairs was chosen that gave problem sets with an average optimal solution of 1.5 to 2, in order to equalize this factor across the density dimension and to make it easier to solve these problems to completion. The expected densities for the different problem sets were 0.1, 0.3, 0.5, 0.7 and 0.9.

Table 1. Average time (seconds) required to find an optimal solution for various problems and walk probabilities.

| | | walk probability | | |
	0.05	0.10	0.15	0.25
		density = 0.10		
M	4.3	2.3	2.2	4.6
SD	7.9	3.4	3.2	8.1
		density = 0.30		
M	6.8	4.1	4.9	17.2
SD	9.7	5.5	7.3	40.0
		density = 0.50		
M	14.9	6.5	7.0	19.3
SD	30.8	11.2	8.5	27.1
		density = 0.70		
M	11.0	5.4	6.4	22.7
SD	24.8	8.1	7.2	36.9
		density = 0.90		
M	28.4	11.3	9.3	29.5
SD	86.5	19.0	11.9	46.4

Notes. M is mean, SD is standard deviation. 30-variable problems with mean optimal distance 1.5-2.

A second series of problem sets was generated with inclusion probabilities for constraint pairs that gave an average optimal distance of about 8.5. The domain size was again fixed at five. The expected densities for different problem sets were 0.1, 0.3 and 0.5. (Problems with higher densities were too difficult for complete methods.) Finally a few problem sets were generated with domain size fixed at 10. Two sets, with expected densities of 0.1 and 0.3, had average optimal distances of 1.5-2. A third set, with expected density of 0.1 had an average optimal distance of about 8.5.

To solve these problems to completion, heuristic methods were first used to find partial solutions with good distances. These distances were then used as initial upper bounds for the branch and bound algorithm. After optimal distances had been verified, this information was used in more systematic experiments with heuristic methods, to determine the average time and number of constraint checks needed to find an optimal solution. In these experiments four values for the walk probability were tested: 0.05, 0.10, 0.15 and 0.25. Each set of 25 problems was run ten times with each walk probability, and the grand average of all these runs was calculated. (These are the means shown in the tables in this section.)

The results of this analysis are shown in Tables 1-3. For the problem sets with domain size of 10, results based on single runs were basically the same. In addition, single runs with walk probability = 0.35 gave results that were markedly worse than those with the 0.25 setting, for all problem sets.

Table 2. Average constraint checks required to find an optimal solution for various problems and walk probabilities.

	walk probability			
	0.05	0.10	0.15	0.25
		density = 0.10		
M	138,465	73,866	68,591	142,202
SD	271,461	118,424	104,170	258,966
		density = 0.30		
M	326,500	176,156	208,794	715,680
SD	475,090	245,456	320,537	1,693,800
		density = 0.50		
M	752,610	356,698	366,596	965,676
SD	1,536,334	611,429	454,577	1,364,675
		density = 0.70		
M	682,131	327,561	379,321	1,346,160
SD	1,583,098	507,676	441,469	2,220,345
		density = 0.90		
M	1,953,711	766,143	617,370	1,920,546
SD	5,976,511	1,294,147	804,505	3,038,159

Notes. M is mean, SD is standard deviation. Same problems as in Table 1.

Tables 1 and 3 both show a curvilinear relation between walk probability and efficiency, with best results for probabilities in the vicinity of 0.10 and poorer performance when the probability is too high or too low. For problems with low optimal distances, the best probability value was always 0.10 or 0.15. For problems with greater optimal distances, the performance curve was shifted downward, with best results for probability 0.10 and next-best at 0.05. In both cases the performance ranking was the same throughout the range of densities tested. However, with increasing density, differences in performance became accentuated. In addition, problem difficulty increased for all settings with increased density, even with average optimal distance controlled; this may simply be due to the number of constraints that must be checked at each step.

Table 2 shows that effects on number of constraint checks required to find an optimal solution vary in the same way as run time. However, the overall rate of constraint checks tends to increase with problem difficulty as measured by run

Table 3. Average time (seconds) required to find an optimal solution for various problems and walk probabilities.

	\multicolumn{4}{c}{walk probability}			
	0.05	0.10	0.15	0.25
	\multicolumn{4}{c}{density = 0.10}			
M	3.4	3.2	6.5	76.2
SD	4.9	3.8	10.9	231.8
	\multicolumn{4}{c}{density = 0.30}			
M	6.6	6.2	12.8	242.8
SD	8.1	8.0	17.6	467.7
	\multicolumn{4}{c}{density = 0.50}			
M	19.5	12.1	22.4	367.4
SD	40.3	15.7	33.1	615.8

Note. 30-variable problems with mean optimal distance about 8.5.

time, from about 30 to 50 thousand constraint checks per second.

The reset strategy was also effective in finding optimal solutions. But even with the best parameter settings (150-250 assignments between resetting at density = 0.10 and 500 at density = 0.50), the mean time to find an optimal solution was 5-6 times as long as that given by the best walk probability values.

The simple version of min-conflicts was tested with the first problem series (one run of 1000 seconds per problem). The number of optimal solutions found was always ≤ 4 in each group of 25 problems.

4.3 Results for larger problems

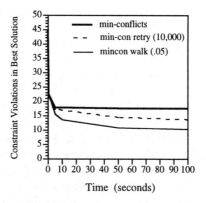

Fig. 2. Averaged anytime curves for heuristic repair algorithms. 100-variable problems, exp. density = 0.10, exp. tightness = 0.25, domain size = 5.

Figure 2 shows performance over time for three heuristic methods, using 100-variable PI problems with expected density = 0.10, domain size fixed at 5, and expected tightness = 0.25. The best average distance after 500 seconds (for walk probability = 0.05) was 9.56 for this set of problems. Based on results with smaller problems, this is probably close to the optimum. The walk and retry results are each part of a series; walk probabilities of 0.10, 0.15 and 0.25 and resetting after every 250, 500, 1000, 5000 and 25,000 tries were also tested. The parameter values shown in the figure were the best that were found in each series. In a further test, the best walk and retry values were combined, but this did not improve on the walk probability shown here.

4.4 Results for MAX-SAT

For 50-variable problems with 225 clauses, GSAT and walk-SAT found optimal solutions for all but one or all problems, respectively, within 50 seconds; both procedures found optimal solutions for most problems within a few seconds. For 300-clause problems the times were somewhat longer (all or almost all optimal solutions being found within 100 seconds instead of 50), but the pattern of results was similar. These results are reflected in the anytime curves, which, like the curves for min-conflicts, show a very steep initial descent, and much more gradual improvement thereafter (Figure 3). These curves are similar to those reported earlier for ordinary SAT [4].

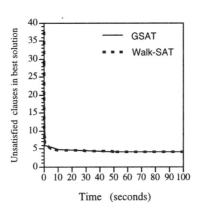

Fig. 3. Averaged anytime curves for heuristic repair algorithms on MAX-SAT problems. 50-variable, 300-clause problems.

5 Conclusions

This work provides a partial answer to the question posed in the Introduction. For MAX-CSPs, the basic min-conflicts procedure does *not* usually find an optimal solution to the maximal constraint satisfaction problem. However, variants which introduce an element of randomization can find optimal solutions with great efficiency, at least for small to medium-sized problems (i.e., up to about 100 variables). For MAX-SAT, GSAT is quite effective in most cases, but here again an element of randomization serves to enhance performance by eliminating some of the longer times to find a solution.

These results also indicate that, for MAX-CSPs, the random walk strategy of [4] is superior to a repeated resetting strategy. Preliminary results did not suggest any improvement when resetting was combined with the walk strategy or when a version of tabu search was used to avoid local minima. For both walk and reset, the effectiveness of the method depends on proper settings of parameter values in the procedure. However, for the walk strategy the best settings were always within a fairly restricted range. In constrast, for the reset strategy the best parameter value for the 100-variable problems was much greater than the best values for the 30-variable problems.

An obvious question is why a particular restricted range of probability values for the walk strategy always gave the best results. A more important question is whether this simple strategy is generally competetive with more complex procedures like tabu search or genetic algorithms in this problem domain, and whether the basis for the superiority of these techniques over the basic min-conflicts procedure can be accounted for in the same terms.

References

1. E. C. Freuder and R. J. Wallace. Partial constraint satisfaction. *Artificial Intelligence*, 58:21–70, 1992.
2. F. Glover. Tabu search: a tutorial. *Interfaces*, 20:74–94, 1990.
3. S. Minton, M. D. Johnston, A. B. Philips, and P. Laird. Minimizing conflicts: a heuristic repair method for constraint satisfaction and scheduling problems. *Artificial Intelligence*, 58:161–205, 1992.
4. B. Selman and H. A. Kautz. An empirical study of greedy local search for satisfiability testing. In *Proceedings AAAI-93*, pages 46–51, 1993.
5. B. Selman, H. Levesque, and D. Mitchell. A new method for solving hard satisfiability problems. In *Proceedings AAAI-92*, pages 440–446, 1992.
6. R. J. Wallace and E. C. Freuder. Comparing constraint satisfaction and Davis-Putnam algorithms for the maximal satisfiability problem. In D. S. Johnson and M. A. Trick, editors, *Cliques, Coloring and Satisfiability: Second DIMACS Implementation Challenge*, (to appear). Amer. Math. Soc., 1996.

Cascaded Directed Arc Consistency and No-Good Learning for the Maximal Constraint Satisfaction Problem *

Richard J. Wallace

University of New Hampshire, Durham, NH 03824 USA

Abstract. This paper describes new branch and bound methods for overconstrained CSPs. The first method is an extension of directed arc consistency preprocessing, used in conjunction with forward checking. After computing directed arc consistency counts, inferred counts are derived for each value, based on the counts of supporting values in future variables. This inference process can be 'cascaded' from the end to the beginning of the search order, to augment the initial counts. The second method is a form of wipeout-driven nogood learning: the method for finding nogoods is described and conditions for the validity of the nogood are established. In tests with random problems, significant improvements in efficiency were found with cascaded DACCs; in contrast, no-good learning did not enhance performance when used alone or in any combination of strategies.

1 Introduction

Constraint satisfaction problems (CSPs) involve assigning values to variables which satisy a set of constraints. Algorithms for ordinary CSPs are designed to return a complete solution, i.e., one that satisfies all of the constraints. But if a problem has no complete solution, these algorithms simply report this.

For problems that have only partial solutions, it may still be useful to have an assignment of values to variables that satisfies the most important constraints or, if constraints have equal weight, one that satisfies as many constraints as possible. (For situations in which good partial solutions to overconstrained problems are useful, see [2] [1] [6] [5].) The latter case, in which an optimal solution is one with a maximal number of satisfied constraints, has been termed the maximal constraint satisfaction problem (MAX-CSP).

In recent years, complete algorithms have been developed for MAX-CSPs that are based on branch and bound methods, using depth-first search [8] [7]. Enhancements have also been developed that use information obtained from local consistency tests carried out before search. This information takes the form of *inconsistency counts* , i.e., tallies for each value a of variable v_i, of the number of domains v_j that do not have any values consistent with a in the constraint

---

* This material is based on work supported by the National Science Foundation under Grant Nos. IRI-9207633 and IRI-9504316.

between v_i and v_j. These counts can be used to compute tighter lower bounds and to order domain values to find solutions with fewer inconsistencies earlier in search. The most powerful techniques of this sort use *directed arc consistency counts* (DAC-counts or DACCs) that are based on an initial variable ordering, and that refer to variables that are before or after the current value in this order [10].

In this paper a further elaboration of directed arc consistency counts is considered. Given a domain some of whose values have DACCs, inconsistency counts can sometimes be inferred for values in neighboring domains. For example, if value a in the domain of v_i is supported by values in the domain of v_j that have counts of 1 or more, then a count can be deduced for a on this basis. If such infererences are carried out systematically, DACCs can be carried forward or backward to one or the other end of the search order. In this way, it may be possible to derive much tighter lower bounds at the beginning of search.

This paper also examines nogood learning in conjunction with algorithms for MAX-CSPs. An appealing feature of this technique in this context is that, once discovered, nogoods are independent of the error factors that go into lower bound calculations. A complication here is that it cannot be assumed that a nogood tuple discovered at one point in search is still no-good at a later point. But, as shown below, a simple test can be done to verify that the tuple is still a valid nogood.

The next section, 2, presents some background for this work. Section 3 describes cascaded DAC procedures. Section 4 describes the no-good learning procedure for MAX-CSPs. Section 5 presents experimental comparisons between these new methods and the best branch and bound algorithms studied in earlier work. Concluding remarks are given in Section 6.

2   Background: Basic Concepts

A constraint satisfaction problem (CSP) involves assigning values to *variables* that satisfy a set of *constraints* among subsets of these variables. The set of values that can be assigned to one variable is called the *domain* of that variable. In the present work all constraints are binary, i. e., they are based on the Cartesian product of the domains of two variables. A binary CSP is associated with a constraint graph, where nodes represent variables and arcs represent constraints.

Branch and bound algorithms associate each path through a search tree with a cost function that is non-decreasing in the length of the path. Search down a given path can stop when the cost of the partial assignment of values to variables is at least as great as the lowest cost yet found for a full assignment. The latter, therefore, sets an *upper bound* on the cost function. In addition to calculating the cost at a particular node, projected cost can be calculated to produce a higher, and therefore more effective, *lower bound*. The present algorithms use the number of violated constraints incurred by the partial assignment of values to variables as a cost function; this is called the *distance* of a partial solution from a complete solution. Maximal solutions are associated with minimum distances.

Of the several branch and bound algorithms that have been derived for MAX-CSPs, those based on prospective strategies have proven most successful. Prospective algorithms compare each assigned value against domains of unassigned (or future) variables to determine which of the values in these domains are consistent with the values already assigned. In this way domains can be successively restricted by successive assignments. The most successful prospective algorithms for MAX-CSPs use the forward checking strategy, in which only the future domains of neighboring variables are checked for inconsistency with the value considered for the current assignment [8] [7]. The basic forward checking algorithm is described in more detail below.

The efficiency of branch and bound algorithms is greatly enhanced by variable and value ordering. Variable ordering heuristics are based on problem parameters such as domain size and degree of a node in the constraint graph. With forward checking, ordering by decreasing degree of a node has been found to be the most effective heuristic for random problems with sparse constraint graphs. For problems with density of 0.3-0.4 or greater, dynamic ordering based on the smallest current domain size is more effective [10]. As already indicated, ordering values by increasing number of inconsistencies also has important effects on search efficiency.

3 Directed Arc Consistency and Cascaded DACCs

3.1 Background: Directed Arc Consistency and Forward Checking for MAX-CSPs

The procedure for deriving directed arc consistency counts is shown in Figure 1 (based on Figure 1 of [10]). A variable search order is established at the outset. Arc consistency checking is then carried out in a single pass through the ordered list of variables. Checking is done in one direction only; in the present work this is in the forward direction (i.e., against future variables in the search order) because forward DACCs can be used with forward checking. At each step, values in a domain are checked against values in the domain of each future variable that shares a constraint with the current variable. If a value in the current domain has no supporting values in a future domain, the DACC for that value is incremented by one.

In forward checking for MAX-CSPs, a value a being considered for assignment to variable u is tested against values in the domain of each uninstantiated (future) variable v, if there is a constraint between these two variables. If b, belonging to the domain of v, is inconsistent with a, then an inconsistency count associated with b is incremented. This forward-checking count (FC-count or FCC) is a kind of backward DACC, which is only valid for the current instantiation of the set of past variables. If the sum of this count and the current distance is as great as the current upper bound, b can be discarded from the future domain, given a partial instantiation that includes value a in the domain of u. In addition, a global lower bound can be calculated, based on the sum of the

Establish search order for variables
Set DAC-count for each domain value to 0

For each variable v_i
    for each value a in domain d_i
        for each variable v_j later than v_i in the
        ordering such that v_i and v_j share
        a constraint
            if there is no value b in domain d_j
            such that (a, b) is in the constraint
            between v_i and v_j
                increment the DAC-count for a

Fig. 1. Directed arc consistency for MAX-CSPs. Here, checking is in forward direction, i. e., each value is tested for support in domains of future variables. Counts (DACCs) derived in this manner can be used with forward checking, since inconsistency counts generated during the latter procedure refer to earlier domains in the search order.

minimum FC-counts for future domains. If used with directed arc consistency, the lower bounds can be increased by adding two more factors: the DACC for b and the sum of the minimum forward DACCs (See Figure 2.)

3.2 Directed Arc Consistency and Cascaded DACCs

In this paper an elaboration of directed arc consistency is described that is intended to enhance the effect of DACCs, especially early in the search order. After carrying out the procedure outlined in Figure 1, better estimates are derived for the minimal nonsupport associated with values earlier in the search order, derived from DACCs associated with supporting and non-supporting values in future domains. These improved estimates are then used during search to enhance lower bounds. The procedure starts at the end of the search order and works backward, so that inferences of eventual nonsupport in the form of DACCs can be carried back to the beginning of search. (For this reason these counts are referred to here as "cascaded" DACCs.)

The procedure is outlined in Figure 3. For each value in the current domain (v_i in Figure 3), the sets of supporting and nonsupporting values in the domain of each future, adjacent (i.e., constraining) variable are checked separately, to find the minimum DACCs for each set. If the set of supporting values is non-empty, the minimum count for nonsupporting values can be incremented, since choosing a value from this set would violate the constraint between v_i and v_j. This would not be included in the original DACCs under these conditions, since nonsupport was tallied with respect to the entire future domain. After this, the minimum DACCs must be checked to see if they share violated constraints in common with the value in v_i (due to previous cascading). For each such match the minimum distance $= 0$, $N = \infty$

While upper bound $N > 0$
    If all values have been tried for first variable chosen
        exit
    else
        Choose a future variable v_i as the current variable
            For each value x in domain d_i
                if (distance $+$ FC-ct$_{ix}$ $+$ Σ min FC-ct$_{jy}$
                      $+$ $\underline{\text{DAC-ct}_{ix} + \Sigma \text{ min DAC-ct}_{jy}}$) $\geq N$
                    try next value
                else if all variables are now instantiated
                    save solution and set N equal to distance
                else if preclude (v_i, x, distance)
                    returns nil, try next value
                else add x to partial solution and exit for-loop
            If all values have been tested
                return v_i to the set of future variables,
                reset future domains and continue
                with previous variable chosen

Subroutine:
preclude (current-variable, current-value, distance)
    for each future variable, v_j, that shares a
    constraint with current-variable
        for each value y in domain d_j
            if y and current-value do not satisfy this constraint
                increment count FC-ct$_{jy}$
                if (distance $+$ FC-ct$_{jy}$ $+$ $\underline{\text{DAC-ct}_{jy}}$) $\geq N$,
                    discard y
        if no values remain in d_j return nil

Fig. 2. Directed arc consistency incorporated into forward checking. 'FC-ct' refers to count produced by forward checking. Inclusions based on DAC-counts are underlined. (Note: since the overhead incurred by using DACCs in the subroutine does not justify the modest improvements obtained, this feature was not used in tests reported here).

DACC must be decremented. Finally, the smallest of the two adjusted minimum DACCs is added to the DACC for the value in v_i.

In the present work, variable ordering was by decreasing degree of the node in the constraint graph. Intuitively, this ordering should be well-suited for pushing cascaded DACCs to the front of the search order, since the initial variables should receive 'streams' of DACCs from a large number of adjacent variables.

Unfortunately, the use of cascaded DACCs, as opposed to ordinary DACCs,

Find DACCs (as shown in Figure 1), associated with
constraints whose violation produces these counts.

For variable $v_i = n - 2$ to 1 with respect to the search order
  For each value a in domain d_i
    For each variable v_j later than v_i in the
    ordering such that v_i and v_j share a constraint
      Let min-suppct = minimum DACC for those values
        in v_j that support a.
      Let min-nonsuppct = minimum DACC for values
        in v_j that do not support a.
      If there are supporting values in v_j, increment
        min-nonsuppct by one.
      Let min-suppct' = min-suppct adjusted for
        redundancies with respect to the DACCs already
        associated with a.
      Let min-nonsuppct' = min-nonsuppct adjusted for
        redundancies with respect to the DACCs already
        associated with a.
      Add min (min-suppct', min-nonsuppct') to
        the DACC for a and update the list of constraints
        on which this DACC is based.

Fig. 3. Generation of cascaded directed arc consistency counts (cascaded DACCs).

involves an important tradeoff. This is because in the former case, the global
calculation involving the sum of the minimum DACCs in each future domain
must be restricted to variables that are not connected to the current variable
with respect to variables between them in the search order. Otherwise this sum
could include counts that are redundant with the cascaded DACC for the current
value. On the other hand, use of one lower bound estimate during search does
not preclude use of the other, although there will be some extra overhead due
to the greater number of calculations and comparisons.

4 Nogood Learning with MAX-CSPs

A nogood is a k-tuple that is not found in any solution for a problem. Nogood
learning can be based on the failure to find any value in a domain that is consistent with previously assigned values; in this case the set of assignments made
before considering the current variable forms a nogood [3] [9]. A more perspicu-
ous method for deriving nogoods under these circumstances is to consider only
the earlier assignments that led to inconsistencies [4]. Both methods have been
called deadend-driven learning by Frost and Dechter. In either case the nogood

is added to the constraint graph and can be used to discard the final nogood value whenever the remaining k-1 values have been assigned.

If forward checking is used, then wipeout in a future domain can serve as the basis for nogood discovery. Then, for ordinary CSPs, the set of assignments to past variables that led to deletions in the wiped-out domain forms a nogood [9]. This may be called wipeout-driven learning, to distinguish it from the situation in which running out of feasible assignments for the current variable leads to nogood discovery. As with deadend-driven learning, the nogood is added to the constraint graph and can be used to discard the final nogood value whenever the remaining k-1 values have been assigned. (In the present implementation these nogoods are associated with the penultimate variable in the nogood, with respect to the search order, and are checked whenever search reaches that variable.)

With MAX-CSPs, deletion of a value in a future domain is not necessarily triggered by finding an inconsistency with a current assignment. Only when the sum, based on the inconsistency counts (here, forward checking counts) for that value and the current distance, is greater or equal to the upper bound can the value be discarded. Under these conditions, if a wipeout occurs in the domain of v_k, then the set of values that produced increments in the inconsistency counts associated with values in domain d_k form a nogood. However, the no-good status of this tuple depends on the current distance, d, and the upper bound, N. If the difference between N and d is greater at a later time in search when this no-good is considered, then the nogood is not necessarily valid, because values that were discarded from d_k under the earlier conditions might not now be discarded. Conditions for the validity of a nogood are established in the following theorem:

Theorem 1. If a nogood tuple is found at some point in search, then this tuple will be a valid nogood at subsequent points in search if the value of (N − d) at these subsequent points is ≤ the value of (N − d) when the nogood was discovered.

Proof. Note, first, that the value of (N − d) when the nogood is discovered (at the point in search when a deletion produces a wipeout) will be ≤ the value of this quantity when any other value was deleted. This is because N will be constant during this time while d will either increase or remain the same. The difference at wipeout will be called the "critical difference" of (N − d). Now, consider the conditions in which a nogood is applicable, because the same values have been assigned to the first k-1 variables in the nogood. If the value of (N − d) at this point is ≤ the original value, then because the inconsistency counts are the same for the values of the domain whose wipeout produced the nogood, and because (N − d) is always nonincreasing over the search order (so it will be ≤ the present value when search arrives at variable v_k), this k-tuple would again produce a wipeout.

Thus, if the critical difference is stored with the nogood, it can be used to test the validity of the nogood at later stages of search.

Since N does not increase during search, this will increase the likelihood

Table 1. Experimental Results with Cascaded DACCS

algorithm		density		
		.06	.18	.29
FCdeg/AC	ccks	15	124	1,614
	time	0.5	4.0	49.7
FCdeg/FC	ccks	22	128	1,901
	time	1.0	4.8	64.5
FCdegDAC/AC	ccks	7	66	773
	time	0.3	2.1	23.8
FCdegDACcasc/AC	ccks	7	57	485
	time	0.3	2.2	18.6
FCdegDACcomb/AC	ccks	6	55	457
	time	0.3	2.0	16.6
mean optimal dist.		3.2	7.2	13.9

Note. Means for 25 problems. Constraint checks (ccks) in thousands. Times are in seconds and include arc consistency preprocessing. FC is forward checking, deg is variable ordering by decreasing degree, /FC is value ordering by increasing forward checking counts, /AC is value ordering by full arc consistency counts, DAC is FC with 'ordinary' directed arc consistency counts, DACcasc is FC with cascaded DACCs, DACcomb is FC using both ordinary and cascaded DACCs in lower bound calculations.

that the test will succeed at times after the nogood is discovered. In addition, the probability of success can be increased by using a value ordering like those described above; in this case d will probably increase more rapidly with deeper search because later values are associated with more inconsistencies.

Obviously, nogood tuples found in this way in MAX-CSPs will tend to be larger than those found for CSPs. Hence, the overhead for checking will be greater and the probability of a match will be smaller. These problems can be ameliorated by starting with a good initial upper bound, because then the number of accumulated inconsistencies required to delete a value will be smaller.

5 Experiments

The new algorithms were evaluated using some of the same random problems that were used earlier to test forward checking with ordinary DACCs [10]. These problems were generated according to a "random parameter value model" modified to give a constant number of constraints, with mean domain size = 4.5 and with contraint graphs varying in density according to the formula $(n-1)+k(n/2)$ for the total number of edges, where n is the number of variables, $n - 1$ edges form a connected graph, and $k \geq 1$. In these problems the average degree of a node in the constraint graph is, therefore, equal to $k + 2$. In this model tightness is allowed to vary between the greatest possible limits and is independent for

Table 2. Experimental Results with No-Good Learning

algorithm		density		
		.06	.18	.29
FCdeg	ccks	21,782	142,981	2,010,824
	time	0.7	4.3	56.2
FCdeg/ng	ccks	21,594	142,859	2,010,793
	ngcks	526	1,127	751
	time	1.1	5.9	76.1
FCdeg/FC	ccks	22,210	127,838	1,900,771
	time	1.0	4.8	64.5
FCdeg/FC/ng	ccks	22,017	127,838	1,900,771
	ngcks	546	659	557
	time	1.5	6.2	83.6
FCdegDACcomb/AC/ng	ccks	6,385	54,593	457,016
	ngcks	18	100	206
	time	0.5	2.6	22.0

Note. Same problems as in Table 1. ngcks are no-good constraint checks. See Table 1 for other abbreviations.

each constraint. Problems are built beginning with a randomly generated spanning tree, and then adding the required number of edges. For each combination of parameter values, a sample of 25 problems was generated.

The main performance measures were number of constraint checks (ccks), i.e., tests that value x of variable u is consistent with value y of variable v, and run time. When preprocessing was done before search, its measures were included in the total. Validity of algorithms was checked by comparing means and standard deviations for optimal distances, and by checking the violations in a solution.

Twenty-variable problems were used for the initial testing reported here, with $k = 1$, 3 or 5 (corresponding densities, in terms of the edges added to the spanning tree, were 0.06, 0.18 and 0.29). Table 1 gives results for both cascaded DACCs and the combination of ordinary and cascaded DACCs. These are compared with results for ordinary DACCs and for forward checking. In all cases, variable ordering was by maximum degree of a node in the constraint graph. Value ordering was by increasing inconsistency, or non-support, counts found either during preprocessing (/AC in Table 1) or by forward checking during search (/FC in Table 1).

For these problems, the basic DAC techniques are very effective, reducing the number of consistency checks by a factor of 2-3, in comparison with forward checking enhanced by the same value ordering. However, cascaded directed arc consistency is even more effective, particularly as the density increases. Thus at density 0.29, there is a reduction of almost 40% in the number of consistency checks, although runtime is only reduced by about 20%. The combination DAC strategy improves consistency checking and runtime a bit more.

Table 3. Experimental Results with an Optimal Initial Upper Bound

algorithm		density					
		.06		.18		.29	
		ub	ub/ng	ub	ub/ng	ub	ub/ng
FCdeg	ccks	7882	7748	76,541	76,474	1,393,153	1,393,144
	ngcks		214		430		193
	time	0.3	0.4	2.3	3.0	37.2	50.2
FCdegDAC/AC	ccks	3260		33,325		638,363	
	ngcks						
	time	0.1		1.0		19.9	
FCdegDACcomb/AC	ccks	2996	2996	23,488	23,481	348,199	348,197
	ngcks		0		29		186
	time	0.2	0.2	1.0	1.2	12.9	16.7

Note. Same problems as Table 1. ub indicates that search begins with an optimal initial upper bound. ub/ng is the optimal upper bound with no-good learning also used. No-good learning was not tested with ordinary DACCs.

In contrast to the results with cascaded DACCS, no-good learning did not improve performance on these problems, in any of the combinations in which it was tested (Table 2). These included combining it with the basic forward checking procedure, or with forward checking enhanced by value ordering, and with directed arc consistency counts. Other tests, with lexical variable ordering (essentially a random ordering with respect to problem characteristics) and with other types of random problems also gave negative results. (For the 20-variable problems with density = 0.06, a 20% reduction in ordinary constraint checks was found when lexical ordering was used, but in this case there was an enormous number of no-good-checks, so the overall efficiency was less than for the basic forward checking algorithm.)

The effectiveness of these techniques in conjunction with a good initial upper bound was tested by setting the initial upper bound to the optimal distance. This is the ideal case in which branch and bound serves only to certify optimality. The two main findings were, (i) using cascaded DACCs gave an improvement over ordinary DACCs that was larger than the improvement found with an initial upper bound of 'infinity', (ii) again, no-good learning was ineffective (see Table 3).

6 Conclusions

By using information gained from preprocessing more systematically, it appears possible to solve certain types of MAX-CSPs more readily on the average. The range of random problems that are especially amenable to this treatment is restricted to low densities, at least for problems with 20 or more variables (cf.

[10]). However, it is encouraging that MAX-CSPs of moderate size can sometimes be solved to completion in a reasonable time.

By using simple inferences about support, it is possible to 'cascade' directed arc consistency counts toward the beginning of the search order. This serves to enhance the calculations of lower bounds at the beginning of search, and, as the experiments show, the payoff can be appreciable. As would be expected, this strategy yields even greater improvements when the initial upper bound is reduced. More extensive analysis of the interaction between lower and upper bounds, as well as the use of heuristic repair techniques to find optimal or near-optimal upper bounds is currently underway.

Unfortunately, the tests of no-good learning have given uniformly negative results. It is always possible that there are special cases where this procedure will be helpful; however, it is clear that this is not a generally effective technique with MAX-CSPs.

References

1. A. Borning, R. Duisberg, B. Freeman-Benson, A. Kramer, and M. Woolf. Constraint hierarchies. In *Proceedings OOPSLA-87*, pages 48–60, 1987.
2. R. Bakker, F. Dikker, F. Tempelman, and P. Wognum. Diagnosing and solving overdetermined constraint satisfaction problems. In *Proceedings IJCAI-93*, pages 276–281, 1993.
3. R. Dechter. Enhancement schemes for constraint processing: backjumping, learning, and cutset decomposition. *Artif. Intell.*, 41:273–312, 1990.
4. D. Frost and R. Dechter. Dead-end driven learning. In *Proceedings AAAI-94*, pages 294–300, 1994.
5. R. Feldman and M.C. Golumbic. Optimization algorithms for student scheduling via constraint satisfiability. *Comput. J.*, 33:356–364, 1990.
6. M. Fox. *Constraint-directed Search: A Case Study of Job-Shop Scheduling*. Morgan Kaufmann, Los Altos, CA, 1987.
7. E.C. Freuder and R.J. Wallace. Partial constraint satisfaction. *Artif. Intell.*, 58:21–70, 1992.
8. L. Shapiro and R. Haralick. Structural descriptions and inexact matching. *IEEE Trans. Patt. Anal. Mach. Intell.*, 3:504–519, 1981.
9. T. Schiex and G. Verfaillie. Nogood recording for static and dynamic constraint satisfaction problems. In *Proceedings TAI-93*, pages 48–55, 1993.
10. R. J. Wallace. Directed arc consistency preprocessing as a strategy for maximal constraint satisfaction. In M. Meyer, editor, *Constraint Processing*, volume 923 of *Lecture Notes in Computer Science*, pages 121–138. Springer-Verlag, Heidelberg, 1995.

Partial Arc Consistency

Nick D. Dendris, Lefteris M. Kirousis, Yannis C. Stamatiou, and
Dimitris M. Thilikos

Department of Computer Engineering and Informatics,
Patras University, Rio, 265 00 Patras, Greece.
e-mail: {dendris, kirousis, stamatiu, sedthilk}@cti.gr

Abstract. A constraint network is arc consistent if any value of its variables is compatible with at least one value of any other variable. Enforcing arc consistency in a constraint network is a commonly used preprocessing step before identifying the globally consistent value assignments to all the variables. Since many constraint networks that arise in practice are overconstrained, any global assignment of values to the variables involved, is expected to include constraint violations. Therefore, it is often necessary to enforce some weak form of consistency, to avoid excessive value elimination due to the existence of many constraints. It is also necessary that this form of consistency be enforced fast. In this paper, we introduce a notion of weak local consistency that we call partial arc consistency. We then give an algorithm that, for any constraint network of n variables, outputs a partially arc consistent subnetwork of it in sublinear $(O(\sqrt{n}\log n))$ parallel time using $O(n^2)$ processors. This algorithm removes at least a constant fraction of the local inconsistencies of a general constraint network, without eliminating any globally consistent assignment of values. To our knowledge, it is the first sublinear-time parallel algorithm with polynomially many processors that achieves this extent of local consistency. Moreover, we propose several approximation schemes to a total solution of the arc consistency problem. We show that these approximation schemes are inherently sequential (more formally, they are P-complete), a fact indicating that the approach of partial solutions, rather than that of approximation schemes, is more promising for parallelism.

1 Introduction

One of the most frequently used methods to reduce the search space of a constraint satisfaction problem (or CSP for short), is to enforce, as a preprocessing step, *local* consistency. This is achieved by eliminating, recursively, the values of each variable (or, more generally, the combinations of $k-1$ values corresponding to each set of distinct $k-1$ variables) that are not compatible with every remaining value in the domain of any other variable. Networks having this local consistency property are called arc-consistent (k-consistent, respectively, in the general case). Efficient (polynomial time) optimal algorithms have been devised for each level of local consistency, in contrast, of course, to the NP-completeness of finding globally consistent assignments of values to the variables of a general CSP. Unfortunately, enforcing arc-consistency, the weakest non-trivial local consistency, is inherently sequential (see [7]), i.e. not likely to be amenable to efficient parallel solutions. By this we mean that it is not likely to find a parallel algorithm that, given a constraint network $\mathcal{N}$ of n variables, utilizes a polynomial in the size of the constraint network number of processors and solves the problem in time much less (exponentially less) than the optimal sequential time. The best known parallel algorithm for the ACP, designed for the PRAM parallel computation model (see [6] for a description of the model), has a running time in the order of n. Parallel algorithms that achieve local consistency in time asymptotically less than the optimal sequential time have been devised (see [3]), but at the price of using exponentially many processors. A different line of approach to solve a CSP, and especially an overconstrained instance, is to look for global assignments that violate a small fraction of the constraints (for a similar approach for the boolean formula satisfiability problem, see [8]). This approach is especially fit for overconstrained networks, since a solution to them is expected to include violations of constraints.

In this paper, we define a notion of partiality not on the global level, but locally. Intuitively, a network is partially locally consistent if it can allow a small fraction of violations of the requirements for local consistency (see below for formal definitions). We propose partial local consistency as a useful preprocessing stage for solving an overconstrained network, or, more generally, for finding a partially consistent global assignment for a CSP. Partial consistency can be achieved in parallel time that is asymptotically better than the optimal sequential time, by using only realistically many processors (e.g., sub-linear time and polynomially many processors for the important case of arc-consistency). This is the first algorithm, to our knowledge, that achieves the elimination of a non-trivial number of local inconsistencies at sublinear parallel time with polynomially many processors.

More specifically, an α-partial solution of the arc consistency problem (ACP) is one where every value of any variable is compatible with values of at least $\alpha(n-1)$ of the other variables ($0 < \alpha \leq 1$, and n is the number of variables in the constraint network). Let Δ be the maximum degree in the graph of constraints (i.e. the graph whose edges correspond to constraints among the variables) and q be the maximum number of values for a variable. In many applications of CSP where the constraint network is heavily constrained, Δ is a constant factor of the number of variables. We show that, for any $\alpha < 1 - \frac{q}{q+1}\frac{\Delta}{n-1}$, there is a parallel algorithm that runs in time $O(\sqrt{nq}\log(nq))$ and produces an α-partial solution for ACP that contains the total ACP solution. Observe that the factor $\frac{\Delta}{n-1}$ reflects the density of the constraints. Roughly, our parallel approach utilizes the following idea: for as long as there are more than $\sqrt{nq}$ values that are compatible with values of less than $n-1$ variables, the algorithm removes them in parallel steps. The target is to capture the steps of the original arc consistency algorithm that offer sufficiently high degree of parallelism. Afterwards, the algorithm removes values that are compatible with less than $\alpha(n-1)$ variables. A combinatorial lemma ensures that these steps will not be more than $O(\sqrt{nq})$. The algorithm requires $O(n^2q^2)$ processors on the Concurrent-Read Exclusive-Write (CREW) shared-memory parallel machine model (PRAM).

A question that arises, since the ACP is P-complete, is whether there are efficient parallel algorithms that *approximate* the exact solution, in the sense that the approximate solutions they produce are predictably close (within a constant factor) to the exact solution. To this end, we define several natural approximation schemes to the exact solution of ACP. Unfortunately, all the proposed schemes are inherently sequential too. Let D be the disjoint union of all domains of a network, D_{AC} be the disjoint union of all domains *after* the application of discrete relaxation, and $R_{AC} = D - D_{AC}$. Our first approximation scheme asks to determine, for a given $\epsilon \in (0,1]$, a set D' such that $D_{AC} \subseteq D' \subseteq D$ and $\epsilon|D'| \leq |D_{AC}|$. We prove that, unless P=NC, no such set D' can be found by a parallel algorithm in NC (for the class NC see [6]). We show the same for the approximation scheme that, given an $\alpha \in (0,1]$, asks for a set R' such that $R' \subseteq R_{AC}$ and $\alpha|R_{AC}| \leq |R'|$. The other two approximation schemes that we introduce are related to the degree of incompatibility that a value must exhibit in order to be removed by a relaxation-type algorithm. We define the support of a value d to be the maximum k such that, when we remove values compatible with less than k variables, d is not removed and no domain is emptied. We also define the elimination degree of d as the least k such that, when we remove values not supported by more than k variables, d is not removed and no domain is emptied (see next sections for formal definitions). We show that approximating any of these parameters is P-complete. The above observations imply that looking for partial solutions, and not for approximation schemes, seems to be the correct way to obtain sublinear-time parallel algorithms that approach the solution of ACP.

For extensions of these results to any degree of local consistency and for constraint networks with constraints of arbitrary arity, see [5].

2 Definitions and Preliminaries

A *binary constraint network* $\mathcal{N}$ consists of a set of variables $X_1, \ldots, X_n$, a set of variable domains $D_1, \ldots, D_n$, and a set of binary constraints $\mathcal{C}$. For $i = 1, \ldots, n$, D_i is the set of permissible values for variable X_i. A constraint $R_{ij} \in \mathcal{C}$ is a subset of $D_i \times D_j$ ($i \neq j$). A value d_i of X_i is *compatible* with a value d_j of X_j ($i \neq j$) iff there is no $R_{ij} \in \mathcal{C}$ such that $(d_i, d_j) \notin R_{ij}$. A *domain-reduced subnetwork* of $\mathcal{N}$ is any constraint network $\mathcal{N}'$ with the same variables, domains $D_1' \subseteq D_1, \ldots, D_n' \subseteq D_n$, and set of constraints $\mathcal{C}'$ containing exactly the restrictions of the elements of $\mathcal{C}$ to the corresponding domains.

The *Constraint Satisfaction Problem* (CSP), for a given constraint network, asks for all the n-tuples $(d_1, \ldots, d_n)$ of values such that $d_i \in D_i$, $i = 1, \ldots, n$, and for every $R_{ij} \in \mathcal{C}$, (d_i, d_j) is in R_{ij}. Such an n-tuple is called a *solution* of the CSP. The decision version of the CSP is to determine if any solution exists. Given an instance Π of the CSP with n variables, the *constraint graph* G^{Π} (or

G when no confusion may arise) of Π has n vertices which correspond to the variables of Π and it contains an edge $\{v_i, v_j\}$ iff the corresponding variables are bound by a constraint.

We also use another graph representation of a CSP instance Π, which we call the *compatibility graph*. Let q denote the maximum number of values that can be assigned to a variable, i.e. $q = \max\{|D_i|, i = 1, \ldots, n\}$. The compatibility graph C^{Π} (or just C, when no confusion may arise) of Π is an n-partite graph. The ith part of C^{Π} corresponds to variable X_i of Π and it has exactly $|D_i|$ vertices, one for each value of X_i. For simplicity, we use the same notation for the vertices of the compatibility graph and their corresponding values (domain-elements) in the network. Two vertices d_i and d_j of C^{Π} are adjacent iff the corresponding values d_i, d_j are compatible. We denote by N the total number of vertices of C^{Π}, i.e. $N = |D|$ where $D = \bigcup_{i=1,\ldots,n} D_i$ and $N \leq nq$. In other words, N is equal to the total number of values in the constraint network (values in the domains of different variables are considered to be different).

A constraint network $\mathcal{N}$ is *arc consistent* if the following holds: for any variable X_i, for any value $d_i \in D_i$, and for any other variable X_j, there exists at least one value assignment $d_j \in D_j$ such that d_i is compatible with d_j. The *Arc Consistency Problem* (ACP) is the problem of finding a maximal (with respect to the domains) arc consistent subnetwork $\mathcal{N}_{\mathrm{AC}}$ of $\mathcal{N}$. Also, $\mathcal{N}_{\mathrm{AC}}$ is unique, for if there was another maximal arc consistent subnetwork $\mathcal{N}'_{\mathrm{AC}}$ with domains $A'_1, \ldots, A'_n$, then the domain-reduced subnetwork of $\mathcal{N}$ with domains $A_1 \cup A'_1, \ldots, A_n \cup A'_n$ would also be arc consistent, a contradiction to the maximality of $\mathcal{N}_{\mathrm{AC}}$ and $\mathcal{N}'_{\mathrm{AC}}$. We call $\mathcal{N}_{\mathrm{AC}}$ the *solution* of the ACP. In terms of the compatibility graph C, the ACP is formulated as follows: find the maximal (with respect to the vertices) induced subgraph C_{AC} of C such that any vertex in any of its parts is adjacent to vertices in $n - 1$ other parts.

All values $d_i \in D_i$, $i = 1, \ldots, n$, which participate in some solution of the CSP belong in the domain A_i of the ACP solution. Since CSP is an NP-complete problem, a common approach for solving it involves a preprocessing step to make the network arc consistent and reduce the variable domains, before employing exhaustive search to determine the actual solutions. Arc consistency is achieved by the procedure known as *discrete relaxation*. The algorithm removes values that are incompatible with the values in the domain of at least one variable, until no such values exist. Discrete relaxation runs in optimal $O(eq^2)$ time ([1]), where e is the number of constraints and q is the maximum number of values in the domain of a variable. Also, in [7] it was proved that the ACP is inherently sequential in the sense that given a constraint network $\mathcal{N}$, it is inherently sequential to decide whether a given value for a variable belongs in the corresponding domain of $\mathcal{N}_{\mathrm{AC}}$.

3 Partial Arc Consistency

In this section we introduce the notion of partiality in arc consistency, and give a sublinear-time parallel algorithm that finds partial solutions to the ACP (for other relevant combinatorial problems that admit to partial solutions see [8]).

A constraint network $\mathcal{N}$ is α-*partially arc consistent* if no variable domain is empty and any value of a variable is compatible with values of at least a constant factor of the other variables, i.e. with values of at least $\alpha(n-1)$ other variables $(0 < \alpha \leq 1)$. The α-*partial arc consistency problem* is the problem of finding a maximal (with respect to the domains) α-partially arc consistent subnetwork $\mathcal{N}_{AC,\alpha}$ of $\mathcal{N}$. When $\alpha = 1$, this problem is the same as the original ACP. In a way similar with the case of $\mathcal{N}_{AC}$, one can see that $\mathcal{N}_{AC,\alpha}$ is unique.

For any value d in a constraint network, the support-degree of d is the number of variables in the network that contain a value compatible with d. In the compatibility graph C, the support-degree of a vertex d in C is the number of parts in C that contain vertices adjacent to d. In terms of the compatibility graph C, the α-partial arc consistency problem is formulated as follows: find the maximal (with respect to the vertices) induced subgraph $C_{AC,\alpha}$ of C such that any vertex has support-degree at least $\alpha(n-1)$ and no part is empty. Such a graph is called α-*partially arc consistent*.

In this paper we introduce a modified version of the discrete relaxation algorithm, which we call α-*discrete relaxation*, that solves the α-partial arc consistency problem. In terms of the compatibility graph C, and given a real value α, $0 < \alpha \leq 1$, the algorithm repeatedly removes vertices in C that have support-degree less than $\alpha(n-1)$ until no such vertex exists. The algorithm outputs a subgraph $C_{AC,\alpha}$ of C that contains C_{AC} as a subgraph. If during the α-discrete relaxation a part is emptied, then the ACP has the empty solution.

The algorithm α-discrete relaxation has a parallel version which we call *parallel α-discrete relaxation*. In this version, each step of the identification and elimination of the vertices to be removed is done in parallel in $O(\log(nq))$ time using $O(n^2q^2)$ processors on the CREW PRAM. Therefore the α-partial arc consistency problem can be solved in in $O(nq\log(nq))$ parallel time using $O(n^2q^2)$ processors on the CREW PRAM.

Note that maximality of an α-partially arc consistent subgraph of C is not crucial, as long as this subgraph contains C_{AC}. We show that, given a constraint network $\mathcal{N}$, its constraint graph C and its compatibility graph G, and for any $\alpha < 1 - \frac{q}{q+1}\frac{\Delta}{n-1}$, we can compute in sublinear parallel time an α-partially arc consistent subgraph of $C_{AC,\alpha}$ that contains C_{AC}. In the above, $\Delta = \Delta(G)$ is the maximum degree in in the constraint graph G. The algorithm for doing this is strongly based on the following lemma, which ensures the existence of a partially consistent subgraph of C whenever C is sufficiently dense:

Lemma 1. *Let $\mathcal{N}$ be a constraint network with n variables each one having at most q values. Let G, C be its constraint and compatibility graph respectively, Δ the maximum degree of G, and l the smallest support-degree of any value in $\mathcal{N}$ (observe that $l = n - 1 - \Delta$). Let also $l_2 < l_1 \leq \Delta$ be two integers such that $l_2 < \frac{l_1}{1+q}$. If C contains at most m vertices with support-degree less than $l + l_1$, then, if we apply $(\frac{l+l_2}{n-1})$-discrete relaxation on C, either no more than $\frac{m(l_1-l_2)}{l_1-l_2(1+q)}$ vertices will be removed, or all the vertices of some part will be removed.*

The lemma implies that if the the number of vertices in C that have low degree is sufficiently small, i.e. C is sufficiently dense, then we can find a bound on the number of vertices that will be removed by the α-discrete relaxation. This, in turn, suggests the following idea: as long as the number of vertices with support-degree is above a threshold, remove them in parallel. When their number (the value m in Lemma 1) falls below this threshold, apply parallel α-discrete relaxation. If we arrange things so that the threshold is sufficiently low, then, from Lemma 1, the application of the parallel α-discrete relaxation (for an appropriate value of α, of course) will remove a bounded number of vertices, thus its worst case running time will be bounded by the same bound. All these considerations are formalized by the following theorem:

Theorem 2. *Let $\mathcal{N}$ be a constraint network with n variables each one having at most q values. Let also G, C be its constraint and compatibility graph respectively. If G has maximum degree Δ, then for any $\alpha < 1 - \frac{q}{q+1}\frac{\Delta}{n-1}$, there exists a parallel algorithm that returns a subgraph of $C_{AC,\alpha}$ which contains C_{AC} and where any vertex has support-degree at least $\alpha(n-1)$. This algorithm executes in $O(\sqrt{nq}\log(nq))$ time and uses $O(n^2q^2)$ processors on the CREW PRAM.*

We give the algorithm, without any proofs (for a more complete exposition, see [5]):

Algorithm fast α-discrete relaxation
Input: An n-partite compatibility graph $C = (V, E)$
Output: An α-partial arc consistent subgraph of $C_{AC,\alpha}$ that contains C_{AC},
        where $\alpha < 1 - \frac{q}{q+1}\frac{\Delta}{n-1}$
begin
   $N \leftarrow |V|$
   while $\exists\, V' \subseteq V : (|V'| \geq \sqrt{N}) \wedge (\forall d \in V' : \text{support-degree}(d) < n - 1)$
   do
      for each $d \in V'$ **in parallel**
      do
         $V \leftarrow V - \{d\}$
      od
      if an empty part appears in C
         then terminate and return the empty graph
   od
   run parallel α-discrete relaxation
end

4 Parallel Intractability Results

In this section, we present several natural approximation schemes for the ACP which are inherently sequential (see, also, [5]).

4.1 Domain Approximations

Let $\mathcal{N}$ be a constraint network with n variables and variable domains $D_1, \ldots, D_n$. Let also $A_1, \ldots, A_n$ denote the domains in $\mathcal{N}_{AC}$. If $D = \bigcup_{i=1\ldots n} D_i$, $D_{AC} = \bigcup_{i=1\ldots n} A_i$ and $R_{AC} = D - D_{AC}$. Then the following hold:

Theorem 3. *For a given constraint network $\mathcal{N}$, and for any $\epsilon \in (0, 1]$, it is P-complete to find a set D' such that $D_{AC} \subseteq D' \subseteq D$ and $\epsilon |D'| \leq |D_{AC}|$.*

Theorem 4. *For a given constraint network $\mathcal{N}$, and for any $\epsilon \in (0, 1]$, it is P-complete to find a set R' such that $R' \subseteq R_{AC}$ and $\epsilon |R_{AC}| \leq |R'|$.*

4.2 Support-degree Approximations

Given a constraint network $\mathcal{N}$ and a value d of variable X_i, we define the *support* of d to be the maximum integer h such that, after the $(\frac{h}{n-1})$-relaxation, d is not removed and no domain is empty. The problem of computing support(d) is P-complete, since if we could efficiently compute the support of any value in $\mathcal{N}$, then, by comparing these values with $n-1$, we would compute all the values in $\mathcal{N}_{AC}$.

Computing support(d) is a maximization problem, so we formulate the following approximation scheme for it, for any constant factor $\epsilon \in (0, 1]$: find an integer support(d)$_{approx}$ such that

$$\epsilon \cdot \text{support}(d) \leq \text{support}(d)_{approx} \leq \text{support}(d)$$

Theorem 5. *It is P-complete to approximate support(d) by a factor $\epsilon > 1/q$, where q is equal to the maximum number of values in the domain of a variable.*

We also define the *elimination degree* of d to be the least integer h such that a $(\frac{n-h}{n-1})$-relaxation does not remove d and does not empty any variable domain.

It holds that support(d)+elimin(d) = n, so it is P-complete to find elimin(d). The computation of the elimination degree of d is a minimization problem, so we consider the following approximation of it, for some constant factor $\lambda \geq 1$: find an integer elimin(d)$_{approx}$ such that:

$$\lambda \cdot \text{elimin}(d) \geq \text{elimin}(d)_{approx} \geq \text{elimin}(d)$$

Theorem 6. *It is P-complete to approximate elimin(d) by any factor $\lambda \geq 1$.*

5 Discussion

In this paper we defined a notion of partiality in relation to the Arc Consistency Problem. Our motivation was the fact that many natural approximation schemes we considered for the ACP are inherently intractable. Aided by Lemma 1, we gave an algorithm that, given a constraint network $\mathcal{N}$, returns a subnetwork that contains the solution to the ACP for $\mathcal{N}$, and with the important additional property that at least a constant fraction of the local inconsistencies present in the initial network, do not exist any more. The subnetwork can be subjected to further processing that can be accomplished more easily, since many inconsistencies have already been eliminated. Partial solutions differ from approximate ones in that there is no guarantee as to how far the partial solution will be from the exact solution, a guarantee that is provided in approximate solutions. However, the approximating schemes we considered are inherently sequential, in constrast to the partial scheme that leads to an algorithm running in sublinear time.

References

1. C. Bessiere, Arc-consistency and arc-consistency again. *Artificial Intelligence*, 65:179–190, 1994.
2. M.C. Cooper, An optimal k-consistency algorithm. *Artificial Intelligence*, 41:89–95, 1990.
3. P.R. Cooper and M.J. Swain, Arc consistency: parallelism and domain dependence. *Artificial Intelligence*, 58:207–235, 1992.
4. R. Dechter, Constraint networks. In: S. Shapiro, ed., *Encyclopedia of Artificial Intelligence* (Wiley, New York, 2nd ed., 1992) 276–285.
5. N.D. Dendris, L.M. Kirousis, Y.C. Stamatiou, and D.M. Thilikos, Partiality and Approximation Schemes for Local Consistency in Networks of Constraints. *Proceedings 15th Conference on Foundations of Software Technology and Theoretical Computer Science*, Bangalore, India (1995) 210–224.
6. R.M. Karp and V. Ramachandran, Parallel algorithms for shared-memory machines. In: J. van Leeuwen, ed., *Handbook of Theoretical Computer Science* (Elsevier, Amsterdam, 1990).
7. S. Kasif. On the parallel complexity of discrete relaxation in constraint satisfaction networks. *Artificial Intelligence*, 45:275–286, 1990.
8. K.J. Lieberherr and E. Specker, Complexity of partial satisfaction. *J. of the ACM*, 28:411–421, 1981.
9. A.K. Mackworth. Constraint satisfaction. In: S. Shapiro, ed., *Encyclopedia of Artificial Intelligence*(Wiley, New York, 2nd ed., 1992) 285–293.
10. A. Samal and T.C. Henderson, Parallel consistent labeling algorithms. *International Journal of Parallel Programming*, 16(5):341–364, 1987.

Dynamic Constraint Satisfaction with Conflict Management in Design

Esther Gelle and Ian Smith

Laboratoire d'Intelligence Artificielle
Ecole Polytechnique Fédérale de Lausanne (EPFL)
IN-Ecublens, CH-1015 Lausanne

Abstract. In this paper, we focus on techniques for constraint-based design tasks with *continuous variables* in dynamic environments. We show how design spaces can be explored within an intelligent CAD system using a dynamic constraint satisfaction framework and a conflict resolution paradigm. Navigation between different design spaces is controlled by assumptions the designer makes in situations of incomplete knowledge. These assumptions divide design constraints into defaults, preferences and fixed constraints. Design strategies associate the constraints with the overall goals that the designer wishes to satisfy and allow for a more detailed ordering on the constraints. Reasoning within different belief sets which correspond to the designer's notion of assumptions is related to normal default logic.

1 Introduction

Synthesis tasks such as configuration and design are characterized by an imprecise, often ill structured initial state. At the time a design problem is specified, requirements are not completely identified. When the system is under-constrained, the designer makes assumptions in order to advance the design process and to focus on interesting design alternatives. In this situation, the designer does not always make optimal decisions. Factors and parameters affecting a decision may interact in such a complex manner that the designer becomes unable to decide the best course of action. He may be unable to process good choices or even to generate all alternative courses of action. The best design process is therefore a *dynamic* one where, at any stage of design development, decisions involved can be changed and intermediate solutions are re-evaluated. These requirements show the need for a framework which generates various design alternatives in order to explore the design space in a rational way.

The paradigm of constraint satisfaction is useful for expressing design tasks and subtasks since it accomodates variables and mathematical relations. However finding the relevant constraints and expressing them is much harder. Therefore a constraint-based framework for design should support the exploration of multiple solution spaces and give the designers the possibility to introduce new constraints and their preferences. To satisfy these requirements our system uses

1. Dynamic constraint satisfaction on continuous and discrete variables. Mittal & Falkenhainer [9] have developed a framework for treating constraints on

discrete variables in a dynamic environment. We show how to extend this framework in order to include continuous variables and constraints. Certain aspects of the design process are related to normal default logic.

2. Default and preference assumptions on constraints in order to guide the search of consistent constraint combinations, especially in early stages of the design process.

3. Conflict management methods for detecting and solving inconsistencies arising in over-constraint situations.

4. A general hierarchy on design criteria inducing an ordering on the constraints important for conflict management.

2 Related Work

Sketchpad [11] was the first constraint-based drawing system using constraint satisfaction and relaxation methods on real values. Borning et al. [2] define hierarchies of constraints in their graphics-based drawing system in order to satisfy constraints on graphical objects in an intuitive manner. The functionalities of such an interactive graphics system allow for a simple classification of constraints into required, preferential and default constraints. Another requirement is that the system's response to user requests should be fast. Different types of comparators are implemented to filter out the best possible solution. In a design system, such a simple hierarchy is no longer useful. Comparators are difficult to define for a given design task because often there is more than one hierarchy of preferences. In this cases, users should be able to define their own hierarchies based on design considerations such as cost or esthetics. They might also be interested in several solution spaces; not just a single solution. Freuder et al. [5] describe a related notion of partial constraint satisfaction on discrete variables. Their branch and bound algorithm finds values for a subset of variables that satisfy a subset of constraints. It uses a metric measuring the distance between the current values and the best solution found so far. Preferences in this system can be expressed by assigning weights to the constraints. However, in a design system it is preferable to define a hierarchy based on design criteria rather than on individual constraints.

In the domain of configuration, Mittal and Falkenhainer [9] have defined a framework for dynamic constraint satisfaction based on discrete variables. For design examples however, their definition of activity and compatibility constraints must be adapted in order to treat continuous as well as discrete variables. Other systems allowing for dynamic introduction of constraints include constraint logic programming (CLP) languages. CLP describes a set of languages based on logic programming with embedded constraints and defined on different domains, such as finite, real or interval domains. A CLP program consists of a set of rules. Each rule has a head, the goal, and a body consisting of subgoals and/or constraints. During the execution of the rules, a goal is blocked until its condition is satisfied by the active constraints. One of the most powerful languages for computing numeric systems is CLP(intervals) [1]. It uses box-consistency, an approximation of arc-consistency, to solve numeric constraint systems. Other CLP($\Re$) systems use the simplex algorithm to treat linear constraint systems and delay nonlinear systems until they become linear. One of the major problems in the domain of numeric constraints is the resolution of nonlinear systems which may comprise inequalities. In order to solve a design problem, a unique solution must be found within the solution space. Haroud et al. [6] have developed an algorithm guaranteeing global consistency for continuous constraint satisfaction problems (CCSPs). The advantage of this algorithm is that feasible regions are represented explicitly in the three-dimensional space. The designer can choose a single feasible solution within the solution space. This algorithm needs to know beforehand the set of constraints involved in a problem, i.e. the problem is *static*. Further work will be necessary to adapt it to a dynamic design framework.

3 Dynamic Constraint Satisfaction

Synthesis tasks can be represented naturally as dynamic constraint satisfaction problems (DCSP). In design and configuration, variables define components and properties of the artifact to be created and constraints represent relationships between objects. The sets of variables and the constraints defined on them change dynamically in response to decisions taken during the design process. Typically, the components of an artifact are interconnected: For instance when configuring a laptop, if an efficient secondary cache is required, a cooling fan has to be built in. Thus, the weight of the laptop will change as well as its price. Consequently, the existence of some variables, such as the fan, or constraints is entirely determined by earlier decisions. Simultaneously considering the entire set of possibly relevant constraints would result in a large number of constraints and variables to treat. A large set of conflicts would arise, which would be difficult to solve. In dynamic constraint satisfaction, the activation of a new variable or constraint is *conditional*. A variable or a constraint is called *active* when it must be part of a solution. Not all variables have to be assigned a value to solve the problem and not all constraints are relevant for a given problem. To determine a variable or constraint's activity the *activation conditions* are matched against the current context. A *design context* is defined as the set of active variables and their values, i.e. a partial solution space.

Activation conditions may split up the solution space into disjoint regions. A further difficulty arises, when activation conditions are defined on continuous variables; the subspaces they define are no longer enumerable. In Figure 1 for example, the width W and the height H of a steel structure element of shape I depend on several constraints, in this case linear inequalities, which define multiple solution spaces (shaded regions). The type of the section, which can be plastic, compact or non-compact, induces three different classes of ratio of H and W, denoted Class 1, 2 and 3. The constraint $W \leq w1$ defines sections of elements that must be enforced by stiffeners, whereas no stiffeners are needed when $W \geq w2$. In this example, $w1$ and $w2$ are constants. In general, a constraint satisfaction problem on continuous variables is can be solved in polynomial time when the regions defined by the constraints are arc-wise connected [6]. In the presence of explicit splits in the problem definition as for example shown in Figure 1, it is wise to combine constraint satisfaction with search. Consistent constraint combinations are found and values are then propagated in each subset.

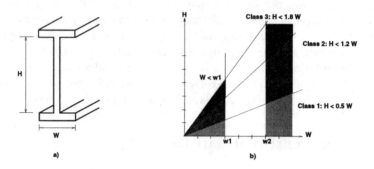

Fig. 1. *a) A structural I element, b) disjoint solution spaces defined by linear constraints on the width W and the height H for a structural element.*

3.1 A Formal Definition of DCSP

In terms of definitions by Mittal et al. [9] *compatibility* constraints define static relations on variables whereas *activity* constraints specify constraints on the activity of variables depending on possible values of other variables. In order to formalize a design problem, we need to enhance the definitions of activity constraints. Not only variables but also constraints may be activated. Consider the following example of calculating the surface of a shape which can be a rectangle or a circle:

R1 $shape = rect \rightarrow active(surface = length * width)$
R2 $shape = circ \rightarrow active(surface = \pi * radius^2)$
R3 $shape = rect \rightarrow active(length, width)$
R4 $shape = circ \rightarrow active(radius)$

The example shows that constraints can be conditional too (R1,R2). In a dynamic CSP, we are given a set of possibly active variables X, a set D defining the union of all domains of X so that each variable $x_i \in X$ has associated a domain $d_i \in D$ and C, the set of potentially active constraints. An instantiation of a set of variables $X' \subset X$ is a tuple representing an assignment for each variable. A solution is a consistent instantiation, so that all the constraints between the variables of X' are satisfied and no more variables and constraints can be activated. A solution space S is a subset of D comprising all solutions. Contrary to a static CSP, we distinguish between active variables and such which are not active. A variable is called *active* when it must be part of the solution. A variable

x_i which is active, has to be assigned values from its domain $d_i \in D$. Similarly, a constraint which is active must be satisfied by a solution s of the solution space S:

$$active(Y_i) \Leftrightarrow \{y_i = d_i | d_i \in D(y_i), y_i \in Y_i\}$$
$$active(constraint(Y_i)) \Leftrightarrow \exists s \in S\{active(Y_i) \wedge satisfies(s, constraint(Y_i))\}$$

We define as activity constraints C^A rules of the form

1. $Y_1 \rightarrow active(Y_2)$ $Y_1, Y_2 \subset X$
2. $predicate(Y_1) \rightarrow active(Y_2)$
3. $predicate(Y_1) \rightarrow active(constraint(Y_2))$

and compatibility constraints C^C

4. $constraint(Y_1)$ $Y_1 \subset X$

More generally, a dynamic constraint satisfaction problem (DCSP) is defined as a set $\langle W, R, X, D \rangle$ with a set of rules R and a set of initial assignments W. R is divided into activity constraints C^A and compatibility constraints C^C. A rule consists of an *activation condition* P_i and its body introducing a new constraint $C_i \in C$ (1). X is the set of variables that may be part of a solution.

$$P_i(Y_1) \rightarrow active(C_j(Y_2)) \Leftrightarrow P_i(Y_1) \overset{act}{\rightarrow} C_j(Y_2) \quad Y_1 \subset X, Y_2 \subset X^1 \quad (1)$$

In its most general formulation, the activation condition P_i is a conjunction of constraints on a subset of variables Y_1 which have to be active. The constraint C_j is a mathematical expression, linear or non-linear, on the subset of variables Y_2. In order to express a compatibility constraint, the consequence of the rule simply evaluates to true. A rule $r \in R$ is satisfied by an assignment s if and only if:

$$active(Y_1) \wedge satisfies(s, P(Y_1)) \wedge active(C_j(Y_2))$$

A solution space S for a given problem $\langle W, R, X, D \rangle$ is a subset of variables $X' \subset X$ each of which has been assigned a range of values consistent with respect to $C^A \cup C^C$ so that $W \subset S$. A solution space S is therefore consistent with respect to all active constraints and no more variables and constraints can be activated.

---

[1] The notation $P_i(Y_1) \overset{act}{\rightarrow} C_j(Y_2)$ is equivalent to IF $P_i(Y_1)$ THEN $C_j(Y_2)$.

3.2 Hierarchy of Constraints

When the given problem is underconstrained, several solution spaces exist in the search space. In order to avoid enumerating all solution spaces, we aim at finding a solution space which fulfills some given design criteria and assumptions of the designer. When the problem is overconstrained, no solution space exists. In this case, we would like to find solution spaces violating as few important constraints as possible. We introduce therefore a hierarchy of constraints, which helps us to guide search in the case of several solution spaces and which aims at relaxing constraints which are the least preferred when the problem is overconstrained.

We define a hierarchy $H : h_1 \ldots h_n$ on the constraints by labeling the rules of R, so that $r_i < r_j$ with $r_i, r_j \in R$ if $r_i \in h_k$ and $r_j \in h_m$ with $k < m$, i.e. r_i is preferable to r_j. Each h_i consists of a set of rules which are at the same hierarchical level.

A solution s in the solution space S is defined to satisfy as many levels of the hierarchy as possible, starting with h_1 being the rules introducing required constraints [2]. The other levels $h_j, j = 2, \ldots n$ are optional and satisfied if possible. The rules are treated in the order of the overall hierarchy H. If a conflict has to be solved between two constraints initiated by two different rules r_i and r_j, the constraint which is less preferable, i.e. the constraint initiated by r_j if $r_i < r_j$, is chosen for relaxation.

3.3 A Comparison with Default Logic - Open Issues

In this section, we want to show how dynamic constraint satisfaction approximates the paradigm of default logic. In a dynamic constraint satisfaction problem, rules add constraints which first have to be checked against the current constraint network. The underlying reasoning process in a DCSP is *non-monotonic*. Suppose that a given precondition P_1 satisfies the current context (the values of variables found so far) and a new constraint C_1 is introduced. Adding a new rule $P_2 \rightarrow C_2$ where P_2 also satisfies the current context does not automatically result in the addition of the constraint C_2. Adding the constraint C_2 may result in a conflict. Hence, the addition of new rules does not automatically mean the addition of more constraints. The general rule structure $P_i(Y_1) \overset{act}{\rightarrow} C_j(Y_2)$ is not just a logical implication but incorporates a consistency check for the relevant constraint. It can be read as:

IF $P(Y_1)$ is satisfied by the current network AND $C(Y_2)$ does not result in a contradiction THEN add $C(Y_2)$ and propagate its results.

The search process within a DCSP augmented with a constraint hierarchy can be compared with techniques used in default logic [10, 8]. A default logic is a proof system $< W, R >$ where W is an initial theory and R consists of default rules of the form

$$\frac{\phi : M\beta_1 \ldots M\beta_n}{\psi} \tag{2}$$

which can be read as *under the assumption that $\beta_1 \ldots \beta_n$, if ϕ then ψ*. Default rules not only consist of a precondition (here the clause ϕ) but also of a justification $\beta_1 \ldots \beta_n$. The role of this justification is that it describes the exceptions blocking the applicability of the default rule in (1). As long as none of the $\neg\beta_i$ belong to the context S, the rule is applicable. S is a belief set, a context, from which so-called *extensions*, i.e. solutions are derived. Such a belief set S determines a subset of the default rules which are applicable. The existence of a proof for a formula f from W using R and S means that f itself is believed and therefore belongs to S. It represents the assumptions under which an extension is derived. Finding consistent solutions in a default theory $< W, R >$ consists of:

1. Guessing a belief set S.
2. Proving that all the consequences derived from W and R with respect to S equal S.

The solutions derived from belief sets using the rules are themselves part of the belief set and can be found using different techniques. One technique generates all elements of the power set of the rule consequences and checks them for consistency with the belief set S and the theory W and R. Another technique uses a well-ordering[2] to treat default rules with given priorities (prioritized defaults). User-defined priorities $\leq$ on the rule set R allow for generating a well-ordering $\preceq$ so that for each $r_i, r_j \in R$ if $r_i \leq r_j$ then $r_i \preceq r_j$. Given the order $\leq$: $r_1 < r_2$ and $r_1 < r_3$ among three rules $\{r_1, r_2, r_3\}$, for example, two possible well-orderings $\preceq$ can be generated, $\{r_1, r_2, r_3\}$ and $\{r_1, r_3, r_2\}$. The rules are processed sequentially with respect to the given well-ordering using one-step operators.

The rules of a DCSP can be modeled in *normal default theory*, a special case of default theory. The rule justifications in a normal default theory coincide with the consequent of the rule. For example:

$$\frac{\phi : M\psi}{\psi} \tag{3}$$

The rule is read as *under normal circumstances, if ϕ then ψ*. The justifications have to be proved within the current belief set. The rules in a dynamic constraint satisfaction can be given the same logic representation:

$$\frac{P(Y_1) : MC(Y_2)}{C(Y_2)} \tag{4}$$

The current belief set is the assumption that the constraints belong to the solution, i.e. they are *active*. In a DCSP, consistency checking among the active constraints provides a proof for believing in a constraint added to the active ones. Our hierarchy defined on rules is used in a similar way as in default logic to derive a consistent solution with respect to the given ordering.

---

[2] A well-ordering is a partially ordered set which is connected and each subset of which has a least element. There is a homomorphism between a well-ordered set and the set of ordinal numbers.

4 Bridge Design

In this section, we describe our implementation in the domain of preliminary bridge design. We sequentially process rules using dynamic constraint satisfaction and an ordering of the rules induced by a hierarchy on design criteria and by assumptions.

4.1 Knowledge Representation

Structural knowledge of the bridge is represented by generic components in the CAD system ICAD[3], which uses a parametric model of the artifact to be created. Design knowledge is represented by design rules each as described in section 3.1. The variables are given an initial range of possible values (intervals in case of real variables). The rules are clustered in consistent rule sets representing different design criteria. Typical criteria are esthetic, structural, economic or safety considerations. They allow the designer to define an ordering on the rules by which the evolution of the design process is controlled. Table 1 shows an example in preliminary bridge design. Initial conditions on the region where the bridge is to be built are entered into the ICAD interface. These include the section of the region, the obstacle over which the bridge is built etc. A geometrical interpretation of the section results in further qualitative information, for example *opposing slopes = similar* or *soil condition = good* illustrated in Figure 3.

4.2 Forward Reasoning with Assumptions

Rule sets are activated in the ICAD model according to the chosen order of design criteria. At each step (corresponding to one criterion), a new rule set is loaded. The design rules that satisfy the current design context are then sequentially processed activating constraints and variables. Each activated constraint is added to the current constraint network and new values are inferred through constraint propagation. If no conflict is detected and the whole rule block has been processed, the ICAD model is updated creating new components if necessary as shown in Figure 2.

A justification-based non-monotonic truth maintenance system (JTMS) reasons explicitly on constraint and variable activity. Variable value pairs are represented as JTMS-nodes and constraints are represented as assumption nodes. If such an assumption is :in the corresponding constraint is active and therefore in the constraint network. There are two types of assumptions: default assumptions and preferences. *Default assumptions* are important in preliminary stages of the design process when very little accurate information is available. *Preferences* control and guide search by focusing on interesting alternatives according to design criteria. *Fixed* constraints represent physical relations which always hold. The first rule of the start-up criterion in Table 1 for instance, introduces the default constraint C_3. The following data is recorded[4] :

---

[3] © Concentra Corporation
[4] The arrow shows data dependencies.

Level	Criteria	Type	Rule
0	physical	fixed	$bridge\ length = environment\ length$ if $nb\ of\ spans = n$ then $\sum_{i=1}^{n} span_i = bridge\ length$
1	start-up	default	if $\frac{1}{10} < aspect\ ratio < \frac{1}{3}$ then $nb\ of\ spans = \frac{environment\ length}{environment\ height}$ if $nb\ of\ spans = n$ then $\forall i(span_i) = \frac{bridge\ length}{nb\ of\ spans}$
2	construction	pref.	if $soil\ condition = good$ then $construction\ span = 55$ and $nb\ of\ spans = \frac{bridge\ length}{construction\ span}$
3	esthetics I	pref.	if $environment\ symmetry = \neg symmetric$ and $centre\ of\ gravity = centre$ then $nb\ of\ spans = 4$ and $inner\ spans = \frac{10}{38} * bridge\ length$ and $outer\ spans = \frac{9}{38} * bridge\ length$
4	cost	pref.	if $angle\ change = abrupt$ then $span\ 1 = point\ x - 15$
5	resistance	pref.	if $environment\ shape = V\ shape$ then $inner\ spans \geq outer\ spans$
6	esthetics II	pref.	if $opposing\ slopes = similar$ then $span_1 = span_n$

Table 1. *Sample of rules used in preliminary bridge design.*

$$default(C_3)\ :in$$
$$(nb\ of\ spans\ [4,5]) \leftarrow (aspect\ ratio\ \tfrac{70}{300}), (default\ (C_3))$$

It depends on the belief in constraint C_3 as well as on the other preconditions of this rule if the inferred values for the spans are :in (or active). Based on new data derived from constraint propagation and justified within the JTMS, the system may activate new rules. When a conflict is detected during constraint propagation, a tms-nogood is installed with the conflicting constraints and conflict management techniques are activated.

A general hierarchy on rule sets is defined in our system. Each rule set is labeled by a design criterion. Rule sets introduced earlier are considered more general than later ones. This order influences the general conflict management strategy and identifies among all the constraints in conflict the one that is preferably relaxed. Earlier default and preference constraints act on a less informed situation than later ones. Therefore the ones earlier introduced are more relaxable. Earlier fixed rules are more resilient. As the rule order proceeds from general to more detailed information later fixed rules are defined by more detailed knowledge. Thus fixed rules introduced later are more relaxable. The hierarchy of design criteria defines the order of activation of the rule sets, as well as the conflict resolution management with defaults, preferences and fixed constraints as explained in section 4.4.

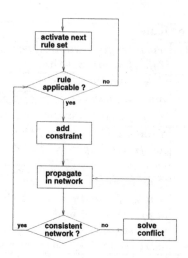

Fig. 2. *General algorithm for DCSP.*

4.3 An Example in Preliminary Bridge Design

The task is:

Given the section of the valley, build a beam bridge in the valley

shown in Figure 3. We will explore two different orders of design criteria presented in table 1: O_1: $0, \ldots, 6$ and O_2: $0, 1, 3, 2, 4, \ldots, 6$. This shows how slightly different rule orderings can result in completely different solutions and how solution spaces can be explored.

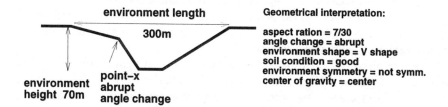

Fig. 3. *The valley entered in the interface of ICAD.*

Solutions for Ordering O_1 In the first ordering O_1 (Figure 4), the start up default rules create a beam bridge with 4 spans equally distributed. $Nbof\,spans = 4$ is chosen from $nb\,of\,spans = [4, 5]$. The construction preference of $nb\,of\,spans = [5, 6]$ overrides the default $nb\,of\,spans = 4$ to 5. The spans are still distributed regularly. The esthetic preference, which has higher priority than the construction preference, weakens [5] $nb\,of\,spans = 5$ again to 4. The spans are distributed so that the inner spans are larger than the outer ones ($outer\,spans = 71$, $inner\,spans = 79$). The cost preference then weakens $span_1$ and $span_2$ to 75 in order to place the first pier near the abrupt angle change in the section. According to the second rule on esthetics, $span_3$ and $span_4$ are weakened to 75 in order to assure that the first and the last span are of equal length.

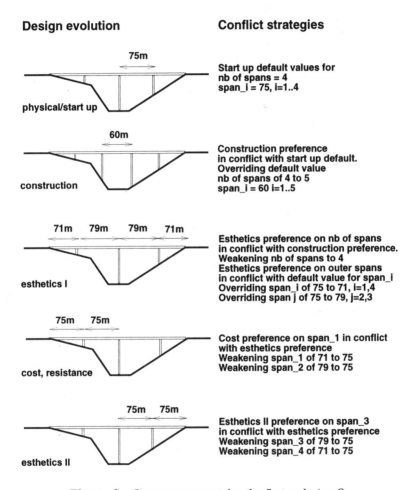

Design evolution

Conflict strategies

75m

physical/start up

Start up default values for
nb of spans = 4
span_i = 75, i=1..4

60m

construction

Construction preference
in conflict with start up default.
Overriding default value
nb of spans of 4 to 5
span_i = 60 i=1..5

71m   79m   79m   71m

esthetics I

Esthetics preference on nb of spans
in conflict with construction preference.
Weakening nb of spans to 4
Esthetics preference on outer spans
in conflict with default value for span_i
Overriding span_i of 75 to 71, i=1,4
Overriding span j of 75 to 79, j=2,3

75m   75m

cost, resistance

Cost preference on span_1 in conflict
with esthetics preference
Weakening span_1 of 71 to 75
Weakening span_2 of 79 to 75

75m   75m

esthetics II

Esthetics II preference on span_3
in conflict with esthetics preference
Weakening span_3 of 79 to 75
Weakening span_4 of 71 to 75

Fig. 4. *Conflict management for the first ordering O_1.*

Solutions for Ordering O_2 In the second solution (Figure 5), the criteria of esthetics I and construction are switched. First, the esthetics I criterion estab-

---

[5] Overriding and weakening are explained in section 4.4

lishes the same results as in ordering O_1 with 4 spans distributed according to the values of $spans = (71, 79, 79, 71)$. As the construction is more important than the start-up criterion $nb\,of\,spans = 4$ is weakened to 5. $Span_1$ is weakened to $[20,51]$ so that the bridge length does not exceed 300 (second rule of level 0): $spans = ([20, 51], 79, 79, 71, [20, 51])$. The cost preference determines $span_1$ to be 75m and in turn $span_2$ is weakened: $spans = (75, [20, 55], 79, 71, [20, 55])$. The resistance preference then weakens $span_3$ in order to accommodate for the resistance criterion and $spans = (75, [75, 114], [20, 60], 71, [20, 60])$. When introducing the rules of esthetics II, the constraint on $span_4$ is weakened in order to be able to change $span_5 = [20, 60]$ to 75. The result of $span_5 = [20, 59]$, however, does not allow for this change: the cost preference of $span_1 = 75$ has to be weakened too. The design ends up with $span_i = [20, 220]$ for $i = 1 \ldots 5$. The only active constraints are then the constraints of the resistance and esthetics II rules. One possible solution respecting all active constraints might be for example $spans = (58.5, 61, 61, 61, 58.5)$

4.4 Conflict Management

Conflicts in design arise as a consequence of conflicting subgoals when assumptions are introduced in situations of incomplete knowledge or when the solution space is infeasible. Conflict management strategies are based on the nature of the constraints (default, preference and fixed) and their importance for the designer (ordering of design criteria). We distinguish two types of conflicts: *assumption conflicts* and *feasibility conflicts* [7]. In assumption conflicts default or preference constraints are involved. Feasibility conflicts arise when all the default and preference information has been retracted and the conflict is still unsolved. They indicate an infeasible region in the solution space. The three conflict management strategies are:

1. **Overriding defaults.** Default constraints are used to start the design process when only incomplete knowledge of the design object is available. When involved in a conflict they are dropped without consequence and information inferred from them is overridden. Example: $nb\,of\,spans$ [4,5] is overriden by the new value for $nb\,of\,spans$ [5,6] when the construction criterion is introduced.

2. **Weakening preferences.** If a conflict exists, after all the default information has been overridden, preference constraints involved in the conflict are examined. These constraints determine local optimal solutions and should not be dropped completely, therefore they are weakened, and can be reactivated at a later stage. Their activation conditions are not revised and search continues in the same space of activation conditions. Example: after the introduction of the esthetic I criterion $span_1 = 71$ is in conflict with the cost preference $span_1 = point\,x - 15 = 75$. It is weakened to 75 meters because of the higher importance of the cost criterion.

3. **Dependency directed backtracking.** If no assumed information is left in conflict, activation conditions have to be reviewed. In this case, the system backtracks to the last activation condition (in the sense of a defined

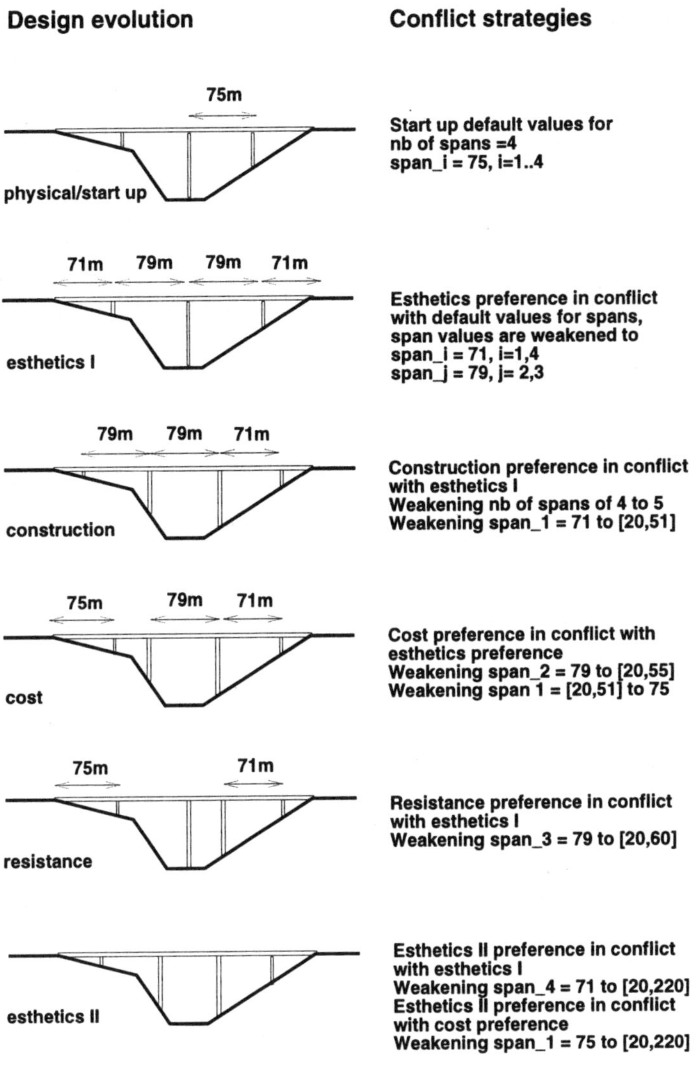

Design evolution   **Conflict strategies**

75m

physical/start up

Start up default values for
nb of spans =4
span_i = 75, i=1..4

71m   79m   79m   71m

esthetics I

Esthetics preference in conflict
with default values for spans,
span values are weakened to
span_i = 71, i=1,4
span_j = 79, j= 2,3

79m   79m   71m

construction

Construction preference in conflict
with esthetics I
Weakening nb of spans of 4 to 5
Weakening span_1 = 71 to [20,51]

75m   79m   71m

cost

Cost preference in conflict with
esthetics preference
Weakening span_2 = 79 to [20,55]
Weakening span 1 = [20,51] to 75

75m   71m

resistance

Resistance preference in conflict
with esthetics I
Weakening span_3 = 79 to [20,60]

esthetics II

Esthetics II preference in conflict
with esthetics I
Weakening span_4 = 71 to [20,220]
Esthetics II preference in conflict
with cost preference
Weakening span_1 = 75 to [20,220]

Fig. 5. *Conflict management for the second ordering O_2.*

order of rule sets) and completely changes the solution path. In our example several solutions exist. If none were feasible, the system would switch to
opposing slopes¬similar in the esthetics II criterion.

4.5 Limits of our Implementation

A designer can explore solution spaces by defining an order on the design criteria.
Limits of this implementation are:

1. When several choices are possible, the system takes the first one in the order. In the second solution, for example, the system continues with $nb\, of\, spans = 5$ chosen from [5,6] according to the construction rule. In order to get a second solution with $nb\, of\, spans = 6$ the system has to backtrack.

2. It can be quite difficult to evaluate in advance the influence of rule ordering on the evolution of the design. The designer has to explore different orderings to get an idea of the solution spaces.

3. Rules acting on a set of similar variables, as spans for example, introduce sets of similar constraints whereas others act on a specific variable, $span_1$ for example. In the example of the second start up rule, spans are distributed at equal distances. Acting only on $span_1$ and $span_2$ when a conflict arises with the cost rule, may lead to a fairly unesthetical design, where the spans are of unequal length.

Currently, we use the Waltz algorithm [3] for processing continuous constraints. It is known that this algorithm is neither complete nor sound [4]. Therefore the entire process of sequential rule processing is only an approximation of the sequential techniques applied in default logic. A theoretical development is under way in order to achieve soundness and completeness. Most of the current methods for nonlinear constraints only apply to local constraint satisfaction. For nonlinear systems, a complete and sound global constraint satisfaction algorithm is also under development [6].

5 Conclusion

We have shown how solution spaces can be explored using continuous constraint satisfaction algorithms in a dynamic environment. Exploration is guided by ordering the constraints into a hierarchy of criteria and by distinguishing between default, preference and fixed constraints. Theoretically, the problem of dynamic constraint satisfaction can be described in terms of normal default logic. Analogue algorithms of sequential rule triggering are used in dynamic constraint satisfaction where different rule orderings lead to different solutions.

Acknowledgments

Funding for this research was provided by the Swiss National Science Foundation. The authors would like to thank Boi Faltings, Djamila Haroud, Rainer Weigel, LIA, EPFL, and Sylvie Boulanger, ICOM, EPFL, for useful discussions. They would also like to thank the anonymous referees for their comments.

References

1. F. Benhamou, D. Mc Allester, and P. Van Hentenryck. Clp(intervals) revisited. *Technical Report CS-94-18*, University of Marseille, France, April 1994 1994.

2. Alan Borning, Bjorn Freeman-Benson, and Molly Wilson. Constraint hierarchies. *Lisp and Symbolic Computation*, 5(3):223–270, September 1992.
3. E. Davis. Constraint Propagation with Interval Labels. *Artificial Intelligence*, 32:281–331, 1987.
4. Boi V. Faltings. Arc-consistency for Continuous Variables. *Artificial Intelligence*, 65:363–376, 1994.
5. Eugene C. Freuder and Richard J. Wallace. Partial Constraint Satisfaction. *Artificial Intelligence*, 58:21–70, 1992.
6. Djamila Haroud and Boi Faltings. Global Consistency for Continuous Constraints. *Proc. of the ECAI-94*, 1994.
7. D. Haroud, S. Boulanger, E. Gelle, and I. Smith. Strategies for Conflict Management in Preliminary Engineering Design. *AIEDAM Special Issue on Conflict Management in Design*, 1995, pages 313–323.
8. V.W. Marek and M. Truszczynski. *Nonmonotonic Logic* Springer-Verlag, 1993.
9. S. Mittal and B. Falkenhainer. Dynamic Constraint Satisfaction Problems. In *Proc. of AAAI-90*, pages 25–32, Boston, MA, 1990.
10. Raymond Reiter. A logic for default reasoning. *Artificial Intelligence*, 13:81–132, 1980.
11. Ivan Sutherland. Sketchpad: A man-machine graphical communication system. In *Proceedings of the Spring Joint Computer Conference*, pages 329–346. IFIPS, 1963.

Specifying Over-Constrained Problems in Default Logic

Abdul Sattar[1], Aditya K. Ghose[2], Randy Goebel[3]

[1] School of Comp. and Info. Technology, Griffith University, Brisbane, Australia, 4111. sattar@cit.gu.edu.au
[2] Knowledge Systems Group, Basser Department of Computer Science, University of Sydney, Syndey NSW 2006, Australia. aditya@cs.su.oz.au
[3] Department of Computing Science, University of Alberta, Edmonton, Alberta, Canada, T6G 2H1. goebel@cs.ualberta.ca

Abstract. In the previous studies, it has been shown that the classical constraint satisfaction problem (CSP) is deductive in nature, and can be formulated as a classical theorem proving problem [1, 10]. Constraint satisfaction problems for which an assignment of values to all variables which satisfy all available constraints is not possible are referred to as *over-constrained problems*. This paper shows how computing partial solutions to over-constrained problems can be viewed as a default reasoning problem. We propose two methods for translating over-constrained problem specifications with finite domains to two different variants of default logic. We argue that default logic provides the appropriate level of abstraction for representing and analyzing over-constrained problem even if other methods are used for actually computing solutions.

1 Introduction

A *constraint satisfaction problem* (CSP) involves a set of *variables*, a *domain* of possible values for each variable, and a set of *constraints*, representing *acceptable* and *non-acceptable* relations over subsets of variables. A *solution* is an assignment of values to variables that satisfy all constraints. Constraint satisfaction problems for which an assignment of values to each variable that satisfies all constraints is not possible are called *Over-Constrained Problems (OCP)*. If the domains of variables in an OCP (resp. CSP) are restricted to be *finite*, we obtain the class of Finite Over-Constrained Problems (FOCP) (resp. Finite Constraint Satisfaction Problems (FCSP))[4].

Over-constrained problems are ubiquitous in AI and arise due to conflicting constraints in a domain of reasoning such as design problems in which conflicting goals have to be achieved, diagnosis in which competing hypotheses explain the set of symptoms, schedule conflicts etc. This problem has been variously termed as *partial constraint satisfaction* [5], reasoning with *constraint hierarchies* [14], reasoning with hard and soft constraints [12], or *constraint relaxation* and preferences over relaxations in resolving conflicting schedules [4].

---

[4] Mackworth [10] specifies a finite constraint satisfaction problem in an identical manner

Logically, the notion of constraint satisfaction has been viewed as a *deductive* reasoning problem [1, 10]. It has been previously shown that classsical CSP can be formulated as a hypothetical reasoning problem using the THEORIST system [13]. We show that formulating constraint satisfaction as a nonmonotonic reasoning problem is general enough to cover OCP's in addition to classical CSP's. Intuitively, conflicting constraints could be treated as conflicting defaults in a default reasoning framework. Thus, finding a solution to a partial constraint satisfaction problem corresponds to computing an extension of a default theory.

A variety of default reasoning systems exist in the literature. In this paper we shall consider a restricted version of Reiter's default logic [11] as well as a recent variant called prerequisite-free constrained default logic (PfConDL) presented by Delgrande, Schaub and Jackson in [3]. We shall present translations from finite over-constrained problem specifications to default theories in each of these formalisms. We have earlier shown how default extension computation can be viewed as solving over-constrained problems [7]. The research presented in this paper thus completes the picture by presenting the reverse translation as well.

We maintain that default logic provides the appropriate level of abstraction for representing and analyzing over-constrained problems even if other techniques are used for actually computing solutions. The ability to specify such problems in a formal language with well-defined semantics has several practical advantages. These include semantically well-founded criteria for defining preference relations on constraints and solutions as well as methods for revising OCP specifications in a principles manner. Further, we believe that complexity results from the default reasoning area can suggest tractable classes of OCP's.

2 Over-constrained problems

Formally, a constraint satisfaction problem (CSP) specification consists of a finite set of variables $Var = \{X_1, \ldots, X_n\}$, each associated with a domain of discrete values, $d_1, \ldots, d_n$, and a set of constraints $Con = \{C_1, \ldots, C_m\}$. Each constraint is a relation defined on some subset of the set of variables. A constraint C_i consists of the *constraint-subset* $S_i = \{X_{i_1}, \ldots, X_{i_{j(i)}}\}$, where $S_i \subseteq X$, denoting the subset of the variables on which C_i is defined and the *relation* rel_i defined on S_i such that $rel_i \subseteq d_{i_1} \times \ldots \times d_{i_{j(i)}}$.

Formally, an *over-constrained problem* is a constraint satisfaction problem for which there is no assignment of values to all variables such that all the constraints are satisfied. The following example from [5] illustrates the idea.

Example 1. Consider a robot seeking to select matching shoes, shirts and slacks while getting dressed. It has two kinds of shoes (cordovans, sneakers), two kinds of shirts (green, white) and three kinds of slacks (denims, dress blue, dress grey). The only allowable combinations are: white shirts and cordovan shoes, cordovan shoes and gray dress slacks, sneakers and denim slacks, green shirts and dress gray slacks, white shirts and denim slacks and white shirts and dress blue slacks.

We shall formulate the problem as a CSP. We consider three variables: Shoes (S_1), Shirts (S_2), and Slacks (S_3). The corresponding domains are:

- $D_1 = \{cordovans, sneakers\}$
- $D_2 = \{white, green\}$
- $D_3 = \{denims, dress\_blue, dress\_gray\}$

The constraints are:

- $C_{S_1 S_2} = (cordovans, white)$
- $C_{S_1 S_3} = (cordovans, dress\_gray)$
- $C_{S_1 S_3} = (sneakers, denims)$
- $C_{S_2 S_3} = (green, dress\_gray)$
- $C_{S_2 S_3} = (white, dress\_blue)$
- $C_{S_2 S_3} = (white, denims)$

The problem is over-constrained since there is no choice of shoes, shirts and slacks which satisfy all the constraints.

In a *finite* CSP, the domain of every variable is finite [10]. In this paper, we shall confine our interest to *finite over-constrained problems* (FOCP's).

Definition 1 (FOCP Specification). An FOCP specification is a 3-tuple (Var, Dom, Con) where:

1. $Var = \{X_1, X_2, \ldots, X_n\}$ is the set of variables.
2. $Dom = \{d_{X_1}, d_{X_2}, \ldots d_{X_n}\}$ where d_{X_i} specifies the domain for variable X_i. Each $d_{X_i} = \{v_i^1, v_i^2, \ldots, v_i^{n_i}\}$ is required to be finite.
3. $Con = \{C_1, C_2, \ldots, C_m\}$ where each C_i is a relation defined on some subset of Var involving values from the appropriate domains.

Over-constrained problems may be solved using partial constraint satisfaction (PCS) techniques [5]. PCS techniques enables us to identify the "best" partial solution, where the notion of "best" can be defined using a variety of metrics. Given the difficulty of obtaining *a priori* guarantees on the existence of solutions that satisfy all constraints, it is clear that PCS techniques have a broader applicability than classical CSP techniques. PCS techniques are also suitable for solving problems in resource-bounded situations, such as when the time available for computing a solution is bounded. PCS techniques can help us identify solutions that are "good enough" or "close enough" to the complete solution in the available time.

The notion of the "best" partial solution can be defined along two different dimensions. Partial solutions can be classified on the basis of whether they are "maximal" with respect to the variables, or with respect to the constraints. Similarly, the notion of "maximality" can be defined on the basis of cardinality or set inclusion. Classifying along the first dimension, two kinds of partial solutions are possible:

- **Var-partial solutions.** These are solutions which assign values to a subset of the set of variables, but satisfy all available constraints.
- **Con-partial solutions.** These are solutions which assign values to all variables, but staisfy only a subset of the set of constraints.

Classifying along the second dimension, two notions of "maximality" are possible:

- **Cardinality-maximal solutions**. Maximality is defined in terms of set cardinality. Thus, a cardinality-maximal var-partial solution is a solution which satisfies all available constraints and assigns values to the largest number of variables. A cardinality-maximal con-partial solution assigns values to all variables and satisfies the largest number of constraints.
- **Inclusion-maximal solutions**. Maximality is defined in terms of set inclusion. Thus, an inclusion-maximal var-partial solution satisfies all the available constraints and assigns values to some subset x of the set of variables Var such that there is no x' where $x \subset x' \subseteq Var$. An inclusion-maximal con-partial solution assigns values to all variables and satisfies some subset c of the set of constraints Con such that there is no c' where $c \subset c' \subseteq Con$.

Observation: *Every cardinality-maximal var-partial (con-partial) solution of an FOCP-specification S is an inclusion-maximal var-partial (con-partial) solution of S.*

3 Frameworks for default reasoning

Reasoning with incomplete information is a crucial component of intelligent systems. Reiter [11] addressed this problem by adding domain specific *default rules* with classical logic to model patterns of inference of the form "in the absence of information to the contrary, conclude that ...". A default rule consists of three components:

1. The *prerequisite* $\alpha(\mathbf{x})$.
2. The *justification* $\beta(\mathbf{x})$.
3. The *conclusion* $\gamma(\mathbf{x})$.

The informal semantics of a default rule of the form: $\frac{\alpha(\mathbf{x}):\beta(\mathbf{x})}{\gamma(\mathbf{x})}$. is: "If, for some set of instances $\mathbf{c}$, $\alpha(\mathbf{c})$ is provable from what is known and $\beta(\mathbf{c})$ is consistent, then conclude by default that $\gamma(\mathbf{c})$." For the rest of this paper, we shall focus on *closed default thoeries*, i.e., default theories which refer only to formulas with no free variables. A *normal* default is one where the justification and consequent are equivalent. A *semi-normal* default is one where the consequent is a logical consequence of the justification. Formally, a default theory is a pair (W, D) where W is a classical theory containing *facts* that must be true and D is a set of default rules as described above. A default *extension* represents a maximal set of beliefs sanctioned by a default theory. A default theory may, in general, have several extensions. Given a default theory (W, D), the set of *generating defaults* of an extension E, given by $GD(E, (W, D))$, is some $D' \subseteq D$ such that every default in D' is applied in computing E. A precise formulation of generating defaults will depend on the version of default logic under consideration. In this section, we shall present Reiter's original formulation of default logic and a variant developed by Delgrande, Schaub and Jackson called *prerequisite-free constrained default logic* (PfConDL) [3]. In the next section, we shall present two translations from over-constrained problem specifications to default theories. The first translation maps an FOCP specification to a PfConDL theory. The second maps an FOCP specification to a default theory in a restricted version of Reiter's default logic.

3.1 Reiter's default logic

We shall first present a fixed point definition of extensions. In the rest of this paper, $Cn(T)$ will refer to the deductive closure of T.

Definition 2. [11] Let (W, D) be a default theory. For any set S of formulas, let $\Gamma(S)$ be the smallest set satisfying the following properties:

1. $W \subseteq \Gamma(S)$.
2. $\Gamma(S) = Cn(S)$.
3. If $\frac{\alpha : \beta}{\gamma} \in D$ and $\alpha \in \Gamma(S)$ and $\neg \beta \notin S$, then $\gamma \in \Gamma(S)$,

then E is an extension of the default theory (W, D) iff $\Gamma(E) = E$.

Thus, we may informally view a default extension as a sanctioned set of beliefs which necessarily contains all the facts, and which contains the consequent of a default rule in case the partial extension constructed at the point when the default rule is considered entails its prerequisite, and its justification is consistent with the final extension.

In this paper, we shall only focus on *prerequisite-free normal default theories* in Reiter's default logic, i.e., default theories in which every default rule is *normal* and has an empty prerequisite. For a prerequisite-free normal default theory (W, D) in Reiter's default logic, the set of generating defaults for an extension E is given by:

$$GD(E, (W, D)) = \{\tfrac{:\beta}{\beta} \in D \mid E \models \beta\}$$

3.2 PfConDL

Prerequisite-free constrained default logic (PfConDL) [3] is a recent variant of Reiter's default logic which addresses several problems with the original version, including situations where Reiter's default logic produces unwarranted conclusions or fails to produce conclusions which are intuitively warranted. As well, it has several useful properties such as the guaranteed existence of extensions, semi-monotonicity, the weak orthogonality of extensions, cumulativity, the ability to reason with *modus tollens* and reasoning by cases. We shall not describe these properties in any greater detail but shall summarize the relevant results here.

In PfConDL, every default rule is prerequisite-free and semi-normal. A *constrained extension* in PfConDL is a pair (E, C) where E corresponds to the set of sanctioned beliefs (and hence, an extension in the sense of Reiter's logic) and C consists of E together with the justifications of all applied defaults (thus C defines the reasoning context in which E is obtained). Following [3], we can characterize constrained extensions of a theory in PfConDL in the following way.

Definition 3. Let (W, D) be a default theory. For any set of formulas T let $\Gamma(T)$ be the pair of smallest sets of formulas (S', T') such that

1. $W \subseteq S' \subseteq T'$,
2. $S' = C_n(S')$ and $T' = C_n(T')$,
3. for any $\frac{:\beta}{\gamma} \in D$, if $T \cup \{\beta\} \cup \{\gamma\} \not\vdash \perp$, then $\gamma \in S'$ and $\beta \wedge \gamma \in T'$

A pair of sets of formulas (E, C) is a constrained extension of (D, W) iff $\Gamma(C) = (E, C)$.

The theorem below provides a simpler view of what a constrained extension stands for. $Justif(D)$ and $Conseq(D)$ refer to the conjunction of the justifications and consequents of all defaults in D.

Theorem 4. [3] *Let (W, D) be a prerequisite-free default theory. Then (E, C) is a constrained extension of (W, D) iff there exists some $D' \subseteq D$ such that:*

$$E = C_n(W \cup Conseq(D'))$$
$$C = C_n(W \cup Conseq(D') \cup Justif(D')),$$

and $C \not\vdash \perp$, but for every $D' \subset D''$, $C \cup \{Conseq(D'')\} \cup \{Justif(D'')\} \vdash \perp$.

Thus, a constrained extension may be viewed as a pair consisting of a sanctioned set of beliefs and the reasoning context in which this set of beliefs is obtained. Constrained extensions can be computed constructively, unlike extensions in the general case of Reiter's default logic (it is constructive in the case of *normal* default theories). In considering whether a default rule is applicable, one checks whether the default consequent is consistent with the current (partial) set of sanctioned beliefs and in addition, whether the justfication together with the consequent is consistent with the current (partial) set of commited assumptions (i.e., the reasoning context).

Reiter's default logic with prerequisite-free normal defaults turns out to be a special case of PfConDL.

Given a theory (W, D) in PfConDL, the set of generating defaults for a constrained extension (E, C) is given by:

$$GD((E, C), (W, D)) = \{\frac{:\beta \wedge \gamma}{\beta} \in D \mid E \models \beta, C \models \beta \wedge \gamma\}$$

3.3 A taxonomy for extensions

For the purpose of establishing equivalences between FOCP specifications and their corresponding translations into default theories, it is useful to consider a richer taxonomy on the class of default extensions. Each of the classes of extensions we will introduce here will have a counterpart in an FOCP specification.

Traditionally, default extensions have been viewed as maximal sets of sanctioned beliefs, where maximality is defined with respect to set inclusion. Thus, in computing an extension, one attempts to apply as many default rules as is consistently possible, and each maximally inextensible set of beliefs is taken as an extension. However, it is also possible to confine one's interest to those sets of beliefs in which the *maximal number* of default rules are applied. We shall refer to such extensions as *cardinality-maximal extensions*. Extensions in the traditional sense shall be referred to as *inclusion-maximal extensions*. Formally, a cardinality-maximal extension e is an inclusion-maximal extensions such that there is no inclusion-maximal extension e' with $|GD(e, \Delta)| < |GD(e', \Delta)|$ where Δ is a default theory and $GD(e, \Delta)$ stands for the set of generating defaults for e given Δ. In general, a default theory can have several cardinality-maximal extensions.

Consider a default theory (W, D) where the set of defaults is finitely partitioned into partitions $D_1, D_2, \ldots, D_n$. An extension e is a *representative extension* for (W, D) iff $GD(e, (W, D))$ contains at most one element of each of $D_1, D_2, \ldots, D_n$. A representative extension e is *perfect* iff $GD(e, (W, D))$ contains exactly one element of each of $D_1, D_2, \ldots, D_n$; otherwise the representative extension is imperfect. If $GD(e, (W, D)) = D$, then e is referred to as a *total extension*.

4 Translations

We shall propose two translations in this section. The first translation maps an FOCP-specification to a default theory in PfConDL. The resulting default theory has a finitely partitioned set of defaults and the representative extensions of this theory provide var-partial solutions to the corresponding OCP. The second maps an FOCP-specification to a normal default theory in Retier's default logic. The extensions in the resulting default theory represent con-partial solutions to the corresponding OCP.

4.1 Trans1

Trans1 views every assignment of a value to a variable as a potentially refutable default, while the constraints and domain specifications are treated as facts. The set of resulting defaults is thus partitioned on the basis of the variables they refer to. Representative extensions of the resulting default theory thus correspond to var-partial solutions to the corresponding FOCP-specification.

Definition 5 (Trans1). Let (Var, Dom, Con) be an FOCP specification. Let (W, D) be a PfConDL theory.
$Trans1(Var, Dom, Con) = (W_{(Var, Dom, Con)}, D_{(Var, Dom, Con)})$ where:

1. $W_{(Var, Dom, Con)} = W_{domains} \cup W_{constraints}$.
    - $W_{domains} = W_{X_1} \cup W_{X_2} \cup \ldots \cup W_{X_n}$ where $W_{X_i} = \{x_i\_domain(v_j) \mid v_j \in d_{X_i}\}$.

$-\ W_{constraints} = \{\forall x_1, \ldots, x_k (c_i(x_1, \ldots, x_k) \leftarrow x_1\_value(x_1) \wedge \ldots \wedge x_k\_value(x_k)) \mid$
$\quad\ c_i(x_1, \ldots, x_k) \in C\}$

2. $D_{(Var,Dom,Con)} = D_{X_1} \cup D_{X_2} \cup \ldots \cup D_{X_n}$ where
$D_{X_i} = \{ \frac{:x_i\_value(v_j) \leftarrow x_i\_domain(v_j) \wedge Comp(C)}{x_i\_value(v_j) \leftarrow x_i\_domain(v_j)} \mid v_j \in d_{X_i}\}$
and
$comp(C) = \{\neg c_i(v_1, \ldots, v_j) \mid$ there exists
no $c_i \in Con$ containing the tuple $< v_1, \ldots, v_j >\}$

Theorem 6 (Trans1-FOCP Equivalence). *Let :*

- (Var, Dom, Con) *be an FOCP-specification.*
- $Trans1(Var, Dom, Con) = (W_{(Var,Dom,Con)}, D_{(Var,Dom,Con)})$.
- $D_{X_1}, D_{X_2}, \ldots, D_{X_n}$ *be a partitioning of* $D_{(Var,Dom,Con)}$ *as defined by Trans1.*
- $EXT1 = \{E \mid (E, C)$ *is a representative constrained extension of*
  $(W_{(Var,Dom,Con)}, D_{(Var,Dom,Con)})$ *given the partitioning of* $D_{(Var,Dom,Con)}\}$
- $EXT2 = \{E \mid (E, C)$ *is a representative cardinality-maximal constrained*
  extension of
  $(W_{(Var,Dom,Con)}, D_{(Var,Dom,Con)})$ *given the partitioning of* $D_{(Var,Dom,Con)}\}$
- $S1$ *be the set of var-partial inclusion-maximal solutions of* (Var, Dom, Con).
- $S2$ *be the set of var-partial cardinality-maximal solutions of* (Var, Dom, Con).

Then:

1. $EXT1 = S1$.
2. $EXT2 = S2$.
3. (Var, Dom, Con) *is not over-constrained iff there exists some* $e \in EXT1$
   such that e *is perfect.*

Example 2. Consider three variables: X_1, X_2 and X_3. The domains are: $\mathcal{D}_{X_1} = \{a, b\}$, $\mathcal{D}_{X_2} = \{e, f\}$, and $\mathcal{D}_{X_3} = \{c, d, g\}$. The constraints are: $c_{X_1 X_2}(b, e)$, $c_{X_1 X_2}(b, f)$, $c_{X_1 X_3}(b, c)$, $c_{X_1 X_3}(b, d)$, $c_{X_1 X_3}(b, g)$, $c_{X_2 X_3}(e, d)$, $c_{X_2 X_3}(f, g)$.

The task is to find one or all possible consistent assignments of values for X_1, X_2 and X_3.

$Trans1(X, \mathcal{D}, C) = (W, D)$, where
$W = \{x_1\_domain(a), x_1\_domain(b), x_2\_domain(e), x_2\_domain(f), x_3\_domain(c),$
$x_3\_domain(d), x_3\_domain(g)\} \cup$
$\{c_{X_1 X_2}(X_1, X_2) \leftarrow x_1\_value(X_1) \wedge x_2\_value(X_2)$
$c_{X_1 X_3}(X_1, X_3) \leftarrow x_1\_value(X_1) \wedge x_3\_value(X_3)$
$c_{X_2 X_3}(X_2, X_3) \leftarrow x_2\_value(X_2) \wedge x_3\_value(X_3)\}$

For compactness of representation, we shall represent each class of defaults referring to a single variable as a default schema. Thus:
$D = \{ \frac{:x_1\_value(X_1) \leftarrow x_1\_domain(X_1) \wedge comp(C)}{x_1\_value(X_1) \leftarrow x_1\_domain(X_1)}$
$\frac{:x_2\_value(X_2) \leftarrow x_2\_domain(X_2) \wedge comp(C)}{x_2\_value(X_2) \leftarrow x_2\_domain(X_2)}$
$\frac{:x_3\_value(X_3) \leftarrow x_3\_domain(X_3) \wedge comp(C)}{x_3\_value(X_3) \leftarrow x_3\_domain(X_3)} \}$

$$comp(C) = \{\neg c_{X_1 X_2}(a, e), \neg c_{X_1 X_2}(a, f), \neg c_{X_1 X_3}(a, c), \neg c_{X_1 X_3}(a, d),$$
$$\neg c_{X_1 X_3}(a, g),$$
$$\neg c_{X_2 X_3}(e, c), \neg c_{X_2 X_3}(e, g), \neg c_{X_2 X_3}(f, c), \neg c_{X_2 X_3}(f, d)\}$$

So in this example, there are two extensions (here we shall refer only to the E portion of a constrained extension (E, C)):

$W \cup \{x_1\_value(b) \leftarrow x_1\_domain(b), x_2\_value(e) \leftarrow x_2\_domain(e), x_3\_value(d) \leftarrow x_3\_domain(d)\}$ and

$W \cup \{x_1\_value(b) \leftarrow x_1\_domain(b), x_2\_value(f) \leftarrow x_2\_domain(f), x_3\_value(g) \leftarrow x_3\_domain(g)\}$.

Both of these extensions are perfect, hence the problem is not **over-constrained**.

4.2   Trans2

Trans2 views each constraint as a potentially refutable default. The extensions of the resulting default theory thus correspond to con-partial solutions of the FOCP specification.

We assume that a function *Flatten* exists which takes a relation and "flattens" it into a formula in disjunctive normal form. Each disjunct in the resulting formula is a conjunction of assignments of values representing a tuple in the relation. Thus, a relation with two tuples, defined on the variables X_1, X_2, X_3, $R(X_1, X_2, X_3) = \{< v1, v2, v3 >, < v4, v5, v6 >\}$ is flattened to produce a formula $(X_1 = v1 \wedge X_2 = v2 \wedge X_3 = v3) \vee (X_1 = v4 \wedge X_2 = v5 \wedge X_3 = v6)$.

Definition 7 (Trans2). Let (Var, Dom, Con) be an FOCP specification. Let (W, D) be a PfConDL theory.
$Trans2(Var, Dom, Con) = (W_{(Var, Dom, Con)}, D_{(Var, Dom, Con)})$ where:

1. $W_{(Var, Dom, Con)} = W_{domains} \cup W_{constraints} \cup$
   EQUALITY_AXIOMS $\cup$ UNIQUE_NAMES_AXIOM.
   - $W_{domains} = W_{X_1} \cup W_{X_2} \cup \ldots \cup W_{X_n}$ where $W_{X_i} = \{x_i\_domain(v_j) \mid v_j \in d_{X_i}\}$.
   - $W_{constraints} = \{c_{i+X_1 \ldots X_n} \rightarrow Flatten(C_i(X_1, X_2, \ldots, X_n)) \mid C_i(X_1, X_2, \ldots, X_n) \in C\}$
2. $D_{(Var, Dom, Con)} = \{\frac{:c_{i+X_1 \ldots X_n}}{c_{i+X_1 \ldots X_n}} \mid C_i(X_1, X_2, \ldots, X_n) \in C\}$.

Theorem 8 (Tans2-FOCP Equivalence). *Let:*

- *(Var, Dom, Con) be an FOCP-specification.*
- *$Trans1(Var, Dom, Con) = (W_{(Var, Dom, Con)}, D_{(Var, Dom, Con)})$.*
- *EXT1 be the set of inclusion-maximal extensions of*
  $(W_{(Var, Dom, Con)}, D_{(Var, Dom, Con)})$.
- *EXT2 be the set of cardinality-maximal extensions of*
  $(W_{(Var, Dom, Con)}, D_{(Var, Dom, Con)})$.
- *S1 be the set of con-partial inclusion-maximal solutions of (Var, Dom, Con).*
- *S2 be the set of con-partial cardinality-maximal solutions of (Var, Dom, Con).*

Then:

1. $EXT1 = S1$.
2. $EXT2 = S2$.
3. (Var, Dom, Con) is not over-constrained iff there exists $e \in EXT1$ such that e is total.

Example 3. To demonstrate *Trans2* we will again consider Example 1. There are three variables: Shoes (S_1), Shirt (S_2), and Slacks (S_3). The corresponding domains are: $D_1 = \{cordovans, sneakers\}$, $D_2 = \{white, green\}$, and $D_3 = \{denims, blue\_dress, gray\_dress\}$. The constraints are:

- $C_{S_1 S_2} = (cordovans, white)$
- $C_{S_1 S_3} = (cordovans, gray\_dress)$
- $C_{S_1 S_3} = (sneakers, denims)$
- $C_{S_2 S_3} = (green, gray\_dress)$
- $C_{S_2 S_3} = (white, blue\_dress)$
- $C_{S_2 S_3} = (white, denims)$

The problem is to find a consistent assignment of values for S_1, S_2 and S_3. It is clear from the above that there is no such assignment. Let (W, D) be the theory resulting from applying *Trans2* to this FOCP-specification. $W = \{s_1\_domain(cordovans), s_1\_domain(sneakers), s_2\_domain(white),$
$s_2\_domain(green),$
$s_3\_domain(denims), s_3\_domain(gray\_dress), s_3\_domain(blue\_dress)\}$
$\cup$
$\{c_{1+S_1 S_2} \rightarrow (S_1 = cordovans \land S_2 = white), c_{2+S_1 S_3} \rightarrow (S_1 = cordovans \land S_3 = gray\_dress), c_{3+S_1 S_3} \rightarrow (S_1 = sneakers \land S_3 = denims), c_{4+S_2 S_3} \rightarrow (S_2 = green \land S_3 = gray\_dress), c_{5+S_2 S_3} \rightarrow (S_2 = white \land S_3 = blue\_dress), c_{6+S_2 S_3} \rightarrow (S_2 = white \land S_3 = denims)\}$
$$D = \{ \frac{:c_{1+S_1 S_2}}{c_{1+S_1 S_2}}, \frac{:c_{2+S_1 S_3}}{c_{2+S_1 S_3}}, \frac{:c_{3+S_1 S_3}}{c_{3+S_1 S_3}}, \frac{:c_{4+S_2 S_3}}{c_{4+S_2 S_3}}, \frac{:c_{5+S_2 S_3}}{c_{5+S_2 S_3}}, \frac{:c_{6+S_2 S_3}}{c_{6+S_2 S_3}} \}$$
Here we have three extensions:
$W \cup \{c_{1+S_1 S_2}\}$, $W \cup \{c_{2+S_1 S_3}, c_{3+S_1 S_3}\}$, and $W \cup \{c_{4+S_2 S_3}, c_{5+S_2 S_3}, c_{6+S_2 S_3}\}$.
None of them is a **total** extension, so this problem is over-constrained problem, and all three extensions correspond to partial solutions of the problem.

5 The utility of specifying OCP's in default logic

In this section, we shall outline what we believe are strong reasons for viewing default logic as the appropriate level of abstraction for specifying OCP's. These motivations stem largely from two observations. First, considerable advantages accrue from being able to specify problems in a formal language in which the meaning of assertions is well-defined. Second, while constraint satisfaction problems have natural translations into classical logic, over-constrained problems must necessarily be translated into some nonmonotonic logic. Reiter's default logic, and its variants turn out to be simple, syntactically-oriented formalisms into which OCP's can be translated in a straightforward manner. In practical terms, the benefits of specifying OCP's in default logic derive from the ability to provide a semantic basis for preference relations on constraints, or OCP solutions, the ability to analyze the problem of updating OCP specifications as a problem of belief revision in default logic and the possibility of using complexity results in default logic to identify tractable classes of OCP's.

Semantically well-founded preference criteria In general, an OCP may have several partial solutions, one of which must be selected. The problem is identical to the theory preference problem in selecting amongst the multiple extensions of a default theory. Typically, systems for solving OCP's use preference criteria with no semantic basis whatsoever. For instance, in the context of hierarchical logic programming, Wilson and Borning [14] define a number of comparators which use somewhat arbitrary arithmetic criteria on a weighted system of potentially conflicting constraints to compute the best solution. They admit to the lack of any well-defined prescription for using one comparator over another. In the context of default reasoning, a variety of meaningful theory preference criteria have been considered [8]; these include specificity, simplicity, nonredundancy among others. Many of these results can be applied to OCP's to generate meaningful ways of identifying preferred solutions.

Revising OCP specifications While belief revision is a well-studied problem in the context of formal logic, the problem of revising OCP specifications has received scant attention in the literature. An existing belief revision system that uses PJ-default logic [2] (a precursor of PfConDL) presented in [6] can be directly used as a system for maintaining OCP specifications over iterated revision steps.

Identifying tractable cases Few results (except for one in [5]) exist on tractable cases for OCP's. We believe that complexity results from default logic (such as those in [9]) can provide useful pointers for identifying tractable classes of OCP's.

Ideally, OCP's should be specified in default logic and solved using partial constraint satisfaction techniques such as those presented in [5]. We have earlier identified translations from default theories to OCP specifications. The utility of the translations we propose here lie not only demonstrating that OCP's can be specified using default theories, but in translating intermediate OCP specifications back into defaul theories for the purpose of debugging.

Acknowledgement

We would like to acknowledge constructive comments/criticism from Peter van Beek, Jia You and Vladimir Alexiev.

References

1. W. Bibel. Constraint satisfaction from a deductive viewpoint. *Artificial Intelligence*, 35(3):401–413, July 1988.

2. J. P. Delgrande and W. K. Jackson. Default logic revisited. In *Proc. of the Second International Conference on the Principles of Knowledge Representation and Reasoning*, pages 118–127, 1991.

3. J. P. Delgrande, Torsten Schaub, and W. K. Jackson. An approach to default reasoning based on a first-order conditional logic: Revised report. *Artificial Intelligence*, 36:63–90, 1988.

4. Fox, M.: *Constraint Directed Search: A Case Study of Job-Shop Scheduling*. Morgan Kaufman, 1987.

5. E. C. Freuder and Richard J. Wallace. Partial constraint satisfaction. *Artificial Intelligence*, 58:21–70, 1992.

6. A.K. Ghose, P. Hadjinian, A. Sattar, J. You, and R. Goebel. Iterated belief change: A preliminary report. In *Proceedings of Australian Joint Conference on Artificial Intelligence*, pages 39–44, Melbourne, Victoria, November 1993. World Scientific Publishing Co.

7. A.K. Ghose, A. Sattar, and R. Goebel. Default reasoning as partial constraint satisfaction. In *Proceedings of Australian Joint Conference on Artificial Intelligence*, Armidale, NSW, November 1994. World Scientific Publishing Co.

8. S. Goodwin and A. Sattar. On computing preferred explanation. In *Proceedings of Australian Joint Conference on Artificial Intelligence*, pages 45–52, Melbourne, Victoria, November 1993. World Scientific Publishing Co.

9. H.A. Kautz and B. Selman. Hard problems for simple default logics. In *Proc. of the First International Conference on the Principles of Knowledge Representation and Reasoning*, pages 189–197, 1989.

10. A. K. Mackworth. The logic of constraint satisfaction. *Artificial Intelligence*, 58(1-3):3–20, December 1992.

11. R. Reiter. A logic for default reasoning. *Artificial Intelligence*, 13(1&2):81–132, 1980.

12. Satoh K.: Formalizing soft constraints by interpretation ordering. In *Proc. of the 9th European Conf. on AI*, pages 585–590, 1990.

13. Sattar A. and Goebel R.G.: Constraint Satisfaction as Hypothetical Reasoning. In *Proceedings of the Vth International Symposium on Artificial Intelligence*, Cancun, Mexico, December 1992. AAAI-Press.

14. Molly Wilson and Alan Borning. Hierarchical constraint logic programming. *Journal of Logic Programming*, 16:277–318, 1993.

Implementing Constraint Relaxation over Finite Domains Using Assumption-Based Truth Maintenance Systems

Narendra Jussien and Patrice Boizumault

École des Mines de Nantes. Département Informatique.
4 Rue Alfred Kastler. La Chantrerie.
F-44070 Nantes Cedex 03, France.
{Narendra.Jussien,Patrice.Boizumault}@emn.fr

Abstract. Many real-life Constraint Satisfaction Problems are over-constrained. In order to provide some kind of solution for such problems, this paper proposes a constraint relaxation mechanism fully integrated with the constraint solver. Such a constraint relaxation system must be able to perform two fundamental tasks: identification of constraints to relax and efficient constraint suppression. Assumption-based Truth Maintenance Systems propose a uniform framework to tackle those requirements. The main idea of our proposal is to use the ATMS to record and efficiently use all the information provided by the constraint solver while checking consistency. We detail the use of ATMS in our particular scheme and enlight their efficiency by comparing them with existing algorithms or systems (Menezes' IHCS and Bessière's DnAC4).

1 Introduction

Many real-life problems are over-constrained (time-tabling, CAD, ...). A constraint relaxation mechanism integrated with the constraint solver becomes necessary. Such a constraint relaxation system must be able to perform two fundamental tasks: identification of constraints to relax and efficient constraint suppression. This research area has been initiated by the Constraint Logic Programming community, from the works of Borning *et al.* [3] in the Logic Programming community and from Freuder's first theoretical framework [11] in the CSP community.

Several systems or algorithms have been proposed to provide a constraint relaxation system over finite domains such as IHCS [14], DnAC4 [1], DnAC6 [7], ... These propositions present two similarities : first, they all deal with a single point of constraint relaxation over finite domains – intelligent identification of constraint(s) to relax in IHCS or efficient constraint suppression in DnAC* – and second, they all record supplementary information provided by the constraint solver. It is worth noticing that this information is the same in all the cited systems – for each removed value in a variable domain, one records the constraint which first removed this value when checking consistency –, just the use of this information differs.

The aim of this paper is to propose a Constraint Relaxation scheme for CLP($\mathcal{FD}$) languages integrating at the same time appropriate relaxation decision, identification of responsibilities and constraint suppression. Assumption-based Truth Maintenance Systems [5] propose a uniform framework to tackle those three requirements. The main idea of our proposal is to use the ATMS to record and efficiently use all the information provided by the constraint solver while checking consistency. The ATMS framework provides a direct management of constraint deletion, and a more precise identification of constraints to relax.

First, we briefly recall some definitions and results about Constraint Relaxation. Then, we present our framework for Constraint Relaxation over Finite Domains and recall basic results about ATMS. We describe the use of ATMS to implement our system and illustrate it using an example. Finally we give results from a first implementation and draw further works.

2 Constraint Relaxation

In order to deal with over constrained systems, various approaches have been developed for a few years. Constraint hierarchies and their use were introduced by Borning *et al.* [3], and Freuder [11] proposed a first theoretical framework. In this section, we briefly recall the basic concepts and review different proposals.

2.1 Basic Concepts

HCLP [3] introduces a **hierarchy** upon constraints. A **weight** is assigned to each constraint. This weight represents the relative importance of the constraint. This weight allows the setting of a partial order relation between constraints. Thus, constraints with no weight (which can be assimilated as a zero weight) are called **required** or **mandatory** constraints and those with a positive weight are called **preferred** constraints. Let us recall that the greater the weight, the less **important** the constraint is.

A substitution (values for variables) which satisfies all the required constraints and which satisfies the preferred constraints in the best possible way with respect to a given comparator is called a **solution**. Thus, a *maximal* (to a given criterion[1]) sub-problem of the initial problem is searched. This sub-problem must have existing solutions.

2.2 Different approaches

A constraint relaxation problem can be handled in different ways.

- The problem can be seen as the modification of an already solved instance using repairing algorithms [12, 16]. This approach needs an existing solution in order to start and needs explicit operations, i.e. we have to tell the system which operation has to be executed (suppressing a constraint, adding a constraint, ...)

---

[1] This criterion is called the comparator in Borning's approach.

- A constraint relaxation problem can be solved using branch and bound methods with preferred constraints in the objective function. Those approaches use known Operations Research results and techniques [8]. In those approaches, the programmer or user cannot specify his labeling strategy and does not control the resolution which is guided by the objective function.
- Fages *et al.* [9] proposed a reactive scheme that can efficiently add and suppress constraints. An abstracted framework is presented in terms of transformations of CSLD trees. In this approach, the constraint(s) to suppress are selected by the user but not automatically detected by the system.
- The Constraint Relaxation problem can also be studied as the handling of a tower of constraints [3, 13]. For example, the HCLP($\mathcal{R}$) system considers that the whole problem does not have any solution. Therefore it tries to find a solution respecting only the required constraints, and then refines the solution adding the preferred constraints level[2] by level until failure. Unfortunately, this approach is no more usable in the Finite Domains paradigm, because consistency techniques are local (eg., all the variables must be instantiated to insure the existence of a solution under the required constraints).

All those systems are not satisfactory for our purpose: providing an automated constraint relaxation system embedded in a Constraint Logic Programming language over Finite Domains.

3 Constraint Relaxation over Finite Domains

First, we review some works about efficient addition or suppression of constraints using dynamic CSP, and then about identifying constraint(s) to relax using intelligent backtracking. We show that even though not sharing the same objectives, they all keep the same *justification* system. Finally, we present our general framework.

3.1 Dynamic Arc Consistency

For a few years, the CSP community has been interested in Dynamic CSP. A few algorithms have been developed by Bessière (DnAC4 [1]) and Debruyne (DnAC6 [7]) to achieve dynamic arc-consistency which allows efficient addition and suppression of constraints.

DnAC4 extends AC4 by recording justifications during restrictions. The justification consists in recording for each removed value in the domain of a variable, the constraint which first removed it.

When deleting a constraint, DnAC4 finds which values must be put back in the domains thanks to the system of justifications. This procedure is more efficient than re-running an AC4 algorithm on the new problem from scratch.

---

[2] A set of constraints with a same weight forms a level in the hierarchy of constraints.

3.2 Identifying constraint(s) to relax

The Logic Programming community has been interested for a few years in Intelligent Backtracking. The results of this study have been embedded in the CLP($\mathcal{FD}$) framework to identify responsibility of constraint for inconsistency and thus to identify constraints to relax.

When achieving arc-consistency, Cousin's approach [4] records the same justification as DnAC4. When the domain of a variable becomes empty, it is then easy to identify the appropriate constraints, but this notion of responsibility is very limited (we just identify an immediate responsible constraint). It is worth noticing that even though the recorded information is the same as DnAC4, Cousin does not make the same usage of this information. Cousin's approach appears as a rudimentary strategy for determining constraints to relax because of its short range view of the problem.

IHCS gives a deeper analysis of the situation. IHCS uses the same justification as DnAC4. Moreover, it defines a dependency relation between constraints: *a constraint C_a depends on a constraint C_b if the constraint C_b modifies the domain of a variable appearing in C_a*. The resulting graph can be used in order to identify responsibility of failures. Thus, when a constraint fails, it is easy to identify a set of constraints who could be responsible for the failure. This set is the transitive closure from the failing constraint in the dependency graph. The problem of the IHCS system is encountered in the fact that all the analysis is done from the *constraints* i.e. a lot of information is lost about *values* in domains of variables, then, the dependency analysis becomes too general and unnecessary possibilities of relaxation are handled.

It is worth noticing that all these orthogonal works lie upon the same notion of justification (the constraint who removed a value from the domain of a variable).

3.3 Our general framework

We state that Constraint Relaxation over Finite Domains relies on three main points which cannot be treated separately:

- *When to relax ?* When exactly during computation do we have to relax a constraint.

  We have to start a relaxation process when a contradiction is raised ensuring the non existence of solution. Exploiting this approach is not so easy in a standard CLP($\mathcal{FD}$) scheme. In such a scheme, there are two kinds of contradictions: a contradiction caused by a wrong value given during the labeling phase, and a contradiction due to the intractability of the problem. Those two possibilities lead to different treatments, in the first case a simple backtrack will lead to another possibility and in the second case, we must start a relaxation process.

  Therefore, we would like to collapse the two possibilities into a single one: start the relaxation process in case of contradiction. This can be achieved by redefining the labeling in terms of adding and suppressing constraints[3].

- *Which constraint(s) to suppress ?* Which constraint or which set of constraints do we need to suppress (relax) in order to obtain a satisfiable subproblem. The previous systems (Cousin's system and IHCS) proposed answers to this question (see section 3.2).

  We would like to do it more accurately exploiting all the information that the solver can gather i.e. the justifications for deletions of values (not only trivial justification).

- *How to delete a constraint ?* How can we efficiently perform the suppression of a constraint. An obvious answer is to use systems that can add **and** retract constraints, i.e. incremental systems as [10, 17].

  Another answer comes from the CSP community. As we stated before, DnAC4 achieves dynamic arc-consistency and thus enforces the maintenance of arc-consistency after the deletion of a constraint.

A Constraint Relaxation system must provide an efficient answer to the *all* three questions. For example, using DnAC4 to perform the effective suppression does not allow an efficient answer to the *who* question. This is the reason why we would like to perform the deletion easily but keeping in mind that we have to answer other questions. So, we would like to use a DnAC4 like method but with no suppression of information (i.e. not really putting back a value but merely make temporarily unbelievable its deletion).

We propose in the following sections a specialization of the Assumption-based Truth Maintenance Systems concepts to implement a constraint relaxation system which fully answers the three basic questions upon which our framework lies.

4 Truth Maintenance Systems

4.1 Overview of Truth Maintenance Systems

As stated by de Kleer [5], most problem solvers search and are constantly confronted with the obligation to select among equally plausible alternatives. When solving problems, one aims to provide a good answer to two questions:

- How can the search space be efficiently explored, or how can maximum information be transferred from one point in the space to another ?
- How, conceptually, should the problem solver be organized ?

---

[3] for example, in order to label the variable X whose domain is $[1, 2, 3]$, the first step is to add the constraint $X = 1$, and if this not succeeds then perform: retract $X = 1$ and add $X = 2$.

First of all, classical backtracking techniques cannot provide an efficient scheme to achieve such tasks. The problems caused by such approaches lead to the development of truth maintenance systems (TMS). These systems make a clean division between two components: a problem solver which draw inferences and another component (namely the TMS) solely concerned with the recording of those inferences (called justifications) (see figure 1).

The TMS serves three roles:

- It maintains a cache of all the inferences ever made. Thus inferences, once made, need not be repeated and contradictions, once discovered, are avoided in the future,
- It allows the use of non-monotonic inferences and thus needs a procedure (called truth maintenance) to determine what data are to be believed,
- It ensures that the database is contradiction free. Contradictions are removed by identifying absent justification(s) whose addition to the database would remove the contradiction. This is called dependency directed backtracking.

The Assumption-based Truth Maintenance Systems (ATMS) extend the TMS architecture, by extending the cache idea, simplifying truth maintenance and avoiding dependency directed backtracking.

In a classical (justification-based) TMS every datum is associated with a status of *in* (believed) or *out* (not believed). The entire set of *in* data defines the current context.

In an ATMS each datum is labeled with the sets of assumptions (representing the *environment*) under which it holds. These environments are computed by the ATMS from the problem-solver-supplied justifications.

Those labels enable an efficient way of shifting context. When shifting contexts, one wants to know how much of what was believed in a previous context can be believed in the new one. This can be efficiently achieved by ATMS using the labels and testing their accuracy in the current context.

Chronological backtracking after failure looses useful information because assertions derived from data unaffected by a contradiction remain useful. Thus, only assertions directly affected by the contradiction causing the failure should be removed. In an ATMS, one can tell easily whether an assertion is affected or not.

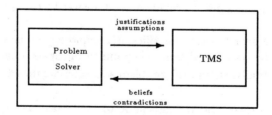

Fig. 1. *TMS and Problem Solver*

4.2 Basic definitions

Here are some basic definitions for ATMS.

- A *node* corresponds to a problem-solver datum.
- An *assumption* designates a decision to assume without any commitment. An assumption is a special kind of node.
- A *justification* describes how a node is derivable from other nodes.
- A *context* is formed by the assumptions of a consistent environment combined with all nodes derivable from those assumptions.
- A *label* is a set of environments associated with every node. A label is consistent i.e. from the label and the set of justifications, one can prove that the current node is to be believed.

  While a justification describes how the datum is derived from immediately preceding antecedents, a label environment describes how the datum ultimately depends on assumptions. These sets of assumptions can be computed from the justifications, but computing them each time would disable the efficiency advantage of the ATMS.

The fundamental task of the ATMS is to guarantee that the label of each node is consistent, sound, complete and minimal with respect to the justifications (see [5] for more details).

5 Constraint Relaxation using ATMS

In this section, we show how ATMS can handle uniformly the three components of a Constraint Relaxation system.

5.1 Architecture

The overall architecture of our system is as stated in figure 2. In this figure, the problem solver is a CLP($\mathcal{FD}$) solver which uses an Arc Consistency algorithm and a labeling procedure coupled with an analyzer which computes constraint(s) to relax from the ATMS information.

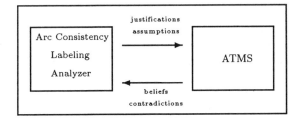

Fig. 2. *ATMS and CLP(FD)*

We give the meaning of the ATMS concepts in our Constraint Relaxation scheme.

- A *node* reflects the suppression of a value in the domain of a variable, which represents the atomic operation of a constraint solver over finite domains. The constraint solver gives a *justification* of this operation. This justification is composed with the *conjunction of constraints* that are responsible for this suppression. This justification is in fact the *environment* explaining the corresponding suppression. This definition of a justification represent the difference between our approach and DnAC4, Cousin's method or IHCS.
- Each constraint in the problem appears as an *assumption*. This assumption can be *in* or *out* whether the constraint is considered *in* the system or *out* of the system.
- The current context is represented by the *in* assumptions i.e. the active constraints.
- The label of a node is a set of environments which indicates the reason why a value has been retracted.

This architecture is very flexible and allows the use of any constraint solver over Finite Domains. The efficiency our system is related to the precision and accuracy of the justifications provided by the constraint solver. A constraint solver that uses a lot of information to remove a value from a domain will give all this information to our system and thus allow an efficient relaxation procedure.

Our system gives a useful help to the constraint solver but does not create information by itself, it can only use the information contained in the solver.

5.2 Utilization

The ATMS as previously described provides a unified framework to answer the three main questions.

- The identification of responsibility will be given by the node label, which answers the **who** question. As we stated before, the label gives all the conjunctions of constraints that are responsible for deletion. We then have to determine a set of constraints whose relaxation will free at least one value back in the domain.
- Contradictions are raised when the domain of a variable becomes empty. When a contradiction occurs (the domain of a variable, the *failing variable*, becomes empty), the current system of constraints is not satisfiable, we start the determination of constraint(s) to relax. We answer here the **when** question.
- Incrementality will be ensured by keeping useful information between relaxation thanks to the use of good labels. This will answer the **how** question. To relax a constraint, one just have to specify that the related ATMS assumptions may not be believed anymore. This is done by switching the status of the assumption (constraint) from the *in* status to the *out* status.

If the problem solver raises a contradiction during the labeling phase (answering the **when** question), responsibility (answering the **who** question) will efficiently be determined by examining the associated labels. These labels contain all the necessary information to enforce this task.

Once responsibility determined using the previously called analyzer, one just need to retract (answering the **how** question) the corresponding assumption(s) and to reexecute the labeling phase in order to ensure a complete exploration of the new search space[4].

6 Example

We describe our constraint relaxation scheme and the use of the ATMS considering the Conference problem [4].

6.1 Stating the Conference problem

Michael, John and Alan must attend work-sessions taking place over four half days.

John and Alan want to present their work to Michael and Michael wants to present his work to John and Alan.

Let Ma, Mj, Am, Jm be the four presentations (Michael to Alan, Michael to John, Alan to Michael, John to Michael respectively). Those variables have the domain $[1, 2, 3, 4]$. Each presentation takes one half-day.

Michael wants to know what John and Alan have done before presenting his work. Michael would like not to come the fourth half-day and Michael does not want to present his work to Alan and John at the same time.

Someone who attends a presentation cannot present something in the same half-day. Two different persons cannot present to the same person at the same time.

We introduce the following constraints:

- $C_1 : Ma > Am$, $C_2 : Ma > Jm$, $C_3 : Mj > Am$ and $C_4 : Mj > Jm$. Those constraints are the most important ones so they are given the weight 1.
- $C_5 : Ma \neq 4$, $C_6 : Mj \neq 4$, $C_7 : Am \neq 4$ and $C_8 : Jm \neq 4$. Those constraints are the least important ones, so they are given the weight 3.
- $C_9 : Ma \neq Mj$. This constraint is quite important, so it is given the weight 2.
- $C_{10} : Ma \neq Am$, $C_{11} : Ma \neq Jm$, $C_{12} : Mj \neq Am$, $C_{13} : Mj \neq Jm$ and $C_{14} : Am \neq Jm$. Those constraints are required, so they are not given any weight.

---

[4] The search space has been modified by the suppression of a constraint.

Variable	Labels				Resulting
	1	2	3	4	Domain
Ma	$[[C_2],[C_1]]$	$[]$	$[]$	$[]$	[2,3,4]
Mj	$[]$	$[]$	$[]$	$[]$	[1,2,3,4]
Am	$[]$	$[]$	$[]$	$[[C_1]]$	[1,2,3]
Jm	$[]$	$[]$	$[]$	$[[C_2]]$	[1,2,3]

Table 1. *Introduction of the first two constraints*

Variable	Labels				Resulting
	1	2	3	4	Domain
Ma	$[[C_2],[C_1]]$	$[]$	$[]$	$[[C_5]]$	[2,3]
Mj	$[[C_4],[C_3]]$	$[]$	$[]$	$[[C_6]]$	[2,3]
Am	$[]$	$[]$	$[[C_3,C_6],[C_1,C_5]]$	$[[C_7],[C_1],[C_3]]$	[1,2]
Jm	$[]$	$[]$	$[[C_4,C_6],[C_2,C_5]]$	$[[C_8],[C_2],[C_4]]$	[1,2]

Table 2. *Introduction of the 14 constraints*

6.2 Solving the Conference problem

The ATMS labels are associated with each removal of a value in a domain, so this label can be attached with the corresponding value instead of creating explicitly an ATMS node.

For example, after the introduction of the first two constraints, we obtain the table 1.

For the variable Ma, the set of labels can be read as:

- The value 1 has to be removed from the domain if the constraint C_2 is to be believed or if the constraint C_1 is to be believed.
- The values 2, 3 and 4 cannot be removed with the current information[5].

After the introduction of the 14 constraints, the labels are as in table 2.
For the variable Am, the set of labels can be read as:

- The values 1 and 2 cannot be removed with the current information.
- The value 3 has to be removed from the domain if the constraints C_3 and C_6 are both to be believed or if the constraints C_1 and C_5 are both to be believed.
- The value 4 has to be removed from the domain if the constraint C_7 is to be believed or if the constraint C_1 is to be believed or if the constraint C_3 is to be believed.

---

[5] We use here arc consistency with AC5 [18]. We can parameterize the real signification of the label with the achieved degree of consistency.

Domain Value	Label
1	$[[C_2, C_5, C_{16}], [C_3], [C_4]]$
2	$[[C_1, C_2, C_3, C_4, C_5, C_6, C_7, C_8, C_9], [C_{16}]]$
3	$[[C_4, C_5, C_6, C_9, C_{16}], [C_1, C_2, C_3, C_4, C_5, C_6, C_7, C_8, C_9]$ $[C_1, C_5, C_9, C_{16}], [C_2, C_5, C_9, C_{16}]]$
4	$[[C_6]]$

Table 3. *The second contradiction*

Considering the context (the active constraints) we find the current domain of the variable.

To ensure the existence of a solution to this problem, we have to enumerate the variables. This is the labeling phase. We choose to first enumerate the variable Am. Giving the value 1 ($C_{15} : Am = 1$ weight[6] 100) to Am leads to a contradiction on the variable Mj so we try to give[7] the value 2 ($C_{16} : Am = 2$ weight 100). This leads to another contradiction for the variable Mj.

Handling the contradictions Contradictions occur when the *current* domain of a variable becomes empty, i.e. all the values have got a non empty label and those labels are valid (i.e. can be believed given the current context).

In order to find the responsible constraint(s) of the a contradiction, we have to *read* the labels for the variable who provoked the contradiction.

In the Conference problem, the labels of the variable Mj after the second contradiction are as in table 3.

We only keep the valid environments (justifications) in these labels. The real labels contained information about the constraint C_{15} which is not valid i.e. relaxed in the current context.

The classical interpretation of the labels leads to following statement:

In order to recover from the inconsistency lying on the variable Mj, we have to free a value in the domain, i.e. we have to make a label invalid by modifying the current context. This modification lies upon the relaxation of a constraint.
In order to recover from the inconsistency, we have to relax:
(C_2 or C_5 or C_{16}) and C_3 and C_4 or (C_1 or ... or C_9 and ... and C_{16}) or ... or C_6

This formulation is not satisfactory because, we have to choose the value that we want to make free. In order to avoid this choice, we transform the preceding disjunction of conjunctions in a conjunction of disjunctions, i.e. :

---

[6] This is an arbitrary great value.
[7] We must first relax the constraint C_{15}.

(C₁ or ... or C₉ or C₁₆) **and** *(C₁ or ... or C₉)*
and ... **and** *(C₄ or C₅ or C₆ or C₉ or C₁₆)*

We then have to make sure that in each conjunction there is at most one true component considering that C_i is true if the constraint is relaxed and false otherwise.

This problem can be seen as the determination of a set covering problem in an hypergraph whose vertices are the different disjunctions and whose edges are the different constraints. An edge represents the sharing of a constraint between different disjunctions. An edge can involve only one disjunction. This set covering must minimize a certain criterion. This criterion defines the nature of the solutions we want to determine to solve our Constraint Relaxation problem.

A criterion could be:

- minimizing the *cost* of the relaxation, i.e. to minimize the sum of the weights of the edges. In our case, the weight would have been the inverse of the weight associated to the constraint. This problem is $\mathcal{NP}$-Complete. In our example, this method leads to the set of relaxable constraints: $\{C_5\}$.
- minimizing the maximum weight of the covering, i.e. ensuring that the more important of the relaxed constraint is the least important one. This can be answered by simply selecting the least important constraint in each disjunction, i.e. selecting for each vertex the better edge (the one with the smaller weight). This can be done in a linear time: $O(m)$ where m represents the size of the formula. In our example, this method leads to the set of relaxable constraints: $\{C_5, C_{16}\}$.

Performing the relaxation We simply have to remove from the context (change their status from *in* to *out*, which is immediate) the constraints determined in the previous step and then reexecute the labeling procedure. As we reexecute the labeling procedure we benefit from the previously done work because the marks did not changed during the relaxation (we did not erase any deduction).

Giving an answer If the labeling procedure can be achieved, the current solution is our solution, and the satisfied constraints are those in the current context. In our example, we reconsider the value 1 for the variable Am which leads to a solution $[Ma, Mj, Am, Jm] = [4, 3, 1, 2]$ which violates only one constraint: C_5.

7   First results

We implemented our system in Prolog. We chose the AC5 algorithm [18] to achieve arc-consistency. The complete system represents 3000 lines of code.

System	Relaxed Constraint	Backtracks
HCLP	1	27
IHCS	3	3
FELIAC	1	3
AC+ATMS	1	2

Table 4. *Complete Results for the Conference Problem*

7.1 Comparing systems on the Conference Problem

We chose to compare our system with the more complete (answering the three basic questions) systems we found in the literature. We did not compare it with incremental algorithms (DnAC∗) since the more important basic questions are *how* **and** *who.*

So, we implemented HCLP and IHCS as presented in the related papers to make comparisons between different systems. We also used the FELIAC system presented in [2]. This system uses known techniques (IHCS, Van Hentenryck's oracles [17]) to create a constraint relaxation system fully but not uniformly answering the three basic questions. This system is interactive, the user must give the constraint to add or retract (FELIAC gives a choice of judicious constraints), it cannot deal with a Prolog program.

We can find the results on the Conference problem for the four systems in table 4.

To obtain a solution in the Conference problem, one needs to relax only one constraint (namely C_5). The HCLP and FELIAC systems gave the *good* answer as well as our system (AC+ATMS). IHCS had to relax three constraints to obtain a feasible solution. Let us recall that IHCS works on the failing *constraint* and not the failing *variable* and so works with incomplete information, that's why it has to relax inaccurate constraints.

When we look at the number of backtracks which reflects the real efficiency of the systems, we can see that HCLP generates a lot of backtracks compared to the other systems. This illustrates the inaccuracy of such a method with incomplete solvers, after each addition of variable, HCLP must ensure that there exist a solution and so has to exhibit one when using arc-consistency.

Finally, our system is more efficient than FELIAC and illustrates the incremental capability of ATMS. In other words, the cache of inferences of the ATMS is more efficient than Van Hentenryck's oracles.

7.2 Complexity

In this section, we give first results about complexity in the worst case of our implemented system. We cannot really compare the complexity of our system with other systems since we propose a complete treatment of the relaxation problem (answering the three basic questions) whereas no other system, as far as we know, proposes it.

Let e the number of constraints in the problem, n the number of variables and d the size of the largest domain. We give here the final results for complexity using AC5 as constraint solver.

Spatial Complexity The space use of our system will increase when adding justifications. We cannot had more justifications than the solver removes values.

A justification can contain at most e constraints. Adding a justification depends on the constraint solver. When using AC5, at each step of the algorithm, a justification can be added to explain the removal of a value in a domain for a variable. Thus, the space complexity would be $O(nd \times e \times ed)$ in the worst case since there are nd labels in the system. This complexity is considered without the ATMS treatment which ensure the minimal size of the label by suppressing redundant justifications in a same label.

For the Conference problem, the number of added justifications would be, in the worst case, $O(ned^2) = O(4 \times 14 \times 4^2) = O(896)$. The final number of justifications for the Conference problem is 61.

Time complexity The time complexity of the constraint solver is not affected by the use of the ATMS system although at each removal the constraint solver must justify his removal to the ATMS recording system (this operation is not significant).

The set of labels for a variable is transformed in a normal conjunctive form at each relaxation. This transformation is achieved in $O(m^2)$ where m is the size of the treated set. This set would contain in the worst case all the added justifications : $O(ned^2)$. For the Conference problem, as stated in table 3 this set contains only 10 justifications.

The selection of the constraints to relax, (see section 6.2) depends on the retained criterion. When using the second criterion, the complexity is: $O(m)$, of course when using the first criterion we try to solve an $\mathcal{NP}$-complete problem and therefore the complexity does not remain polynomial.

8 Conclusion and further works

Using ATMS for constraint relaxation provides a uniform framework to answer the three basic questions (when, which and how). This framework is very flexible and can be used with any constraint solver (domain reduction) to provide an automated constraint relaxation system over finite domains.

The IHCS system and DnAC4 were only focused in a single aspect of the problem. Our results enlight the utility of tackling the three questions in the same framework.

Some further works are to be done:

- First, to evaluate the average behavior of our approach, in particular in terms of space complexity. We plan to study how to retain only useful information when adding a justification.
- Our research area lacks for significant benchmarks. We plan to study how to generate random instances of intractable problems avoiding simple contradictions.
- De Kleer [6] shows how CSP can be mapped to ATMS by encoding each domain and constraint as boolean formulas, and then achieving arc-consistency by defining inference rules. It would be interesting to extend this scheme to Dynamic CSP, giving a semantics for constraint deletion.
- Saraswat *et al.* [15] initiated works to actively use the ATMS labels as boolean formulas. This leads to the integration of those formulas inside the constraint language. This interesting approach would worth a study for finite domains.

References

1. Christian Bessière. Arc consistency in dynamic constraint satisfaction problems. In *Proceedings AAAI'91*, 1991.
2. Patrice Boizumault, Christelle Guéret, and Narendra Jussien. Efficient labeling and constraint relaxation for solving time tabling problems. In Pierre Lim and Jean Jourdan, editors, *Proceedings of the 1994 ILPS post-conference workshop on Constraint Languages/Systems and their use in Problem Modeling : Volume 1 (Applications and Modelling)*, Technical Report ECRC-94-38, ECRC, Munich, Germany, November 1994.
3. Alan Borning, Michael Maher, Amy Martindale, and Molly Wilson. Constraint hierarchies and logic programming. In Giorgio Levi and Maurizio Martelli, editors, *ICLP'89: Proceedings 6th International Conference on Logic Programming*, pages 149–164, Lisbon, Portugal, June 1989. MIT Press.
4. Xavier Cousin. Meilleures solutions en programmation logique. In *Proceedings Avignon'91*, 1991. In French.
5. Johan de Kleer. An assumption-based tms. *Artificial Intelligence*, 28:127–162, 1986.
6. Johan de Kleer. A comparison of ATMS and CSP techniques. In *IJCAI-89: Proceedings 11th International Joint Conference on Artificial Intelligence*, pages 290–296, Detroit, 1989.
7. Romuald Debruyne. DnAc6. Research Report 94-054, Laboratoire d'Informatique, de Robotique et de Micro-électronique de Montpellier, 1994. In French.
8. François Fages, Julian Fowler, and Thierry Sola. Handling preferences in constraint logic programming with relational optimization. In *PLILP'94*, Madrid, September 1994.
9. François Fages, Julian Fowler, and Thierry Sola. A reactive constraint logic programming scheme. In *International Conference of Logic Programming, ICLP'95*, Tokyo, 1995.
10. Bjorn Freeman-Benson, John Maloney, and Alan Borning. An incremental constraint solver. *Communications of the ACM*, 33(1):54–63, January 1990.

11. Eugene Freuder. Partial constraint satisfaction. In *IJCAI-89: Proceedings 11th International Joint Conference on Artificial Intelligence*, pages 278–283, Detroit, 1989.

12. ·Alois Haselböck, Thomas Havelka, and Markus Stumptner. Revising inconsistent variable assignments in constraint satisfaction problems. In Manfred Meyer, editor, *Constraint Processing: Proceedings of the International Workshop at CSAM'93, St. Petersburg, July 1993*, Research Report RR-93-39, pages 113–122, DFKI Kaiserslautern, August 1993.

13. Michael Jampel and David Gilbert. Fair Hierarchical Constraint Logic Programming. In Manfred Meyer, editor, *Proceedings ECAI'94 Workshop on Constraint Processing*, Amsterdam, August 1994.

14. Francisco Menezes, Pedro Barahona, and Philippe Codognet. An incremental hierarchical constraint solver. In Paris Kanellakis, Jean-Louis Lassez, and Vijay Saraswat, editors, *PPCP'93: First Workshop on Principles and Practice of Constraint Programming*, Providence RI, 1993.

15. Vijay Saraswat, Johan de Kleer, and Brian Williams. ATMS-based constraint programming. Technical report, Xerox PARC, October 1991.

16. Gilles Trombettoni. CCMA*: A Complete Constraint Maintenance Algorithm Using Constraint Programming. In Manfred Meyer, editor, *Constraint Processing: Proceedings of the International Workshop at CSAM'93, St. Petersburg, July 1993*, Research Report RR-93-39, pages 123–132, DFKI Kaiserslautern, August 1993.

17. Pascal Van Hentenryck. Incremental constraint satisfaction in logic programming. In *Proceedings 6th International Conference on Logic Programming*, 1989.

18. Pascal Van Hentenryck, Yves Deville, and Choh-Man Teng. A generic arc-consistency algorithm and its specializations. *Artificial Intelligence*, 57(2–3):291–321, October 1992.

Experiences in Solving Constraint Relaxation Networks with Boltzmann Machines

Rolf Weißschnur[1], Joachim Hertzberg[1], Hans Werner Guesgen[2]

[1] German National Research Center for Computer Science (GMD), AI Research Division, Schloss Birlinghoven, D-53754 Sankt Augustin, Germany
[2] Computer Science Department, University of Auckland, Auckland, New Zealand

Abstract. Earlier, Guesgen and Hertzberg have given a theoretical description of how to implement constraint relaxation in terms of combinatorial optimization using the concept of Boltzmann Machines. This paper sketches some lessons that an implementation of this idea has taught us about how to tailor the translation from constraint networks to Boltzmann Machines such that the resulting implementation be efficient.

1 Background and Overview

Usually, a constraint satisfaction problem (CSP) is defined as follows: Given a set of variables over some arbitrary domains and a set of constraints, each constraint ranging over a subset of the variables, find an assignment of values to the variables such that the constraints are satisfied.

Solving a CSP using the usual constraint satisfaction methods assumes that it is solvable in the first place. However, experience tells that many real-world problems, when formulated as CSPs, do not have a solution. Nevertheless, human beings are often able to handle them; a natural way for doing so is trying to find an approximation (or almost solution) for the problem, which satisfies our needs in the best possible way. If it is not possible to satisfy the entire set of constraints, the over-constrained problem can be relaxed by switching off or weakening some constraints such that the relaxed problem is close to the original problem and has a solution.

There are several approaches to constraint relaxation, ranging from theory as in [10, 6] to practical applications as in [3, 5]. All these approaches have in common that they attack a given CSP by finding a solution of a relaxed CSP that differs only minimally from the original one. The difference is expressed in terms of a metric.

There are various ways to define a metric on a given CSP, i.e., to state how different a relaxed CSP is from the original CSP and how far away the approximate solution is from the ideal one. In [9], for example, we have used the concept of penalties. Values not included in the original constraint relations are marked by natural numbers greater than 0. More recently, fuzzy set theory has been used to capture the idea of constraint relaxation [4, 7, 11, 13].

Since the aim is to find a relaxed CSP whose distance to the original CSP is minimal, constraint relaxation can be seen as an optimization problem. In [8, 9], we have proposed to formulate this optimization problem in terms of *Boltzmann Machines* (BMs) [1]. While BMs are a little exotic an optimization tool, they offer a number of nice properties that makes considering them plausible:

- The behavior of BMs is governed by a simulated annealing algorithm. Given that the very use constraint relaxation presupposes that there is no sharp concept of solution of a constraint network, the statistical flavor of simulated annealing leads to a nice *anytime* [2] solution behavior: given short run time bounds, reasonable solutions can be expected; given no or large time bounds, one is guaranteed to get a globally optimal solution.
- BMs offer the potential of being implemented directly on massive parallel hardware. Given that constraint problems, by their very nature, have a flavor of distributedness, massive parallel implementation is a point to consider, even if todays standard hardware is of, or close to, von Neumann type.

This paper sketches a number of ideas we found useful when actually implementing constraint relaxation in terms of BMs. The reader is referred to [8, Ch. 4] for a more detailed description of constraint relaxation; to [1] for an excellent introduction to BMs; to [8, Ch. 8] or [9, Sec. 5] for a description of the basic idea how to translate relaxation networks into BMs; and to [14] for details of the work presented here. Our implementations uses a sequential BM simulator [12] on SUNs.

This is the plot of the paper: Section 2 recapitulates the essentials of translating relaxation networks into BMs, including the most basic BM notions. The key section of the paper is Sect. 3, describing our findings how to best tailor a BM corresponding to the original constraint problem. Section 4 concludes.

2 The Basics of the Experiments

2.1 Boltzmann Machines: The Most Basic Notions

In this section, which is drawn from [9], we recapitulate the most basic BM notions that are necessary to understand what follows. Any reader who is knowledgeable about BMs may safely skip this section.

A BM is a graph $B = (U, K)$, where U is the finite set of *units* and K is a set of unordered pairs of elements of U, the *connections*, each connection written as $\{u, v\}$, for $u, v \in U$. K includes all *bias connections*, i.e., connections connecting a unit with itself. In short: $\{\{u, u\} \mid u \in U\} \subset K$. The units are binary valued, i.e., they are either *on* or *off*, which is represented as 1 and 0, respectively. A *configuration* of a BM is a 0-1-vector of length $|U|$ describing the state of each unit. If k is a configuration, $k(u)$ denotes the state of u in k.

Connections are either active or passive. A connection between u und v is *active* in a configuration k if both connected units are on, i.e., if $k(u) \cdot k(v) = 1$; else it is *passive*. Every connection $\{u, v\}$ has an associated *connection strength* $s_{\{u,v\}} \in \mathbb{R}$, to be interpreted as the desirability that $\{u, v\}$ be active. The strength of a bias connection $\{u, u\}$ is called the *bias* of u.

The desirability of a whole configuration k, expressed in terms of a *consensus function* is the sum of the strengths of all active connections; hence the consensus function $\mathcal{K}(k)$ looks as follows:

$$\mathcal{K}(k) = \sum_{\{u,v\} \in K} s_{\{u,v\}} \cdot k(u) \cdot k(v).$$

A BM is "run" by generating new configurations from given ones by toggling units; then the consensus value of these new configurations is determined. The objective of a BM is to find a global maximum of the consensus function by generating new configurations. In general, a BM can be run sequentially, where units are allowed to change their states only one at a time, or in parallel. For simplicity, we here use the sequential mode of operation.

The idea of how to arrive at a global maximum is: take an arbitrary configuration as the recent one; generate a neighboring one (by changing the state of one unit); accept it as the recent configuration with some probability depending on the difference of consensus compared to the recent configuration and on the number of previous iterations of the procedure; continue.

To explain this briefly, given a configuration k, we define a *neighboring configuration* k_u as the one obtained from k by changing the state of the unit u. The *difference in consensus* between k and k_u, is defined as

$$\Delta\mathcal{K}_k(u) := \mathcal{K}(k_u) - \mathcal{K}(k)$$

A configuration k is *locally maximal* if $\Delta\mathcal{K}_k(u) \leq 0$ for all units u, i.e., if its consensus is not increased by a *single* state change.

If, while in search for a globally maximal consensus value, the BM would accept only configurations as the new recent configuration that have a consensus higher than the one of the old recent configuration, then it would perform hill climbing, which would only be guaranteed to find a *locally* maximal consensus. To avoid that, from time to time worse configurations are accepted as the new recent configuration, or better configurations are rejected, where the probability of doing so decreases with with the run time of the procedure and the difference in consensus between the old recent configuration and the newly generated one. In more detail: Given a configuration k, assume that neighboring configurations k_u are generated with equal probabilities for all u. The *acceptance criterion* to accept k_u as the recent configuration is

$$\frac{1}{1 + \exp\frac{-\Delta\mathcal{K}_k(u)}{c_t}} > random[0,1),$$

where $c_t > 0$ and converges to 0 for increasing run time t. It is known that under certain conditions, e.g., concerning the convergence behavior of c_t, the BM is guaranteed to converge towards a globally optimal configuration.

2.2 Issues in Translating Constraint Networks into Boltzmann Machines

In general, a constraint represents an arbitrary relation on variables over arbitrary domains; in this paper we assume that this relation is finite, hence consists of finitely many relation elements specifying the combinations of values that the variables are allowed to take.

To introduce our demo constraint network for this paper, consider you want to draw a graphics consisting of two boxes B_1, B_2 connected by an arrow. The boxes shall be equally sized and horizontally aligned. You have six places for positioning each of the boxes, say, $a, \ldots, f$. Figure 1 sketches the grid of positions and the intended mini-graphics. To translate this into a constraint problem, we

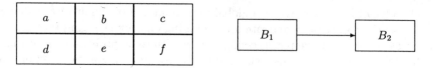

Fig. 1. The demo problem: the grid for positioning the boxes, and the desired appearance of the graphics.

need the following relations and—corresponding to them—constraints:

- $Position(B_1)$ enumerates all positions (initially) possible for B_1; it is initialized with $\{a, \ldots, f\}$.
- $Position(B_2)$ analogously to $Position(B_1)$.
- $Arrow(B_1, B_2)$ enumerates all pairs of positions that can be connected with an arrow spanning one position, possibly crossing levels; it is initialized with $\{(a, c), (c, a), (a, f), \ldots\}$.
- $Same\text{-}Level(B_1, B_2)$ would normally enumerate all (different) positions on the same level, i.e., $\{(a, b), (b, a), \ldots, (f, e)\}$.

Note that all information is expressed in terms of relations here; while the information about possible positions of the boxes (relations $Position(B_i)$) could have been represented by domains of the respective variables. For technical convenience, we assume that all information is expressed in terms of relations; a constraint formulation that guarantees this, while being as expressive as the usual relation-and-variable formulation, can be found in [8].

This problem has obvious solutions, e.g., put B_1, B_2 on positions a, c, respectively. However, to make it more interesting and to satisfy possible other constraints on the boxes that might forbid these positions, let us relax the $Same\text{-}Level(B_1, B_2)$ constraint: accepting some *penalty* of, say, 2 points, pairs of positions crossing levels (i.e., the pairs (a, f), (f, a), (d, c), and (c, d)) would also be acceptable; the "really" same-level pairs obtain zero penalty. To represent this in a constraint network, we increase the arity of the $Same\text{-}Level(B_1, B_2)$

constraint by one, now representing position pairs plus a penalty value, and we introduce a new relation, *SL-Penalty*(P), on an additional variable P, providing the two possible penalty values of 0 and 2. The respective constraint network is shown in Fig. 2.

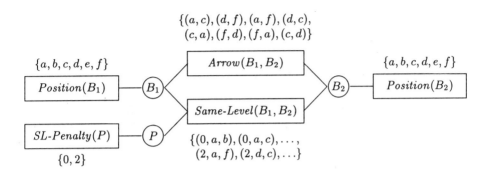

Fig. 2. Constraint network representing the graphics demo problem. Boxes represent constraints; circles respresent variables. Each constraint is annotated with its initial relation elements.

Here then is the basic idea of mapping constraint networks to BMs, the details of which are described in Section 3. In principle, the idea is the same as in [8], but we are aiming here at a mapping yielding more efficient BMs.

Relation elements are mapped to BM units; e.g., we obtain 8 units from translating the $Arrow(B_1, B_2)$ constraint, labeled with $(a, c), (d, f), \ldots$, respectively. Naturally, BM configurations with a high consensus are intended to correspond to solutions of the corresponding constraint network. Consequently, we have to take care that no two units are *on* that represent different elements of the same relation, and that only units are *on* that represent fitting elements of relations of different constraints. E.g., if a unit is *on* that represents $Position(B_1)$ being a, then only a unit should also be *on* that represents a relation element of the $Arrow(B_1, B_2)$ constraint with a in its first argument. So there are certainly "good" links in a BM connecting units that should be *on* together, and "bad" links connecting other pairs of units. Moreover, we have to represent different qualities of solutions in the BM; this will be done by translating the penalty values from the constraint networks into BM biases.

Finally, as the point here is to find a "useful" translation of a constraint network into a BM, we have to state a measure telling which translation is to be preferred to another. Components of such a measure might be

- time for transforming a constraint network into the corresponding BM,
- average or expected run time of the BM,
- average or expected ratio of finding an optimal solution,
- average quality of the solutions found after some fixed run time.

These components are clearly not independent. For example, BM run time can be bought for transformation time: Just find an optimal solution for the problem by exhaustive search, represent it in a one-unit-BM, and solve this one breathtakingly quickly by switching *on* the one unit—to give an extreme example. Consequently, any judgement of the practical value of transformations has to respect the trade-off between the different criteria.

In the following section, we consider in more detail how to find a smart translation. The presentation addresses in turn the different components of the to-be-developed BM, which are units, connections, and weights.

3 Tuning the Transformation

We now discuss in more detail our findings in search for an effective transformation schema. The presentation addresses in turn the different components of the to-be-developed BM, which are units, connections, and weights. Measurements of the performance of our test implementation on the graphics demo problem are summarized and discussed in 3.4 at the end of this section.

3.1 The Unit Set

The straightforward idea of transforming constraint networks into BMs is to generate one unit for each relation element. However, this idea turns out to be impracticable: the configuration space of a BM corresponding to a constraint network with n relation elements consists of 2^n different configurations, n being 52 for the mini-network in Fig. 2. So a practical transformation should save units whenever possible. However, there is the trade-off just mentioned between the effort spent in minimizing the unit set and the effort used for calculating the units: structurally "efficient" BMs take time to be calculated.

Luckily, there is a cheap way to find a potentially large reduction. Take for example the constraint network in Fig. 2. It includes constraints representing only domains that get used in turn to assemble the "proper" relations—the $Position(B_1)$ constraint representing the 6 possible positions is an example. These constraints state essentially no restrictions that are not present in the "higher" constraints, e.g., the *Arrow* constraint, given that their relation elements involve no other than the values sanctioned by the "lower" ones. So we can skip the respective units in the corresponding BM—the units representing $Position(B_1)$, in our example—and generate units only for those relation elements that don't stem from a domain binding constraint.

3.2 The Connection Pattern

To differentiate between arbitrary BM configurations and configurations that correspond to solutions of the constraint network, the BM's consensus function must be such that solution configurations get high values, and all other configurations get low ones. To model this difference in the BM, we introduce the following two types of connections between units.

Firstly, positive connections connect locally consistent units. Two units are locally consistent if they represent relation elements sharing variables and agreeing on all assignments of shared variables. To give an example, the units representing $Arrow = (a, c)$ and $Same\text{-}Level = (0, a, c)$ are locally consistent, sharing the assignment to $Arrow$ and agreeing that it be (a, c). For example, assuming that $Position(B_1)$ and $Position(B_2)$ were not deleted from the BM according to the unit-saving approach just described, they do not share variables; hence, the units representing them are not connected by positive connections. Positive connections get a positive strength to give the respective units a tendency to be *on* in accepted BM configurations.

Secondly, negative connections connect locally inconsistent units. Two units are locally inconsistent if they represent different assignments to one and the same variable. This may be the case for conflicting elements of different relations, e.g., $Arrow = (a, c)$ and $Same\text{-}Level = (0, a, b)$; moreover, this is *always* the case for units representing different elements of the same relation, like $Arrow = (a, c)$ and $Arrow = (a, f)$. Giving negative connections a negative strength yields a tendency of accepted configurations to have at least one of the respective units *off*. In our previous work, we have been somewhat reluctant to introduce negative connections, reserving them for connecting units that represent different values of the same relation. However, introducing additional ones between conflicting elements of different relations should allow solution configurations to be more easily separated from other ones. Our experiments confirmed this guess. ·

While classifying connections into positive and negative allows to tell configurations corresponding to solutions from those that do not, it does *not* allow to qualify BM configurations as corresponding to good or bad solutions. Following the lines of our previous work, this is done by associating a bias connection with every unit, i.e., a unary connection from a unit to itself, allowing to associate a value with this one unit being *on*. How to set these is described below.

Figure 3 shows the structure of the BM corresponding to the constraint network in Fig. 2.

3.3 The Weights

Having described the structure of a BM corresponding to a constraint network, we must now be more specific about the weights. The ideal weights would be such that they allow a BM, firstly, to converge to a configuration with a high value that is sure to correspond to a solution of the constraint network and, secondly, to do so quickly. A large part of the experiments described in [14] dealt with finding a reasonable compromise between these inconsistent desires.

To start with something that does *not* work, a suggestive idea for helping the BM stumble over solutions is to increase the quality differences between configurations that represent solutions and those that do not. In our experiments, this had just a bad effect: The BM did not find solutions significantly more

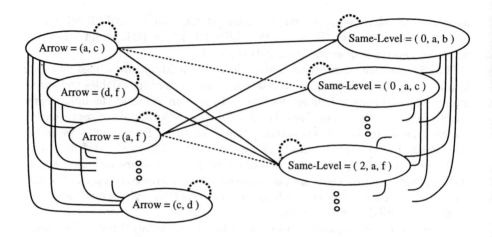

Fig. 3. A sketch of the Boltzmann Machine corresponding to the graphics problem constraint network. Positive connections are represented by dashed lines, negative connections by solid lines, and bias connections by bold dashed semicircles.

quickly, but on the average, they got significantly worse. An explanation for the deterioration is that the quality difference between different solutions gets lower if the average value of an solution increases. Consequently, to find not only solutions, but good solutions, the share of the bias weights within the value of a solution configuration must be made as large as possible.

Our original schema for setting the bias values [8, 9] was to associate a nonzero bias with every value of a variable that could be relaxed, i.e., that allowed to formulate solutions of different qualities. Moreover, it was assumed that BMs corresponding to a CSP would have to have a particular property, namely, be *feasible,* meaning that all configurations with a consensus above a certain known threshold correspond to solutions of the CSP. To guarantee that, bias values of relaxable constraints could contribute only very little to the overall consensus of a configuration in order to guarantee thet they cannot make a nonsolution a seeming solution.

Now, given the different connection pattern with a higher number of negative connections that tend to favor solution configurations automatically, there is somewhat more liberality. We stick to the idea of giving a 0 bias to all units corresponding to nonrelaxable constraints. However, we have experimented with slightly nonfeasible bias functions accounting for a larger percentage of the overall consensus. (Exact definitions would require some overhead in terms of definitions—please consult [14] for that.) The idea is that the generation of a good—rather than just any—solution becomes more likely.

We have not experienced any problems with the theoretical lack of feasibility. And we suspect that this is no serious practical problem anyway: A BM is never guaranteed to find a solution in finite time, even if one exists. So, whether or not a BM practically converges to a configuration that corresponds to a solution does not depend on the feasibility of the consensus function. Feasibility just guarantees the existence of a quick check whether a BM solution represents a CSP solution—no more, no less.

3.4 Experimental Results on the Graphics Demo Problem

Measure	Problem Transformation Variant					
	Basic procedure	Introduce negative connections	Increase weight of negative connections	Decrease weight of positive connections	Maximize bias weights	Reduce number of units
Transform. (sec)	0.20000	0.66666	0.19666	0.18666	0.19666	0.08333
Runtime (sec)	1.68819	0.39234	0.41226	0.42749	0.38942	0.17246
Total time (sec)	1.88819	1.05894	0.60892	0.61414	0.58608	0.25579
Hit rate (%)	47.77	92.86	89.71	89.82	90.48	91.90
Variance (%)	1.70	0.44	11.54	15.55	15.44	30.22

Table 1. Measurements for different transformation variants of the graphics demo CSP (Fig. 2). See text for explanations.

Given all these theoretical considerations, what is their gain in terms of run time and solution quality? To provide an impression of that, we finally discuss a number of transformation variants of the graphics demo CSP. Numerical values are summarized in Table 1, whose structure we have to explain first.

Its six columns correspond to different problem transformation variants. The first variant ("Basic procedure") shows the respective measures for the straight-forward transformation as described in [8, Sec. 8.2]; it is included here as a point of reference for the other variants. The other transformations are cumulative from left to right: The transformation in the n-th column *includes* the technique in the $(n-1)$-th column, adding other features. The most dramatic saving comes from reducing the number of units; it is done last here to keep the BMs at more significant sizes for the other transformation variants.

The overall run times of the different BM variants are shown in the middle row ("Total time"). It consists of the time for the respective transformation plus the run time of the BM. As the transformation may be done off-line and just once, i.e., it may be considered less interesting, we give the individual values for transformation and run times in rows No. 1 and 2. Given that BMs are stochastic devices, the value for the run time has to be an average value; we have averaged over as many runs as necessary for the respective BMs to terminate 50 times in a solution configuration (and possibly in nonsolution configurations in other cases). The measurements were done on a Sun 4 workstation, using the PARABOL BM simulator [12].[3]

The hit rate (4th row) describes the ratio of BM runs terminating in a solution configuration (50 runs in our case) compared to the overall number of runs.

For the runs terminating in a solution configuration, we computed the value shown in the last row ("Variance") as follows: Let q be the average consensus value of all BM configurations that correspond to a solution, and q^* the average consensus value of all 50 runs terminating in a solution configuration. Then the variance is the percentage of $q^* - q$ in q, i.e., $\frac{q^*-q}{q} \cdot 100$. Note that this value could, in principle, be negative. However, it is positive in all columns of Table 1, signalling that all BM variants tend to find such solution configurations (if any) that are better than average.

We will now discuss the results. Starting with the connections, let us first remark that the straightforward translation of the graphics demo CSP in Fig. 2 *does* already contain all positive connections, so this would bring no enhancement. Introducing negative connections—second column—slows down the transformation process considerably: There is just more work to be done. On the other hand, it cuts the run time for the BM by about 4, and almost doubles the hit rate—practically a clear advantage.

The next three variants tune the weights. Increasing the weight of the negative connections (third column) has the desired and anticipated effect of tuning the solution quality: The hit rate does not change significantly, but the solution quality is far better than before (variance 11.54 vs. 0.44). The value for the transformation time deserves additional mentioning: In the previous variant, the strength of the newly-introduced negative connections had to be set according to the BM structure in order to guarantee feasibility of the solutions, as previously discussed. As we do not want to guarantee this now, this quite cumbersome computation can now be saved, setting the weight to some standard value that is local for each relevant constraint. This accounts for the dramatic saving in transformation time wrt. the previous variant; but note that this saving is not intrinsic in the idea of increasing the weight of negative connections. As suspected, the run time does not change significantly.

The next two variants continue along this line, the essence being to favor finding a good solution rather than just any solution. This is shown by another significant increase in the variance value in the fourth column (15.55 vs. 11.54, all other values staying nearly constant). Further maximizing the bias weights (fifth column) does not yield much in this situation.

The final blow comes from reducing the number of units (last column), as described in Sect. 3.1. As expected, all times drop significantly, due to the re-

---

[3] Although the transformation is a deterministic process, transformation times do vary in our implementation environment, due to the unpredictability of the workload of a machine in the network, and due to a timing routine in our programming language (Standard ML) implementation that seems to be unable to measure proper process time. Consequently, the transformation time values shown here are averaged over 3 transformation runs.

duction in size of the corresponding BM. There is again a significant increase in the variance value (30.22 vs. 15.44), but this has not occured in all test domains that we have used, and there is no reason why this effect should result in general. On the other hand, other domains showed a significant increase in hit rate after the reduction of the unit set. So, the definite result is: First, and theoretically obvious: This transformation reduces significantly the transformation and run times. Second, as a matter of empirical evidence: The overall solution behavior of the respective BMs gets better.

For a more elaborate description and discussion of the results, consult [14].

4 Conclusion

In this paper, we have sketched some results about obtaining efficient Boltzmann Machines for solving constraint networks. Theoretical consideration suggests that a number of improvements are possible with respect to our original method of transforming CSPs into BMs [8, 9]. The essence of these improvements is

- Reducing the number of units.
- Introducing negative connections.
- Raising the relative contribution of bias values in the overall consensus.

These considerations could be validated in a number of experiments.

References

1. E. Aarts and J. Korst. *Simulated Annealing and Boltzmann Machines.* John Wiley & Sons, Cichester, England, 1989.
2. T.L. Dean and M. Boddy. An analysis of time-dependent planning. In *Proc. AAAI-88*, pages 49–54, St. Paul, Minnesota, 1988.
3. Y. Descotte and J.C. Latombe. Making compromises among antagonist constraints in a planner. *Artificial Intelligence*, 27:183–217, 1985.
4. D. Dubois, H. Fargier, and H. Prade. Propagation and satisfaction of flexible constraints. Rapport IRIT/92-59-R, IRIT, Toulouse Cedex, France, 1992.
5. B.N. Freeman-Benson, J. Maloney, and A. Borning. An incremental constraint solver. *Communications of the ACM*, 33:54–63, 1990.
6. E.C. Freuder. Partial constraint satisfaction. In *Proc. IJCAI-89*, pages 278–283, Detroit, Michigan, 1989.
7. H.W. Guesgen. A formal framework for weak constraint satisfaction based on fuzzy sets. In *Proc. ANZIIS-94*, pages 199–203, Brisbane, Australia, 1994.
8. H.W. Guesgen and J. Hertzberg. *A Perspective of Constraint-Based Reasoning.* Lecture Notes in Artificial Intelligence 597. Springer, Berlin, Germany, 1992.
9. H.W. Guesgen and J. Hertzberg. A constraint-based approach to spatiotemporal reasoning. *Applied Intelligence (Special Issue on Applications of Temporal Models)*, 3:71–90, 1993.
10. J. Hertzberg, H.W. Guesgen, A. Voß, M. Fidelak, and H. Voß. Relaxing constraint networks to resolve inconsistencies. In *Proc. GWAI-88*, pages 61–65, Eringerfeld, Germany, 1988.

11. A. Philpott. Fuzzy constraint satisfaction. Master's thesis, University of Auckland, Auckland, New Zealand, 1995.
12. J. Prust and W. Vonolfen. Dokumentation PARABOL 1.0. Unpublished documentation document, GMD, 1993.
13. Z. Ruttkay. Fuzzy constraint satisfaction. In *Proc. FUZZ-IEEE'94*, Orlando, Florida, 1994.
14. R. Weißschnur. Die Projektion von Constraint-Satisfaction-Problemen auf Boltzmann-Maschinen. Master's thesis, Universität Bonn, Institut für Informatik, May 1994.

Solving Over-Constrained CSP Using Weighted OBDDs

Fabrice Bouquet and Philippe Jégou

{bouquet,jegou}@lim.univ-mrs.fr
LIM - URA CNRS 1787
CMI - Université de Provence
39, rue Joliot Curie
13453 Marseille Cedex 13, FRANCE

Abstract. In Artificial Intelligence, for practical applications, we often have to manage over-constrained systems of constraints. So, a model based on the formalism of finite Constraint Satisfaction Problems (CSPs) [14] has been proposed with Dynamic CSPs (DCSPs) to handle this kind of problems [10][11]. Some classical techniques defined in the field of CSPs are usable in DCSPs, but the management of *over-constrained system with* DCSPs induces new problems. The purpose of this paper is to introduce an efficient way to solve DCSPs based on a logical approach. We use Ordered Binary Decision Diagrams (OBDDs) [3] and propose a particular coding for dynamicity. We show that our approach allows to solve some major questions in the field of DCSP, particularly consistency maintenance. This kind of problems is naturally expressed as a problem of optimal path computing in weighted graphs. Moreover, we shall see that the problem of finding optimal solutions can be solved easily and efficiently by our approach. One important problem in OBDD is the amount of memory required to represent the OBDD. In the worst case, this amount is in $O(2^N)$ where N is the number of propositional variables for static CSPs. We prove here that, if the number of dynamic constraints is m and if n is the number of variables in the problem, the size of OBDD is bounded by $O(m \times 2^n)$. First experimental results attest the interest of the approach.

1 Introduction

In Artificial Intelligence, we often have to manage dynamic environments of constraints. So, a model based on the formalism of finite Constraint Satisfaction Problems (CSPs) [14] has been proposed with Dynamic CSPs [10][11]. Some classical techniques defined in the field of CSPs are usable in DCSPs, but the management of over-constrained system with DCSPs induces new problems. For example, achieving consistencies in DCSPs must rather be considered as a problem of consistency maintenance than as a filtering problem.

Informally, we can consider a DCSPs as a sequence of CSPs such that each element of the sequence differs from the previous one by the set of constraints defining it: either a new constraint appears or a constraint disappears. At each time, only one CSP (static) is considered, but an efficient manner to solve the DCSP consists in exploiting previous reasonings. To realize that, we can consider all problems already appeared in the sequence, and implicitly, all their sets of solutions.

The purpose of this paper is to present an efficient way to manage sets of constraints, that is sets of problems and their associated sets of solutions, in a dynamic environment. The most important characteristic of dynamic environement is in any step its sets of solutions evolve, the best set is not the best last. The principle is based on a data structure defined in the field of boolean equations : the Ordered Binary Decision Diagrams (OBDDs) [3]. Intuitively, we can consider that OBDDs permit to memorize efficiently the decision tree associated with a system of Boolean equations. Consequently, when a new constraint is added, this kind of data structure allows to compute only a part of the new decision tree associated with was the new system. Nevertheless, two important problems appears if we want to handle DCSPs using OBDDs. On one hand, the formalism, and on the other hand, dynamicity.

Exploiting OBDDs in the field of reasoning about multiple contexts has been proposed in [8], but they used OBDDs to build an assumption based truth maintenance system (ATMS); our purpose is different here.

In this paper, the considered DCSPs will always be defined in the propositional formalism. It is well known that CSPs can be expressed in the propositional formalism using clausal formulas (see for example [9]). Dynamicity is harder to handle. OBDDs give a good framework particularly well adapted to dynamicity when dynamicity is limited to addition of constraints; constraint deletion is not possible with this kind of approach. So, in this paper, we propose a formulation of DCSPs allowing deletion of constraints in OBDDs, *in other words solving over-constainted system*. In our approach, the OBDD codes the static CSP of the current step of the DCSP, but also allows to reason on all possible sub-problems that can be defined with respect to the current CSP. Suppose that we want to add a new constraint C defined by a logical formula f. A boolean variable c associated with the constraint C will be defined, and the formula actually added will be the formula $c \rightarrow f$ where $\rightarrow$ is the logical implication. So, if the value of the variable c is true, the constraint defined by the formula f must be satisfied, and if the value of the variable c is false, the formula f can be satisfied or not. This coding of constraints allows to act as a deterrent to deletion of constraints. Now suppose that we want to absolutely satisfy the constraint C; it is sufficient to add the unary formula c. Now suppose that we want to remove the constraint C. It is sufficient to add the unary formula $\neg c$. This approach is based on the fact that, generally, the addition of constraint is easier to handle than the deletion of constraints. So, with our coding, a deletion is processed applying an addition of constraint (formula $\neg c$).

During the processing of a DCSP, the current state (a static CSP) can be inconsistent, i.e. the current set of constraints induces a CSP without solution. So, we must be able to restore consistency achieving deletion of one or several constraints. We can formulate this problem defining an optimization problem :

what set of constraints must be deleted to restore consistency, tacking account a measure of optimization ? Different measures of optimization can be considered. For example, we can consider the number of deleted constraints. We shall see that our approach allows to solve this kind of problems naturally, expressing it as a problem of optimal path computing in weighted graphs. Moreover, we shall see that the problem of finding optimal solutions can be solved easily and efficiently by our approach.

One important problem in OBDD is the amount of memory required to encode the OBDD. In the worst case, this amount is in $O(2^n)$ where n is the number of proposionals variables. Since, for all dynamic constraints, we introduce a new propositional variable c, if the number of dynamic constraints is m, the size of OBDD is theorically bounded by $O(2^{n+m})$. We give here a better bound of this size : the size of OBDD coding dynamicity is $O(m \times 2^n)$.

This paper is organized as follows. Section 2 introduces DCSPs and related problems. In the next section, we present OBDDs and its interest to solve DCSPs. The section 4 describes the approach, and presents the solutions to the different questions pointed; experimental results are presented in section 5. In the conclusion, we express the relationship between the processing of DCSPs and the problem of knowledge or belief base revision.

2 Preliminaries

In this paper, a *static and finite Constraint Satisfaction Problem* (CSP) involves a set X of propositional variables $\{x_1, ..., x_n\}$, and a set of constraints expressed by a set (a conjunction) of propositional formulas $F = \{f_1, ..., f_m\}$ defined on $X = \{x_1, ..., x_n\}$. The decision problem associated with consistency checking (satisfiability checking) is well known to be NP-Complete. Moreover, every instance of a classical CSP (in the sense of [14]) can be expressed in a logical formalism, the translation between the two formalisms being linear in the size of the CSP (if the CSP is defined by compatibility relations given by tables of tupples or matrix) [9]. So we can work using propositional logic without a loss of generality with respect to the field of CSPs.

The model of *Dynamic CSPs* (DCSP) proposed by [10] formalizes the processing of CSPs in the case of dynamic environments of constraints. In [11], this model has been proposed to assist design process when a human operator interactivally adds or removes constraints. But in many fields of applications, this approach is justified and seems to be sufficiently general to be studied : this approach can be used to solve problems defined by constraints with uncertain, incomplete or imprecise informations. We define the formalism of DCSPs in its logical formulation.

To defined DCSP, we consider a pair $P = (X, F)$, such as X is a finite set of propositional variables and F a set (a conjunction) of propositional formulas defined on X. The formulas in F correspond to all the formulas that can appear during the processing of the DCSP. Nevertheless, if neither X nor F are initially known *a priori*, the approach is the same. The processing of the DCSP considers a sequence of static CSPs, $P_0, P_1, ..., P_k$ such as :

- $P_0 = (\emptyset, \emptyset)$.
- P_{i+1} is obtained starting from P_i by adding or removing a constraint (a formula in the set F). Generally, a constraint is added to restrain the current state (the current problem P_i), when this problem possesses too many solutions, while a constraint is deleted when this problem is inconsistent (no solution)
- P_k possesses only optimal solutions, given a particular measure of optimality.

The sequence of static CSPs, $P_0, P_1, ..., P_k$ associated with $P = (X, F)$ defined the DCSP. Solving a DCSP consists in finding and solving the last CSP P_k of the sequence. Achieving P_k being realized step by step by successive refinements.

Given a current state of the processing, i.e. an element P_i of the sequence, several questions can be considered. We focus on two of them that seem to be the most important :

- **Consistency** : Is P_i consistent, i.e. does P_i admit a solution ? Give a solution.
- **Optimal consistent sub-problem** : If P_i is not consistent, find a sub-problem of P_i (defined by an optimal subset of constraints) that is consistent, the measure of optimality being given.

Related works in the field of DCSPs principally studied partial consistency maintenance [1] or global consistency maintenance for particular instances of CSPs (acyclic CSPs in [10]). More recently, an approach based on "nogood recording" has been proposed to handle DCSPs [16].

3 Ordered Binary Decision Diagrams

A detailed presentation of OBDDs is given in [3][4]. In this section, we only present the basic principles of OBDDs. An Ordered Binary Decision Diagram represents a Boolean function using a labeled acyclic directed graph with one root. This structure codes concisely the decision tree associated with the represented function. So a OBDD is defined w.r.t. an ordered set of propositional variables denoted X, and w.r.t. the coded Boolean function denoted F. In a OBDD, there are two classes of vertices s. On one hand two terminal vertices labeled 0 or 1, and on the other hand, non-terminal vertices labeled by variables of X obtained by $var(s)$. Each non-terminal vertex s has two children : left child denoted $lc(s)$ corresponding to the case where the variable is assigned 0, and right child denoted $rc(s)$ corresponding to the case where the variable is assigned 1. So, the coding principle is the same than for decision trees : for a given assignment of variables, the value yielded by the function is determined by the label of the last vertex in the corresponding path in the graph, 0 or 1.

The main advantage of OBDDs with respect to decision trees concerns the concision of the representation. While all possible assignments are explicitly represented in a decision tree, generally, all possible assignments are only implicitly represented in a OBDD (see figure 1). The concision of the representation is achieved by different reduction operations in the graph. These reductions are obtained either by the deletion of redundant vertices, or in grouping vertices possessing the same status in the graph (isomorphic subgraphs). So, for a boolean formula F coding a boolean function, and for an ordering on the set of variables X, the corresponding OBDD gives a canonical representation. The size of the OBDD is not necessarily exponential in the number of variables (remember that the number of vertices in a decision tree is exactly $2^{|X|+1} - 1$). Moreover, there are many practical cases such as the size of the OBDD reasonably grows w.r.t. the number of variables (see [4] for details).

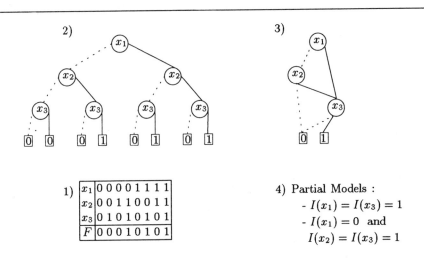

1)

x_1	0 0 0 0 1 1 1 1
x_2	0 0 1 1 0 0 1 1
x_3	0 1 0 1 0 1 0 1
F	0 0 0 1 0 1 0 1

4) Partial Models :
- $I(x_1) = I(x_3) = 1$
- $I(x_1) = 0$ and
  $I(x_2) = I(x_3) = 1$

Fig. 1. *This example is given by Bryant. (1) truth table defining the boolean function F. (2) associated decision tree; a dotted line denotes the case where the variable assignment is 0, and a solid branch 1. (3) OBDD corresponding to the function F; the OBDD is achieved in reducing the decision tree. (4) Model with partial assignment of problem's variables.*

If a single path can code several assignments (its different extensions — see the example in figure 1), there is only one path to code a particular assignment. Finally, an assignment I of X satisfies F if and only if there is a path in the OBDD from the root to the terminal vertex labeled 1 such as I is an extension of the assignment defined by this path. Note that we can say that a path satisfies a formula F if the associated assignment satisfies F. If the OBDD is not reduced to a single vertex, the number of arcs in the graph denoted $|A|$ satisfies the equality

$|A| = 2 \times (|S| - 2)$ since each non-terminal vertex possesses exactly two children.

The interest of OBDDs is related to the concision of the representation, in particular, all the possible assignments are implicitly represented, and it is easy to find an assignment satisfying the represented function. Moreover, though the theoretic complexity remains exponential (in time and memory size), a large number of industrial applications shows the practical interest and the efficiency of the approach (see [4]). Another important advantage of OBDDs concerns the dynamicity in the processing of OBDDs. Consider two logical formulas f and g. Achieving the OBDD representing the formula $f < op > g$ where $< op >$ is a binary boolean operator, is possible in grouping the OBDDs associated with f and g. The time complexity is then $O(|S_f| \times |S_g|)$ where $|S_f|$ and $|S_g|$ respectively code the number of vertices in the OBDDs associated with f and g. Consequently, the addition of constraint (formula f) to a set of constraints (set F) can efficiently be processed computing the BBDs corresponding to $F \wedge f$.

In the field of DCSPs, we point out two major advantages of this approach. On one hand, a dynamic approach is possible, only if we consider the addition of constraints (not the deletion). On the other hand, all the possible assignments are implicitly represented : satisfying assignments and falsifying assignments. Nevertheless, we must observe two drawbacks of great importance : Firstly, constraints are not explicitly represented, and secondly, constraints deletion cannot be realized in OBDDs.

In the next section, we propose a coding of DCSPs that allows to solve the two problems listed above.

4 Handling Dynamicity

The approach we describe below allows to solve the problems defined in the section 2 (consistency checking and search for optimal consistent sub-problems). We also give a general framework to manipulate dynamicity, i.e. addition and deletion of constrain. These tasks can be achieved using an appropriate coding of the constraints. The idea is based on two principles :

- constraints are explicitly represented in the OBDD
- the processing realized for addition of constraints realizes a preventive processing for the deletion of constraints.

4.1 Coding Dynamicity

Consider $P = (X, F)$ and suppose that we add a constraint C_i defined by a logical formula f_i. The next state is the problem $P' = (X \cup \{\text{variables in } f\}, F \cup \{f_i\})$. As a matter of fact, the constraint really added to the OBDD will be the logical formula $c_i \rightarrow f_i$ where c_i is a new variable called *Dynamic Variable*, associated with the new constraint C_i. So, the fact that the constraint appears in P' will be explicitly represented in the OBDD. Consequently, the OBDD associated with P' codes a (Y', G') such that:

- $Y' = Y \cup \{c_i\} \cup \{\text{variables in } f_i\}$
- $G' = G \cup \{c_i \rightarrow f_i\}$
- (Y, G) is associated with P.

So, for each assignment I of the variables in Y', we have two cases :

- $I(c_i) = 1$; in this case, each model of G' satisfies the constraint C_i since the satisfaction of the formula $c_i \rightarrow f_i$ enforces the satisfaction of the formula f_i.
- $I(c_i) = 0$; in this case, it is possible that a model of G' violates C_i.

Later on, we call Dynamic OBDD the OBDD using the coding of constraints above-mentioned, i.e. the OBDD coding (Y, G) is associated with the problem $P = (X, F)$. So, given a CSP (X, F) and the OBDD coding (Y, G), the set of dynamic variables will be denoted $X_{Dyn} = Y \backslash X$. The next property shows the desirable effects of the coding.

Proposition 1. *Given a CSP (X,F) encoded by a pair (Y,G), then a subset $\{f_1, ..., f_m\}$ of constraints of F is satisfiable iff there exists a model I of (Y,G) such that $I(c_i) = 1$, for i = 1,..., m.*

Proof. — $\Rightarrow$ Let I' be a model of $\{f_1, ..., f_m\}$. Consider now the assignment I defined by $I(x) = I'(x)$ for all $x \in X$; then $I(f_i) = 1$, for i = 1,..., m. Since the variable c_i appears only in the formula $c_i \rightarrow f_i$, we establish $I(c_i) = 1$, for i = 1,..., m and $I(c_j) = 0$, for all other formulas f_j in F. Then we have :
  - $I(c_i) = 1$, for i = 1,..., m
  - $I(f_i) = 1$, then $I(c_i \rightarrow f_i) = 1$ for i = 1,..., m
  - $I(c_j) = 0$, then $I(c_j \rightarrow f_j) = 1$ for all other formulas f_j in F.

  Consequently, all formulas $(c \rightarrow f)$ in G are satisfied by I, and therefore, I is a model of (Y, G) such that $I(c_i) = 1$, for i = 1,..., m
- $\Leftarrow$ If I is a model of (Y,G) such that $I(c_i) = 1$, for i = 1,..., m, necessarily, we have $I(f_i) = 1$ to satisfy formulas $(c_i \rightarrow f_i)$ in G. Consequently, I is a model of $\{f_1, ..., f_m\}$. $\square$

We remark that a dynamic variable c_i allows to express the presence of the associated constraint C_i (if the assignment $I(c_i) = 1$), or its absence (if the assignment $I(c_i) = 0$). Therefore, a constraint C_i is now explicitly represented in the OBDD. Moreover, the coding allows to represent the sub-problems of the current problem since all the possible assignments of the variables of $\{c_1, ..., c_m\}$ are implicitly coded, each one corresponding to a problem such that the constraints C_i defining it correspond to the assignments of c_i to true. Moreover, all the solutions of all these problems are implicitly represented since all the possible assignments of the variables of X are implicitly coded in the Dynamic OBDD.

4.2   Exploiting Dynamic OBDDs

The purpose of this section is to show how the questions induced by the processing of DCSPs (see section on preliminaries) can be solved exploiting Dynamic OB-DDs.

First, if we want to satisfy the constraint C_i, it will be sufficient to add the unary formula c_i. To remove the constraint C_i, it will be sufficient to add the unary formula $\neg c_i$, the formula $c_i \to f_i$ being satisfied independently of the satisfaction of the formula f_i.

To operate addition or deletion of constraints, we just realized the addition of formulas. The time complexity cost of these processing is $O(|S_G| \times |S_{(c_i \to f_i)}|)$ for adding the constraint C_i, and $O(|S_{G'}|)$ for removing the constraint C_i later, since the size of the added formula $\neg c_i$ is constant.

$$C_1 : x_1 \vee x_2 \quad C_2 : \neg x_1 \vee x_3 \quad C_3 : x_1 \vee \neg x_2 \vee x_3 \quad C_4 : \neg x_1 \quad C_5 : \neg x_3$$

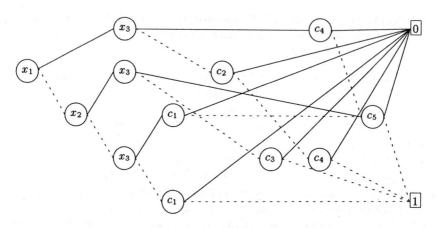

Fig. 2. *A set of constraints and the associated OBDD. The set of constraints $\{C_1, C_2, C_3\}$ expresses the boolean function given in figure 1. The path labeled by $[x_1, x_3, c_2, c_4, 0]$ represents the sub-problem defined by the constraints $\{C_1, C_3, C_4, C_5\}$. This path also represents all the assignments extensions of the assignment $(x_1 = 1, x_3 = 0)$; since the last vertex of this path is labeled 0, all these assignments falsify at least one of the constraints in $\{C_1, C_3, C_4, C_5\}$. The path labeled by $[x_1, x_2, x_3, c_1, c_5, 1]$ represents the sub-problem defined by the constraints $\{C_2, C_3, C_4\}$. This path also represents the assignment $(x_1 = 0, x_2 = 0, x_3 = 1)$; since the last vertex of this path is labeled 1, this assignment is a model for the constraints $\{C_2, C_3, C_4\}$.*

To check consistency in a OBDD, it is sufficient to verify if the OBDD is not reduced to the vertex labeled 0. To verify if an assignment is a model is achieved using a graph algorithm that consists in starting from the root of the OBDD and using branchings defined by the arcs and the assignments (0 or 1) of the propositional variables. On the other hand, the search of a model is possible using a graph algorithm looking for a path from the root to the terminal vertex labeled 1. It is also in exploiting this graph that all desired processing will be realized. The basic question is formulated by the first question, i.e. problem consistency : Is the current state $P = (X, F)$ consistent, that is, does P possess a solution ? This question can be handled applying a graph algorithm finding path in the OBDD. Indeed, the next property justifies this approach :

Proposition 2. *Consider $P = (X, F)$ and the Dynamic OBDD coding (Y, G) associated with (X, F). P is consistent if and only if there is a path from the root to the vertex labeled 1 using no arc corresponding to the left child (a 0 assignment) of a vertex labeled c_i such as $f_i \in F$.*

Proof. Trivial by applying Proposition 1. □

Consistency check can be realized by verifying if a such path exists in the graph. Indeed, consider the graph associated with the Dynamic OBDD. To solve this problem, a graph-searching algorithm can be used. Starting from the root, the algorithm visits the vertices following the arcs in the graph, except prohibited arcs. The prohibited arcs are the arcs (u, v) such that $var(u) = c_i$ and $v = lc(u)$. If the vertex labeled 1 is reached, then P is consistent, else P is inconsistent. When consistency check is associated to a model search, it is sufficient to memorize the assignment of the propositional variables x_k along the good path. So, we obtain an assignment (possibly a partial assignment) that constitutes a model for P. The time complexity for consistency checking and model search is linear in the size of the Dynamic OBDD, i.e. in $O(|S_G|)$ using a classical depth-first search. The figure below shows an example of such searched path.

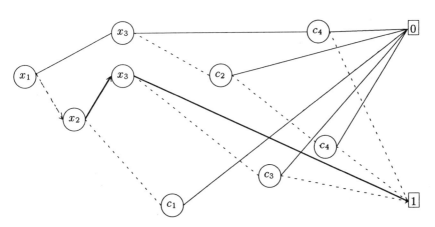

Fig. 3. *The Dynamic OBDD considered is the Dynamic OBDD coding the problem defined by the constraints $\{C_1, C_2, C_3, C_4\}$ given in the example of the figure 2. The path represented with bold lines labeled by $[x_1, x_2, x_3, 1]$ ensures the consistency of the current problem. Moreover, this path also represents the assignment $(x_1 = 0, x_2 = 1, x_3 = 1)$; it is a model for the constraints $\{C_1, C_2, C_3, C_4\}$.*

The method proposed for consistency check also gives a solution to obtain models. This method also gives an approach to solve the second question. Given an inconsistent state in the DCSP related to the CSP P, find a sub-problem of P defined by an optimal subset of constraints that is consistent. The measure of optimality considered here is based on a weighting of constraints : a weight w_i is associated with each constraint C_i. So, the weight of a problem $P = (X, F)$ where $F = \{f_1, f_2, ..., f_m\}$ is exactly equal to the sum of the weights of its constraints, i.e. $\sum_{i=1}^{m} w_i$. This second request can be solved using an algorithm that computes optimal paths from a single source in acyclic weighted directed graphs (an OBDD is a directed acyclic graph with a single source). The weight we use is defined by the weight of constraints. Consider an arc (u,v) in the graph :

- if $var(u) \in X$ or if $v = rc(u)$ then $weight(u, v) = 0$
- if $var(u) = c_i \in X_{Dyn}$, and if $v = lc(u)$ then $weight(u, v) = -w_i$

In the first case, either the arc is not related to a constraint of P_i or it is an arc arriving from a variable c_i (related to a constraint C_i of P) through the right child, i.e. c is assigned 1. So the constraint must be satisfied. In the second case, it is an arc associated with a constraint of P that is not satisfied; its weight must be deducted from the sum. Finally, the total cost of a path is exactly the reverse of the sum of the weights of the violated constraints by the path assignment (and its extensions). A path with weights sum equal to zero is related to a model for the problem P since no one constraint is violated. Next property justifies our approach :

Proposition 3. *Consider $P = (X, F)$, a weight function w on constraints and the corresponding labeling on the Dynamic OBDD coding (Y, G) associated with (X,F). A path from the root to the vertex labeled 1 such that the weight W_{OPT} is minimum defines a sub-problem of P, which is an optimal and consistent sub-problem, which the weight is $\Sigma + W_{OPT}$ where Σ is the sum of the weight of all constraints in P.*

Proof. We just give here the idea of the complete proof. To prove this proposition, it is sufficient to consider first an optimal sub-problem, and to associate it a path. Then, it sufficients to consider an optimal path and see that the weight of this path is equal to the weight of an optimal sub-problem. Finally, we see that every optimal path is associated with an optimal sub-problem. □

The problem of computing one optimal path from a single source in directed acyclic graphs can be solved using a variant of Bellman's algorithm (see [6] page 536). Its time complexity is linear in the size of the graph, consequently, in the number of vertices in a Dynamic OBDD : $O(|S_G|)$. Finding a consistent similar sub-problem of P is then possible with a similar approach than the one used to find models in consistency checking. Indeed, the constraints defining the considered sub-problem are the constraints of P excepted the constraints associated with negative arcs on the optimal path. Finally, finding a model for this optimal sub-problem is possible using the same method than for consistency checking.

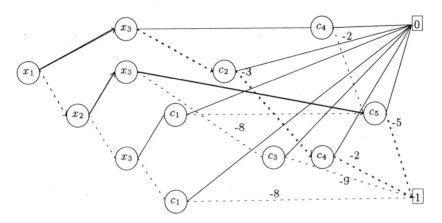

Fig. 4. *The Dynamic OBDD considered is the Dynamic OBDD coding the problem defined by the constraints $\{C_1, C_2, C_3, C_4, C_5\}$ given in figure 2. In this example, the weight of constraints are : $p_1 = 8, p_2 = 3, p_3 = 9, p_4 = 2$ and $p_5 = 5$. The negative values are wrote near the arcs; these negative values correspond to the cost of violating the coresponding constraints. Two paths are represented with the bold lines; these paths correspond to 2 optimal paths. One path is labeled by $[x_1, x_3, c_2, c_4, 1]$; its cost is -5, the sum of (-2) and (-3). This path also represents the partial assignment $(x_1 = 1, x_3 = 0)$; each extension of it being a model for the constraints $\{C_1, C_3, C_5\}$. The second path is labeled by $[x_1, x_2, x_3, c_5, 1]$; its cost is -5 while the associated assignment $(x_1 = 0, x_2 = 1, x_3 = 1)$, is a model for the constraints $\{C_1, C_2, C_3, C_4\}$.*

4.3 A good variables ordering

In the field of OBDDs, the problem of variables ordering is fundamental. The efficiency of OBDDs is strongly related to this parameter (see for example [4]). For Dynamic OBDDs, we represent the set of all sub-problem of a given problem : its size depends exponentially on the number of constraints. We know that the size of a Dynamic OBDD coding (Y,G) is bounded by (2^{n+m}) if $n = |X|$ and $m = |X_{Dyn}|$. Nevertheless, we will see that a good variables ordering can bound the explicit representation of dynamic constraints. Consider an ordering with propositional variables associated with constraints appearing at the end of the ordering, the ordering $(x_1, ..., x_n, c_1, ..., c_m)$; the size of the Dynamic OBDD is bounded by $(m \times 2^n)$. For example, the Dynamic BBD given in figure 2 admits exactly 12 non-terminal vertices while the same set of constraints coded in a Dynamic OBDD using the variables ordering $(c_1, c_2, c_3, c_4, c_5, x_1, x_2, x_3)$ requires 56 non-terminal vertices. Consider the set of variables $Y = X \cup X_{Dyn}$, an *ordering XC* denoted $<_{XC}$ is an ordering on Y satisfying: $\forall x \in X, \forall y \in X_{Dyn}$ then $x <_{XC} y$. We call *Dynamic OBDD coding (Y,G) w.r.t. an ordering XC*, a Dynamic OBDD such that the ordering on Y is an ordering XC.

Proposition 4. *Consider a Dynamic OBDD coding (Y,G) w.r.t. an ordering XC. For all paths $[s_1, ..., s_k]$ of the OBDD from the root to a terminal vertex, we have: if there is a node s_j on this path such that $var(s_j) = c_i$ where $c_i \in X_{Dyn}$, then the partial assignment associated to the path $[s_1, ..., s_j]$, denoted $I_{[s_1..s_j]}$, satisfies $I_{[s_1..s_j]}(f_i) = 0$ and $var(rc(s_j)) = 0$.*

Proof. We remark that the variable c_i appears only in the formula $c_i \to f_i$. So, its assignment does not modify the assignment of f_i. Consider now a path $[s_1, ..., s_k]$ from the root to a terminal vertex in the OBDD that is labeled by c_i. There are two possibilities for $I_{[s_1..s_j]}$:

- $I_{[s_1..s_j]}(f_i) = 0$: If we consider the extension I of $I_{[s_1..s_j]}$ such that $I(c_i) = 1$. Then $I(c_i \to f_i) = 0$ and therefore $I(G) = 0$ and then, $var(rc(s_j)) = 0$.
- $I_{[s_1..s_j]}(f_i) = 1$: In this case, the restriction of the boolean function G induced by the assignment of the variables labeling the partial path $[s_1, ..., s_j]$ which is denoted $G|_{I_{[s_1..s_j]}}$, is equal to the function $G|_{I_{[s_1..lc(s_j)]}} = G|_{I_{[s_1..rc(s_j)]}}$ since $I_{[s_1..lc(s_j)]}(c_i \to f_i) = I_{[s_1..rc(s_j)]}(c_i \to f_i) = 1$. Consequently, subgraphs rooted in $lc(s_j)$ and $rc(s_j)$ encode the same boolean function $G|_{I_{[s_1..s_j]}}$. Applying properties of OBDD (see [Bryant 86]), these subgraphs are isomorphic and then only one can appears in the OBDD. Therefore, we have $lc(s_j) = rc(s_j)$, and then a such node cannot appear in an OBDD. Finally, it can not exist a path labeled c_i such that $I_{[s_1..s_j]}(f_i) = 1$.

Only the first case with $I_{[s_1..s_j]}(f_i) = 0$ is possible. $\square$

We can observe the applying of this property in figure 2: every right child of a node labeled c_i is the terminal node labeled 0. Proposition 5 allows us to bound the size of a Dynamic OBDD:

Proposition 5. *Consider a Dynamic OBDD coding (Y,G) w.r.t. ordering XC where $|X| = n$ and $|X_{Dyn}| = m$. Its number of nodes is bounded by $(m+1).2^n+1$.*

Proof. In a Dynamic OBDD coding (Y,G) w.r.t. ordering XC, the worst case appears when the nodes labeled by variables in X is a decision tree. Consequently, we have at most :

- $(2^n - 1)$ nodes labeled by variables in X
- on the level $(n - 1)$, there are at most $(2n - 1)$ nodes labeled x_n
- on the level n, there are at most 2^n nodes labeled $c_i \in X_{Dyn}$
- by Proposition 4, every node labeled $c_i \in X_{Dyn}$ possesses at most one child labeled in X_{Dyn}. So, the number of nodes 0labeled in X_{Dyn} appearing in the OBDD is bounded by $(m \times 2^n)$ since the maximal number of nodes labeled c_i is 2^n.

Finally, we have at most $2^n - 1$ nodes labeled in X, $(m \times 2^n)$ nodes labeled in X_{Dyn} and two terminal nodes, that is $((m + 1) \times 2^n + 1)$ nodes at most in the OBDD. $\square$

5 Experimental results

The significance of Dynamic OBDDs is naturally related to experimental results; the purpose of this section is to give some results on this topic. We began to experiment our approach using an academic benchmark, the Amazons problem that is defined by the Queens problem with additional constraints defined by Knights. The principle of experiments is based on practical features concerning real life problems. In many real applications, problems are defined by two sets of constraints. On one hand, validity constraints, that must necessarily be satisfied, and preference constraints, that must optimally be satisfied. So, solutions of these problems must satisfy validity constraint and not necessarily preference constraints, but we are interested to find the best solutions, solutions that satisfy optimally preference constraints. The approach of Dynamic CSP can be used to solve this class of problems [11]. In our experiments, the Amazon problem (Table 1), we consider the Queens constraints as validity constraints, and Knights constraints as preference constraints. It is known that the Queens problems with additional constraints such that Knights constraints define an over-constrained CSP. Consequently, Queens constraints are naturally expressed (using clausal form) while Knights constraints are given using dynamicity coding. For our experiments, we apply a variables ordering XC. The OBDD package used for experiment has been realized by A. Rauzy [15]. We experimented dynamicity in three steps. First, we build the OBDD defining Queens problem. In a second step, each preference constraint has been added to the Dynamic OBDD one by one. The third step corresponds to deletion or enforcing preference constraints, to achieve only one solution (actually one class of solutions). All running times are given in seconds.

- n : number of queen
- $|X|$: number of propositional variables for validity
- $|Q|$: number of formulas (clausal formulas) to define queens problem
- $|S_Q|$: number of vertices in OBDD representing queens problem (X, Q)
- Time Q : CPU time required to compute OBDD representing (X, Q)
- $|K|$: number of dynamic constraints (Knights constraints)
- $|f_K|$: number of clausal formulas per dynamic constraints
- $|S_K|$: number of vertices in Dynamic OBDD representing amazons problem
- Time K : average CPU time required to add one dynamic constraint
- Time D : average CPU time required to enforce or delete one dynamic constraint

Now, we use the OBDD for solving random CSPs (Table 2). Each CSP is build in three steps. The first step consists in building the OBDD using the propositional coding of the generated CSP, i.e. domain clauses, negative binary clauses to express the unique value of each variable in its domain and the CSP constraints themselves. The second and third, for manipulation of Dynamic constraints in the Dynamic OBDD, are the same as in the first experimentation in Figure 6. In the whole experimentation, we didn't try to find an optimal insertion order for dynamic and validity variables, we just used the lexicographic order. We preprocessed the problems by a path-consistency filtering.

n	$\|X\|$	$\|Q\|$	$\|S_Q\|$	Time Q	$\|K\|$	$\|f_K\|$	$\|S_K\|$	Time K	Time D
6	36	296	133	0.64	9	8	144	0.01	0.01
7	49	483	1 103	1.29	11	10	1 524	0.06	0.02
8	64	728	2 455	3.96	13	12	4 121	0.14	0.08
9	81	1 065	9 561	16.71	15	14	18 376	0.70	0.39

Table 1. *Applying the Dynamic OBDD approach to solve the Amazon problem with preference constraints. Time results have been obtained on a SUN SPARC 470.*

- $|X|$: number of variables for validity
- $|D|$: domain size for variable X
- $|C|$: number of constraints without definition domain and disjonction value
- $|S_c|$: number of vertices in OBDD representing the CSP
- Tightness : The average percentage of forbidden tuples in each constraint
- Time C : CPU time required to compute OBDD representing CSP
- $|K|$: number of dynamic constraints
- $|S_K|$: number of vertices in the Dynamic OBDD representing the DCSP
- Time K : average CPU time required to add one dynamic constraint
- Time D : average CPU time required to enforce or delete one dynamic constraint

$\|X\|$	$\|D\|$	$\|C\|$	$\|S_C\|$	Tightness	Time C	$\|K\|$	$\|S_K\|$	Time K	Time D
10	10	13	306	80	5.03	13	925	0.25	0.12
10	10	13	7 705	70	12.94	13	137 091	1.35	1.21
10	10	16	1 155	70	16.92	16	8 240	0.35	0.22
10	10	27	4 505	50	47.76	13	10 658	0.43	0.27
10	10	27	109	60	46.81	16	175	0.21	0.07
10	10	27	429	60	13.54	16	1 661	0.28	0.11
12	10	19	4 067	70	7.43	19	157 426	1.73	1.75
15	10	23	7 690	70	47.10	23	148 231	1.97	1.64
15	10	23	6 605	70	66.37	23	135 201	2.02	1.61

Table 2. *Applying Dynamic OBDD approach to solve DCSP. Time results have been obtained on a SUN SPARC station IPX.*

We observe that the hard part is to build the OBDD defined by validity constraints whereas the dynamic constraints insertion seems to be faster. We experimented on random problems to have an idea of the average case, but a study is still to be done on real life problems such as scheduling problems, knowledge bases revision, control systems breakdown detection...

Finally, note that we did not experiment the search for an optimal consistent sub-problems. These experiments will be realized in the next months. We just give the size, i.e. number of vertices $|S_K|$ of the associated OBDDs since the computation time is in $O(|S_K|)$. It seems that this approach may be less efficient than classical CSPs algorithms using a branch and bound approach to find optimal solutions. Nevertheless, Dynamic OBDDs allow to maintain more informations than branch and bound methods and moreover, after the first step that computes the OBDD related to hard constraints next operations and queries can be realized more efficiently.

6 Conclusion and future prospects

In this paper, we have presented an approach to solve Dynamic Constraints Satisfaction Problems. This approach is based on the utilization of a logical constraints solver, Ordered Binary Decision Diagrams, and an appropriate coding of dynamicity. So, we defined Dynamic Ordered Binary Decision Diagrams as OBDDs representing Dynamic Constraint Satisfaction Problems. We shown that this approach allows to handle problems related to dynamicity, as consistency checking and search for optimal consistent sub-problems. The worst case complexity, given by proposition 5 and first experimental results attest the interest of the approach.

Future prospects seem to be interesting. First, we envisage to experiment the approach using an extension of OBDDs to finite domains and classical expression of CSPs (as Toupie [7]). In the same spirit, another approach is possible using finite automata as proposed by Vempaty in [17] to solve static CSPs. The coding of dynamicity for classical constraints given by tables is possible using the same principle than the coding proposed here for dynamicity. We can use 0/1 variables associated with each constraint. If the value is 0 (the constraint is inactive), all possible assignments permit to satisfy the constraint, and if the value is 1 (the constraint is active), only tuples belonging to tables permit to satisfy the constraint. A second interesting topic consists in extending experiments to real applications.

Finally, the question below seems to be more interesting. Can Dynamic CSPs be used to realize non-monotonic reasoning ? The field of non-monotonic reasoning we can consider is clearly related to revision in knowledge or belief bases including preferences. The approach we described in this paper can handle inconsistency in belief bases, in proposing syntax-based revision procedures. This question we first asked in [12] was partially answered in [5]. These authors used our approach showing that the managment of inconsistent propositionnal belief bases defined as sets of prefered subbases (see [2] and [13]), is naturally realized using Dynamic OBDDs with weighting functions on formulas.

References

1. C. Bessière. Arc-consistency for dynamic constraint satisfaction problems. *AAAI*, Anaheim, USA, 1991.

2. G. Brewka. Prefered subtheories: An extended logical framwork for default reasoning. *Proceedings of IJCAI 89, Detroit, USA*, pages 1043–1048, 1989.

3. E. Bryant. Graph-based algorithhms for boolean function manipulation. *IEEE Transactions on computers*, C-35:677–691, 1986.

4. E. Bryant. Symbolic boolean manipulation with ordered bdd. *ACM Computing Surveys, Vol 24 No.3*, September 1992.

5. C. Cayrol, M.C. Lagasquie-Schiex, and T. Schiex. Non-monotonic reasoning: from complexity to algorithms. *4th International Symposium on Artificial Intelligence and Mathematics, Fort Lauderdale, USA*, 1996.

6. H. Cormen, C. Leiserson, and R. Rivest. *Introduction to algorithms*. MIT Press-McGraw-Hill, 1991.

7. M. Corsini and A. Rauzy. Toupie user's manual. Technical report, LABRI, Université de Bordeaux I, France, 1993.

8. O. Coudert and J.C. Madre. A logically complete reasonning maintenance system based on a logical constraint solver. *IJCAI91*, I:294–299, 1991.

9. J. De Kleer. A comparaison of atms and csps techniques. In *IJCAI89*, volume Detroit, USA, pages 290–296, 1989.

10. R. Dechter and A. Dechter. Belief maintenance in dynamic constraint networks. In *AAAI*, volume Saint Paul, USA, pages 37–42, 1988.

11. P. Janssen, P. Jégou, B. Nougier, M.C. Vilarem, and B. Castro. Synthia : Assited design of peptide synthesis plans. *New Journal of Chemistry*, 14-12:969–976, 1990.

12. P. Jégou. Using Binary Decision Diagrams to solve Dynamic CSPs : Preliminary Report. In Constraint Satisfacton issues raised by practical applications Workshop, editor, *ECAI*, 1994.

13. D. Lehmann. Another perspective on default reasoning. Technical report, Leibniz Center for Research in Computer Science. Hebrew University of Jerusalem. Israel, 1992.

14. U. Montanari. Networks of constaints : fundamental properties and applications to picture procesing. *Information Sciences*, 7:95–132, 1974.

15. A. Rauzy. Cedre version 0.2 : user's guide. Technical report, LaBRI URA CNRS 1304, 1994.

16. T. Schiex and G. Verfaillie. Nogood recording for static and dynamic csps. In IEEE, editor, *5th IEEE International Conference on tools with Artificial Intelligence*, 1993.

17. N.R. Vempaty. Solving constraint satisfaction problems using finite state automata. *AAAI*, San Jose, USA:453–458, 1992.

Author Index

Springer-Verlag
and the Environment

We at Springer-Verlag firmly believe that an international science publisher has a special obligation to the environment, and our corporate policies consistently reflect this conviction.

We also expect our business partners – paper mills, printers, packaging manufacturers, etc. – to commit themselves to using environmentally friendly materials and production processes.

The paper in this book is made from low- or no-chlorine pulp and is acid free, in conformance with international standards for paper permanency.

Lecture Notes in Computer Science

For information about Vols. 1–1029

please contact your bookseller or Springer-Verlag